D1559088

AN INTRODUCTION TO ORGANOSULFUR CHEMISTRY

Biographical Details

Professsor R. J. Cremlyn, BSc, PhD(Wales), PhD(Cantab.), DSc(Wales), CChem, FRSC

Richard Cremlyn was educated at Queen Elizabeth Grammar School, Carmarthen, South Wales; Swansea University College; and Trinity Hall, Cambridge. He graduated in Chemistry from Swansea in 1950 and stayed on there to carry out research in steroid chemistry under Professor C. W. Shoppee, FRS, obtaining his PhD degree in 1953. He was awarded a University of Wales Fellowship for postdoctoral research at Cambridge in nucleotide chemistry under Professor Lord Todd, FRS, and obtained the Cambridge doctorate in 1956. Dr Cremlyn then joined ICI (Plant Protection Division) at Jealott's Hill Research Station, where he was involved in the synthesis of a wide range of organic compounds as potential pesticides and in structure–activity studies. In 1959, he was appointed Lecturer in Organic Chemistry at Brunel College of Technology (now Brunel University) and in 1961 became Senior Lecturer in Organic Chemistry at Hatfield Polytechnic (now the University of Hertfordshire), being promoted Principal Lecturer (1964), Reader (1980), Professor of Organic Chemistry (1987), and Head of Department (1990).

Professor Cremlyn's major research interests are the chemistry and biological activity of organosulfur and organophosphorus compounds, especially as crop protection agents. He has published 150 research papers and 24 review articles on the mode of action of pesticides, together with five books (three on organic chemistry and two on pesticides).

Professor Cremlyn was awarded the DSc degree by the University of Wales in 1981, and has acted as External Examiner for the BSc degree in Agricultural Chemistry at Wye College (London University) and for numerous PhD degrees at Universities and Polytechnics.

AN INTRODUCTION TO ORGANOSULFUR CHEMISTRY

R. J. Cremlyn
University of Hertfordshire, Hatfield, Hertfordshire, UK

JOHN WILEY & SONS
Chichester • New York • Brisbane • Toronto • Singapore

Baffins Lane, Chichester,
West Sussex PO19 IUD, England

National 01243 779777
International (+44) 1243 779777

e-mail (for orders and customer service enquiries): cs-books@wiley.co.uk.
Visit our Home Page on http://www.wiley.co.uk.
or http://www.wiley.com

Other Wiley Editorial Offices

John Wiley & Sons, Inc., 605 Third Avenue,
New Your, NY 10158-0012, USA

Jacaranda Wiley Ltd, 33 Park Road, Milton,
Queensland 4064, Australia

John Wiley & Sons (Canada) Ltd, 22 Worcester Road,
Rexdale, Ontario M9W 1L1, Canada

John Wiley & Sons (Asia) Pte Ltd, 2 Clementi Loop #02-01,
Jin Xing Distripark, Singapore 129809

Library of Congress Cataloging-in-Publication Data

Cremlyn, R. J. W. (Richard James William)
An introduction to organosulfur chemistry / R. J. Cremlyn.
p. cm.
Includes bibliographical references (p. –) and index.
ISBN 0 471 95512 4 (hc : acid-free paper)
1. Organosulphur compounds. I. Title.
QD305.S3C74 1996
547′.06—dc20 96-25732
CIP

British Library Cataloging in Publication Data

A catalogue record for this book is available from the British Library

ISBN 0 471 95512 4

Typeset in 10/12pt Times by Thomson Press (INDIA) Ltd., New Delhi
Printed and bound in Great Britain by Biddles Ltd, Guildford and King's Lynn
This book is printed on acid-free paper responsibly manufactured from sustainable forestation, for which at least two trees are planted for each one used for paper production.

CONTENTS

PREFACE

There are currently several excellent advanced texts on organosulfur chemistry; however, there is no introductory book on the subject. It therefore appeared timely to write an introduction to organosulfur chemistry because of the tremendous expansion in this area over recent years. Organosulfur reagents have become of increasing significance in synthetic organic chemistry, and the commercial applications of organosulfur compounds have also greatly expanded. This book first outlines the role of sulfur in biological systems and deals with the special characteristics of organosulfur compounds and the major methods available for their synthesis. Later chapters are concerned with the chemistry of the major classes of organosulfur compounds and their uses in synthetic organic chemistry. The book ends with a chapter on the major industrial applications of organosulfur chemistry, especially in the production of medicinal and pest control chemicals and artificial sweeteners.

It is hoped that this book will provide a stimulating insight into this rapidly developing area of organic chemistry. The book should prove useful to undergraduates in colleges and universities who are taking degrees in chemistry, applied chemistry, biochemistry or pharmacy, and to research workers in industry and universities who are involved in the use of organosulfur reagents.

Thanks are due to Miss Elspeth Tyler for typing the manuscript, and Miss Vanessa Pomfret and the staff of John Wiley & Sons Ltd for their help during the passage of this book through the press.

R. J. Cremlyn
January 1996

LIST OF ABBREVIATIONS

a	axial bond
Ac	acetyl (MeCO)
Ad	adenosyl
Alk	alkyl
6-APA	6-aminopenicillanic acid
aq.	aqueous
Ar	aryl
b.p.	boiling point
BTSP	bis(trimethylsilyl) peroxide
Bu^n	butyl (also Bu^t for tertiary butyl)
Chol	cholesteryl
conc.	concentrated
CSI	chlorosulfonyl isocyanate
DCC	dicyclohexylcarbodiimide
dil.	dilute
DMAP	dimethylaminopyridine
DMF	dimethylformamide
DMSO	dimethyl sulfoxide
e	equatorial bond
E	electromeric electronic effect
e.e.	enantiomeric excess
Et	ethyl
equiv.	equivalent
FVP	flash vacuum pyrolysis
g.	gaseous
Hal	halogen
HMPA	hexamethylphosphortriamide
$h\nu$	ultraviolet radiation
I	inductive electronic effect
IR	infrared
LDA	lithium diisopropylamide
liq.	liquid

MCPBA	*m*-chloroperbenzoic acid
Me	methyl
Men	menthyl
MoOPH	diperoxohexamethylphosphoamidomolybdenum (molybdenumperoxide–pyridine–HMPA complex)
m.p.	melting point
NMR	nuclear magnetic resonance
Nu	nucleophile
PABA	*p*-aminobenzoic acid
Ph	phenyl
Pr	propyl (also Pr^i for isopropyl)
R	any organic radical
Ra Ni	Raney nickel
RT	room temperature
s.	solid
THF	tetrahydrofuran
TMEDA	tetramethylethylenediamine
Tri	trityl (Ph_3C)
Ts	*p*-toluenesulfonyl (or tosyl)
TsMIC	tosylmethyl isocyanide
UV	ultraviolet
X	a halogen atom, unless defined otherwise

INTRODUCTION

An organosulfur compound may be defined as a molecule containing one or more carbon–sulfur bonds.

Sulfur itself has been known since antiquity and, because of its associations with volcanic eruptions, was referred to in the past as 'brimstone' (burning stone). In the third century, Chinese alchemists used a mixture of sulfur with saltpetre (KNO_3) as a primitive gunpowder. Elemental sulfur is widely distributed in nature and also as compounds such as hydrogen sulfide, sulfur dioxide, sulfates, e.g. those of calcium and magnesium, and sulfide ores.

Elemental sulfur deposits occur in many parts of the world, e.g. Texas, Alaska (USA), Sicily, Japan, Mexico and South Africa. Many of these deposits are in areas of high volcanic activity or are found in association with petroleum. Subterranean deposits only became readily available in the twentieth century by the Frasch process in which high pressure steam at 165 °C is forced down to melt the sulfur which is then blown up to the surface. This process yields a high quality product, ≃99% pure sulfur. Over the last 25 years, increasingly large amounts of sulfur have also been obtained from hydrogen sulfide in natural gas and by high temperature roasting of pyrites (metal sulfides). The annual production of sulfur is $\simeq 60 \times 10^6$ tonnes.

Sulfur was recognised as a common element in proteinaceous material in the early eighteenth century, and the disulfide cystine (**1**) was probably the first sulfur amino acid to be discovered (Mörner, 1899); its structure was elucidated by Friedmann (1903).

$SCH_2CH(NH_2)CO_2H$
|
$SCH_2CH(NH_2)CO_2H$

(**1**)

$MeSCH_2CH_2CH(NH_2)CO_2H$

(**2**)

SH
|
CH_2
|
$CH(NH_2)$
|
CO_2H

(**3**)

Scheme 1

Methionine (**2**) was isolated by Mueller (1922) and it was then that the nutritional significance of the relationship of (**1**), (**2**) and (**3**) (Scheme 1) emerged; pure cysteine (**3**) is difficult to isolate owing to its facile oxidation to cystine (**1**). The interrelationship of the amino acids (**1**)–(**3**) is shown in Chapter 4 (p. 49).

Sulfur is essential to the life and growth of all organisms from microbes to man. Plants and microorganisms have the capacity to utilise inorganic sulfur, since they can reduce oxidised forms of sulfur to the sulfur amino acids (**1**) and (**2**). Mammals, on the other hand, cannot utilise inorganic forms of sulfur and consequently have to obtain the essential sulfur amino acids from plant sources. The sulfur amino acids (**1**)–(**3**) are used by living organisms as the source of a vast array of biochemically vital organosulfur compounds, e.g. proteins, glutathione, coenzyme A, the vitamins biotin and thiamine (part of the vitamin B complex), lipoic acid, and plant and fungal metabolites like penicillin.

The relationship between elemental sulfur and the oxidised and reduced forms in nature is depicted in the sulfur cycle (Figure 1). This demonstrates that sources of oxidised sulfur (e.g. inorganic sulfates) and reduced sulfur (e.g. thiols) in the biosphere can be interconverted in living organisms; the special chemistry of sulfur allows these oxido-reductive transformations to occur relatively easily. Thiols (RSH) are easily oxidised under mild conditions to yield disulfides (RSSR), and this process can be reversed by reduction with, for instance, zinc–dilute acid. The reversible disulfide formation from thiols is an important biological process, since many proteins and peptides possess free thiol (SH) groups that form bridging disulfide links. This mechanism is exploited in nature to achieve the intermolecular and intramolecular joining together of amino acid chains (Scheme 2). Thiols

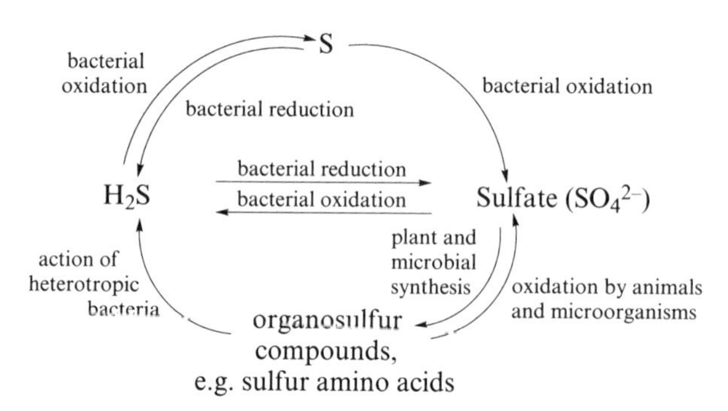

Figure 1 The sulfur cycle in nature.

$$2RSH \underset{\text{reduction }(Zn/H^+)}{\overset{\text{oxidation }(I_2)}{\rightleftharpoons}} RSSR + 2HI$$

SH, SH — amino acid chain —(conformational change in the amino acid chain)→ —SH HS— —(enzymic oxidation)→ —S—S— (disulfide bridge)

Scheme 2

are of considerable biological significance since many vital enzymes in plants, animals and fungi contain the SH group.

The formation of ether by the reaction between ethanol and sulfuric acid was investigated as early as 1540, but the intermediate, ethyl hydrogensulfate, was first observed by Dabit in 1820. Carbon disulfide was originally prepared by Lampadius (1796); this was a very significant discovery since carbon disulfide can be used as the starting material for the synthesis of many organosulfur compounds. The xanthate reaction (Scheme 3) was discovered by Zeise (1815) and really marks the beginning of organosulfur chemistry. The xanthate reaction provides an easy preparative route to dithio acids (**4**) (Chapter 8) and was of major industrial importance in the manufacture of rayon and cellophane, which are obtained by treatment of cellulose with carbon disulfide and alkali.

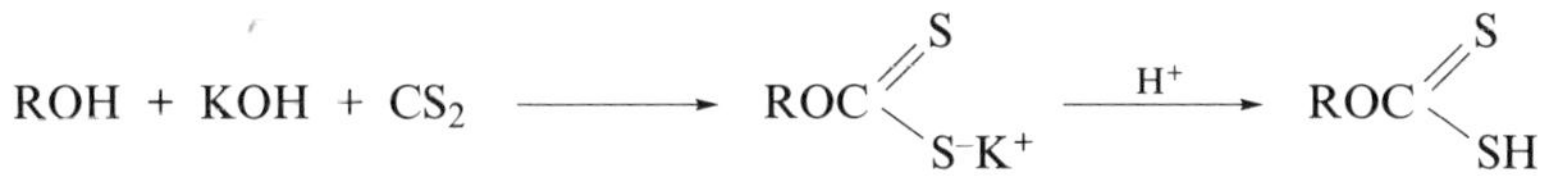

$$\mathrm{ROH} + \mathrm{KOH} + \mathrm{CS_2} \longrightarrow \mathrm{ROC}(=\mathrm{S})\mathrm{S^-K^+} \xrightarrow{\mathrm{H^+}} \mathrm{ROC}(=\mathrm{S})\mathrm{SH}$$

(**4**)

Scheme 3

Uses of sulfur and organosulfur compounds

The major uses of sulfur are in the heavy chemical industry; some 90% of all sulfur produced is used in the production of sulfuric acid. Sulfur dioxide is extensively employed as a preservative and antioxidant in the food industry; wine casks have been sterilised by fumigation with sulfur dioxide for thousands of years. The vulcanisation of natural rubber, discovered by Goodyear in 1839, involves heating rubber with sulfur to increase the sulfur bridges between the hydrocarbon chains to give a product with improved mechanical properties.[1] More recently, several organosulfur compounds, e.g. dithiocarbamates, have been used instead of sulfur. The vulcanisation process to some extent mirrors the biological function of sulfur amino acids (see p. 2). Elemental sulfur has a long history of application in medicine as a result of its scabicidal, insecticidal, fungicidal and purgative properties—sulfur-containing lozenges and ointments are still sold. Sulfur was mentioned as a fumigant by Homer and has been used as an agricultural fungicide for many years. In the nineteenth century, sulfur was increasingly used against powdery mildews and a mixture of sulfur and lime (lime sulfur) (Weighton, 1814) was shown to be effective against apple scab fungus. In 1882, Millardet discovered that a mixture of copper sulfate and lime (Bordeaux mixture) controlled potato blight and vine downy mildew, and carbon disulfide has been used since 1854 as a fumigant for control of insects and nematodes. Sulfur is an essential nutrient for plants, and over the past 30 years an ever-increasing array of synthetic organosulfur compounds have been introduced as pesticides and medicines.

Sulfur sometimes plays a role as propesticide in compounds in which biological activity is enhanced by enzymic oxidation; thus, for the insecticide malathion (**5**) (Figure 2), potency is increased by the *in vivo* conversion of the thiophosphoryl (P=S) into the phosphoryl (P=O) group in the insect.

Of the current agrochemicals, some 30% contain sulfur in a wide variety of oxidative states, e.g. sulfides, thiols, sulfoxides and sulfones. Activity may be related to the special characteristics of the carbon–sulfur bond and the increasing polarity of sulfur–oxygen bonds. Some important examples of biologically active organosulfur compounds include malathion (**5**), the selective sulfonylurea herbicide chlorsulfuron (**6**), the antiulcer drug ranitidine (**7**) and the β-lactam antibiotics such as benzylpenicillin or penicillin G (**8**) (R = $PhCH_2$) (Figure 2).

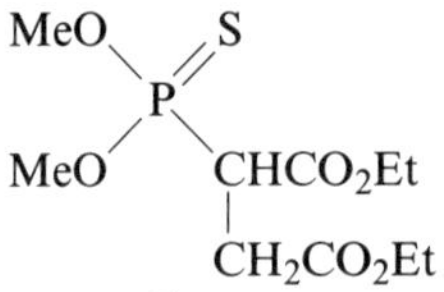

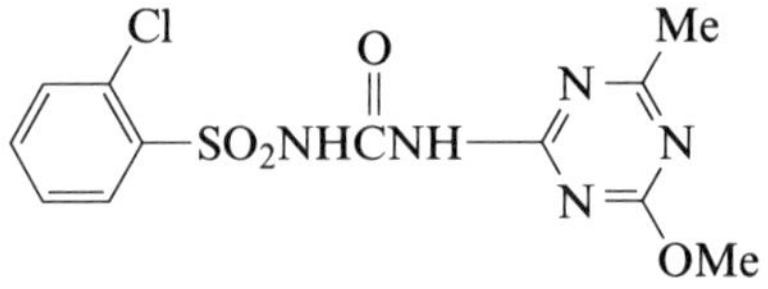

Figure 2 Some biologically active sulfur compounds.

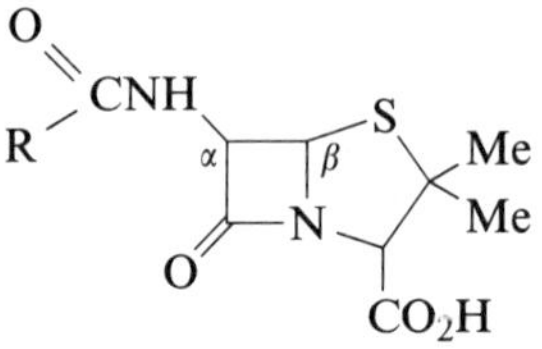

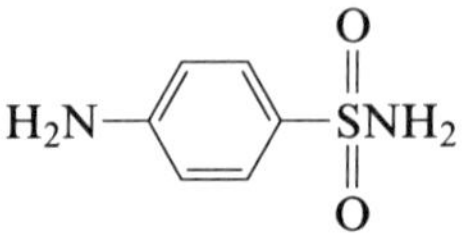

The early introduction of organosulfur compounds as pharmaceuticals derived from the sulfonic acid dyes, one of which, prontosil, showed antibacterial activity in mice. Domagk (1935) demonstrated that the activity was due to the formation of the metabolite sulfanilamide (**9**) (see Chapter 11), and this discovery led to the introduction of a large number of antibacterial sulfonamide drugs which were later modified to give different types of activity, e.g. diuretics and antimalarials. The broad spectrum antibiotic penicillin (**8**) was discovered accidentally by Fleming in 1929. He observed that several cultures of *Staphylococcus* on the laboratory bench had become contaminated by atmospheric microorganisms causing the formation of a green mould, *Penicillium notatum*, and the *Staphylococcus* in its vicinity was being destroyed. The substance causing the antibiotic action was termed penicillin; the discovery was one of the milestones in medicinal chemistry. Pure penicillin was not available until some 10 years later—it was first made commercially in 1945—and since then a large range of penicillins (**8**) have been synthesised containing different R groups.

Several sulfonic acids (RSO_3H) as their sodium salts are important synthetic detergents. For maximum efficiency, R is a highly lipophilic group, e.g. an alkyl-arene moiety as shown in the structure (**11**). The detergents are prepared by sulfonation of the parent hydrocarbon (**10**) by reaction with concentrated sulfuric acid (Scheme 4). Such sulfonate detergents have the advantage compared with ordinary soaps, e.g. sodium stearate, $C_{17}H_{35}CO_2Na$, that they do not produce scums when used in hard water because the calcium and magnesium sulfonates are more soluble.

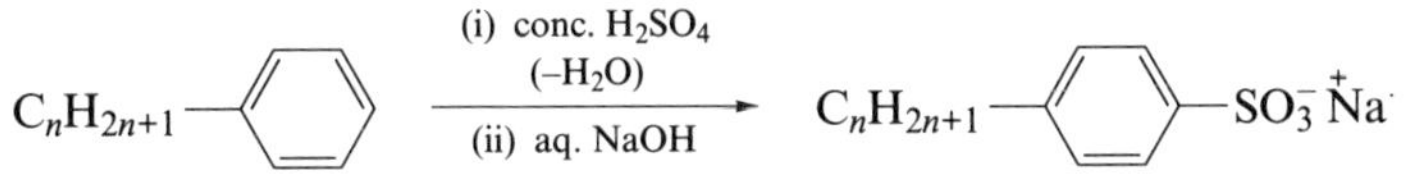

$$C_nH_{2n+1}-C_6H_5 \xrightarrow[\text{(ii) aq. NaOH}]{\text{(i) conc. } H_2SO_4 \; (-H_2O)} C_nH_{2n+1}-C_6H_4-SO_3^-\,\overset{+}{Na}$$

(**10**) (**11**) $n = 9$–15

Scheme 4

Many commercial dyes contain one or more sulfonic acid groups to confer water solubility to the dye and assist in binding the dye to the polar fibres in the textile (cotton, nylon, silk, wool, etc.). An example is Congo Red (**12**) (Figure 3). This is red in alkaline solution; thus, the sodium salt will dye cotton red, but it is very sensitive to acids and on acidification the colour changes from red to blue, and so this compound is also used as a type of indicator. Vat dyes, known as sulfur dyes, can be prepared by heating various organic compounds, e.g. amines, aminophenols, and nitrophenols, with sodium polysulfide.

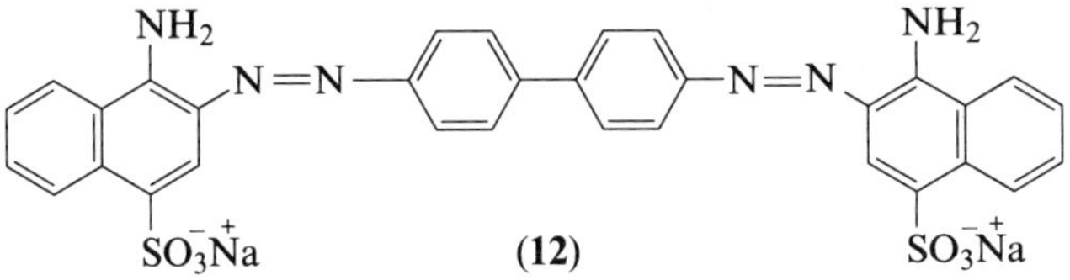

NH_2 N=N N=N NH_2

$SO_3^-\overset{+}{Na}$ (**12**) $SO_3^-\overset{+}{Na}$

Figure 3 An organosulfur dye, Congo Red.

In the food industry, sulfur dioxide and sulfites are extensively employed as preservatives to inhibit microbial spoilage and increase the storage life of foods. In addition, several sulfamic acid derivatives like saccharin (**13**) (1878), cyclamate (**14**) (1937) and acesulfame potassium (**15**) (1973) are important artificial sweeteners (Figure 4).

There are several excellent advanced texts 1–8 on organosulfur chemistry to which the reader is referred for further details of many of the topics mentioned in this book.

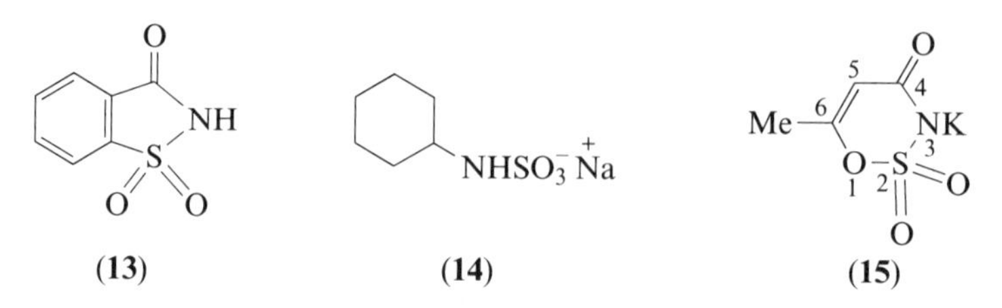

Figure 4 Some synthetic organosulfur sweeteners.

References

1. *The Chemistry of Organosulfur Compounds* (Ed. L. J Belenkii), Ellis Horwood, Chichester, 1990.
2. *Organic Chemistry of Sulfur* (Ed. S. Oae), Plenum Press, New York, 1977.
3. E. Block, *Reactions of Organosulfur Compounds*, Academic Press, New York, 1978.
4. C. M. Suter, *The Organic Chemistry of Sulfur*, Wiley, New York, 1944.
5. F. Bernardi, I. G. M. Csizmadia and A. Mangini, *Organosulfur Chemistry*, Elsevier, New York, 1985.
6. M. J. Janssen, *Organosulfur Chemistry*, Wiley, New York, 1967.
7. S. Oae, *Organic Sulfur Chemistry: Structure and Mechanism*, CRC Press, Boca Raton, FL, 1991.
8. *Organic Sulfur Chemistry: Biochemical Aspects* (Eds S. Oae and T. Okuyama), CRC Press, Boca Raton, FL, 1992.

1 STRUCTURE AND BONDING IN SULFUR AND ORGANOSULFUR COMPOUNDS

Elemental sulfur is a bright yellow crystalline solid, it is the second element of group VI in the periodic table of the elements, just below oxygen, and has the electronic configuration $1s^2 2s^2 2p^6 3s^2 3p^4$.

Sulfur shows a marked tendency to react with itself (catenation) and consequently can exist in a large number of acyclic and cyclic S_n species. For the cycles $n = 2$–20, and n can be >20 for the acyclic chains; this behaviour accounts for the complexity of the physical and chemical behaviour of sulfur.

All the chain and ring forms of sulfur are thermodynamically less stable than cyclooctasulfur (S_8) at 25 °C. Cyclooctasulfur is the most common form of sulfur and this occurs in the crown conformation (Figure 1). The sulfur–sulfur bond in elemental sulfur is probably a resonance hybrid (Figure 1; 1 Å = 0.1 nm), so the S_8 ring exists in the energetically favoured crown shape. Such electronic delocalisation would also account for the colour of sulfur and the special features of Sulfur–Sulfur bonds in the ultraviolet spectrum.

Figure 1 The S_8 ring as found in orthorhombic (S_α) sulfur.

Cyclooctasulfur exists as three main crystalline forms (allotropes); the most stable allotrope at 25 °C is orthorhombic sulfur (S_α) (m.p. 112.8 °C), found as large yellow crystals in volcanic deposits. When S_α is heated slowly to 95.5 °C, it is converted to the high temperature monoclinic form (S_β) (m.p. 119.5 °C), which also contains S_8 rings but differently arranged. Another monoclinic form of sulfur (S_γ) (m.p. 106.8 °C) can be prepared by the slow crystallisation of sulfur from ammonium polysulfide. When sulfur melts, a yellow, mobile fluid is first formed, but on further heating up to 200 °C this becomes increasingly brown and viscous. Above 200 °C, the viscosity of the liquid decreases until at the boiling point of 444.6 °C sulfur becomes a mobile red liquid. The S_8 rings persist in the liquid up to ≃195 °C.[1,2] The viscosity changes on heating are as a result of ring cleavage, chain formation and the presence of various other S_n ring species (where $n = 6, 7, 12, 18$ and >20). The acyclic sulfur species reach their maximum average chain length ($5–8 \times 10^5$ atoms) at 200 °C. The red colour of molten sulfur at temperatures >250 °C is associated with the formation of S_3 and S_4 species. Sulfur vapour contains a mixture of S_n species ($n = 2–10$) in a temperature dependent equilibrium; at approximately 600 °C S_8 predominates, while above 720 °C S_2 is the main species. Sulfur atoms (S_1) only become predominant at very high temperatures (more than 2200 °C) and low pressures.[2] The large stability range of S_2 molecules is a reflection of the strength of the S=S bond (422 kJ mol^{-1}) based on the triplet ground state and analogous to molecular oxygen (O_2).

When diatomic sulfur (S_2) is dissolved in a polar solvent like methanol or acetonitrile, an equilibrium is established in which ≃1% of the sulfur exists as the S_6 and S_7 ring forms. Since these are much more reactive than the normal S_8 rings, they may provide a pathway for the reactions of sulfur in polar solvents. Cyclic sulfur allotropes also dissolve in carbon disulfide, benzene and cyclohexane. Sulfur reacts with amines, e.g. piperidine, giving coloured solutions containing polythiobisamines (**1**) (Scheme 1).

$$2RR'NH + S_8 \longrightarrow \underset{(\mathbf{1})}{(RR'N)_2S_7} + H_2S$$

Scheme 1

Sulfur reacts with many organic molecules, and in such sulfur–sulfur bond-breaking reactions free radicals may be involved; the reactions are often catalysed by amines and Lewis acids. Amines and other bases activate sulfur by formation of nucleophilic sulfur species, while Lewis acids cleave the sulfur–sulfur bond to give electrophilic sulfur moieties.

Saturated hydrocarbons can be dehydrogenated by heating with sulfur; thus, cyclohexane (**2**) is converted to benzene (**3**) (Scheme 2).

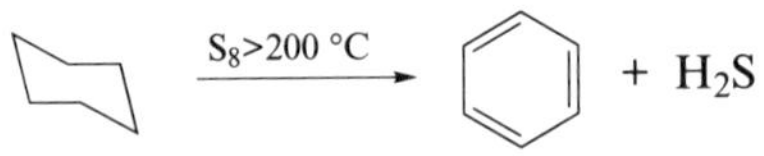

(**2**) $\xrightarrow{S_8 > 200\ °C}$ (**3**) + H_2S

Scheme 2

In some cases, organic compounds may cyclise on heating with sulfur to give thiophenes (**4**) (Scheme 3).

In the presence of bases, mild thiolation of organic compounds by treatment with sulfur provides an efficient route to organosulfur compounds. The reaction

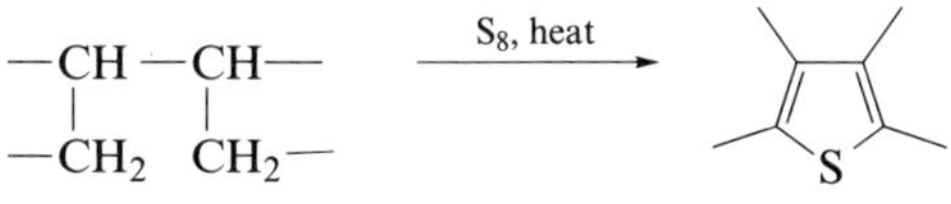

$$\begin{array}{cc} -CH-CH- \\ -CH_2\ \ CH_2- \end{array} \xrightarrow{S_8,\ heat} \textbf{(4)}$$

Scheme 3

with alkenes is of great industrial importance in the vulcanisation of rubber. Many common reactions of sulfur can be interpreted as a nucleophilic attack on the sulfur–sulfur bonds; for instance, the reaction with cyanide anion is shown in Scheme 4.[3a] The initial ring opening (i) is followed by a series of nucleophilic reactions on the Sulfur–Sulfur bonds (ii) with displacement of the thiocyanide anion leading to the overall reaction (iii).

(i) $S_8 + {}^-CN \longrightarrow {}^-S{-}S_6{-}SCN$

(ii) $S_6^-{-}S{-}SCN + CN^- \longrightarrow S_6^-{-}SCN + SCN^-$ and so on

(iii) $S_8 + 8CN^- \longrightarrow 8SCN^-$

Scheme 4

The special characteristics of organosulfur compounds

Since sulfur lies just below oxygen in group VI of the periodic table, the chemistry of organosulfur compounds should parallel that of the oxygen analogues; indeed, there are many similarities, e.g. between alcohols and thiols and ethers and sulfides, because both elements possess the same outer electronic configuration (s^2p^4). However, the prediction is not fully realised since many factors serve to differentiate sulfur from oxygen.[4]

(i) Sulfur (electronegativity 2.44) is appreciably less electronegative than oxygen (3.5), and this lessens the ionic character of organosulfur compounds in comparison with the analogous oxygen derivatives and decreases the importance of hydrogen bonding.

(ii) Sulfur, like most elements in the second and higher rows of the periodic table, is reluctant to form normal π-double bonds; thus, thiocarbonyl (C=S) compounds are comparatively rare and are usually unstable with a tendency to polymerise. This is a result of the relatively low effectiveness of $p\pi$–$d\pi$ bonding involving lateral overlap of the $3p$-orbitals and arising from the larger size of the sulfur atom as compared with carbon. For sulfur, unlike oxygen, the diatomic S_2 molecule is relatively unstable; the most stable form of sulfur is the cyclic S_8 molecule known as cyclooctasulfur (see p. 7).

(iii) The sulfur atom (atomic radius 1.02 Å; 1 Å = 0.1 nm) is larger than oxygen (atomic radius 0.73 Å), and its outermost electrons are therefore more shielded from the attractive force of the positive nucleus. Sulfur is consequently more polarisable than oxygen; the sulfur lone pairs of electrons are better nucleophiles but weaker bases in reactions with acids. The outer electronic shell in sulfur contains not only *s*-electrons and *p*-electrons but also vacant 3*d*-orbitals which can be utilised in bonding. The valency of sulfur, therefore, unlike that of oxygen, is not limited to two. Sulfur can expand its octet and form many hypervalent compounds of higher oxidation states, e.g. SO_2, SO_3, SF_4, SF_6, organic sulfoxides, sulfones and sulfonic acids. The bonding in such four-valent and six-valent sulfur compounds may involve promotion of 3*p*-electrons and 3*s*-electrons into the 3*d*-orbitals by formation of either four sp^3d or six sp^3d^2 hybrid bonds; however, a three-centre four-electron bond, termed a hypervalent bond, is now generally favoured (see Chapter 3, p. 36).

As previously mentioned, although normal $p\pi$–$p\pi$ double bonds are not generally found in sulfur compounds, another type of π-bonding arising from overlap involving the vacant sulfur *d*-orbitals is often very important. The possibility of such π-bonding arises in compounds containing sulfur attached to an element like oxygen, in which, in addition to a π-bond between the atoms, a π-bond may result by utilising an unshared electron pair on the oxygen atom and the vacant sulfur *d*-orbitals. The resultant 'double bond' is correctly described as involving the $p\pi$–$d\pi$ overlap and differs from a π-double bond because it involves expansion of the valence shell of the sulfur atom to accommodate more than eight electrons. A typical example is dimethyl sulfoxide (DMSO) (**5**) which may be regarded as a resonance hybrid of (**5a**) and (**5b**) (Figure 2).[3b]

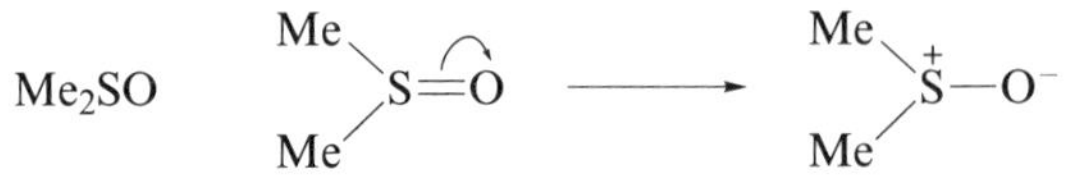

Me_2SO (**5**) $Me_2S{=}O$ (**5a**) ⟶ $Me_2\overset{+}{S}{-}O^-$ (**5b**)

Figure 2

Both structures appear significant; thus, aliphatic sulfoxides possess high dipole moments (3.9 D; 1 D = 3.336×10^{-30} C m) indicative of high polarity in the sulfur–oxygen bond. On the other hand, the bond distances and the infrared stretching frequencies are in agreement with partial double-bond character; thus, the observed length of the sulfur–oxygen bond in DMSO is 1.47 Å, which is in much better agreement with the calculated length (1.49 Å) of the S=O bond than with the calculated length for the S—O bond (1.69 Å). The structure (**5a**) depends on the capacity of sulfur to accommodate more than eight electrons, but this is clearly demonstrated by the existence of stable compounds like SF_4 and SF_6. In this book, the double-bonded structure, i.e. (**5a**), will be used for sulfoxides and analogous compounds, but it must be emphasised that such sulfur–oxygen double bonds do not always correspond to normal carbon–oxygen or nitrogen–

oxygen double bonds with $p\pi$–$p\pi$ overlap (see Chapter 5, p. 63). For instance, S=O bonds involving $p\pi$–$d\pi$ overlap do not require coplanarity of the groups attached to the sulfur atoms as distinct from normal double bonds, so DMSO (**5**) differs from acetone in possessing a non-planar pyramidal configuration. Divalent sulfur compounds, because of the available lone electron pairs and vacant *d*-orbitals which permit expansion of the sulfur valence shell, are readily oxidised to a wide variety of compounds differing in the oxidation state of the sulfur atom. In such hypervalent compounds, the oxidation site is sulfur and not carbon; for example, thiols (**6**) and sulfides (**7**) are oxidised to sulfonic acids (**8**), sulfoxides (**9**) or sulfones (**10**) (Scheme 5).

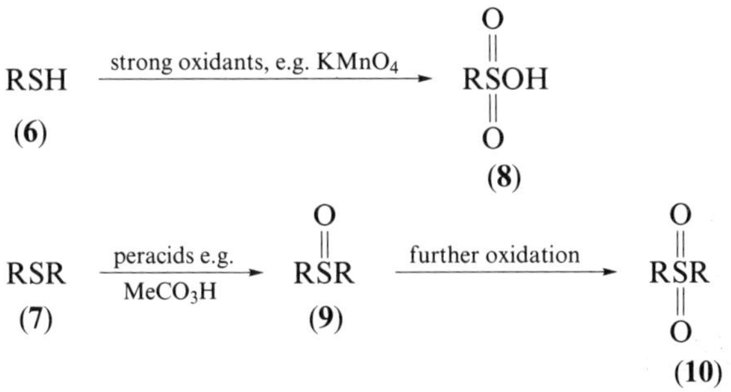

RSH (**6**) —strong oxidants, e.g. $KMnO_4$→ $RS(=O)_2OH$ (**8**)

RSR (**7**) —peracids e.g. $MeCO_3H$→ $RS(=O)R$ (**9**) —further oxidation→ $RS(=O)_2R$ (**10**)

Scheme 5

The sulfur–oxygen bonds are polar, so oxidation results in changes in the chemical behaviour of the parent compounds. On oxidation, the nucleophilicity is decreased but acidity, polar character, water solubility and leaving group capacity all increase. These changes are important in the metabolism of organosulfur compounds; the higher oxidation state compounds like sulfones can function as electrophiles and are good leaving groups in nucleophilic displacement reactions.

Nomenclature of organosulfur compounds[3c,5]

The nomenclature is illustrated by Table 1, which gives the names of the major types of organosulfur compounds. The original name 'mercaptan' for thiols is no longer recommended, except that the term 'mercapto' is retained for the unsubstituted SH group. In the table, an asterisk indicates the International Union of Pure and Applied Chemistry (IUPAC) recommended name.

Thioalcohols (RSH) are known as thiols; otherwise they are named similarly to alcohols. For example,

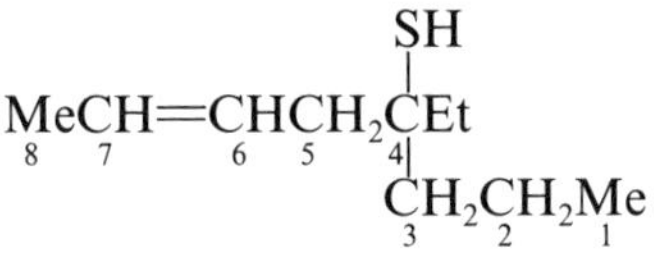

$$\underset{8}{\mathrm{Me}}\underset{7}{\mathrm{CH}}{=}\underset{6}{\mathrm{CH}}\underset{5}{\mathrm{CH_2}}\underset{4}{\mathrm{C}}(\mathrm{SH})(\mathrm{Et})\underset{3}{\mathrm{CH_2}}\underset{2}{\mathrm{CH_2}}\underset{1}{\mathrm{Me}}$$

is 4-ethyloct-6-ene-4-thiol.

Table 1 The nomenclature of some organosulfur compounds.

Structure	Name
(i) Divalent sulfur compounds	
RSH	thiol (mercaption)
e.g. EtSH	ethanethiol (ethyl mercaptan)
RSR′	sulfide
e.g. Et_2S	diethyl sulfide (ethylthioethane*)
RSSR′	disulfide
e.g. PhSSPh	diphenyl disulfide (phenyldithiobenzene*)
RSOH	sulfenic acid
e.g. MeSOH	methanesulfenic acid
(ii) Sulfur compounds of higher valencies 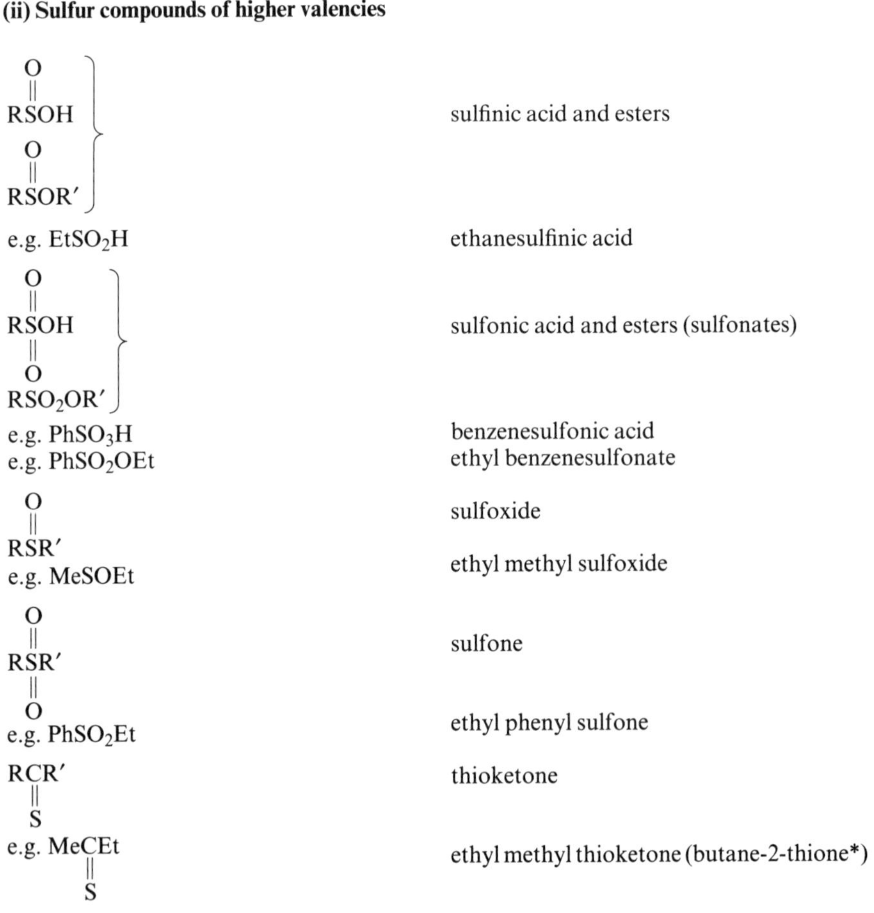	
RS(=O)OH, RS(=O)OR′	sulfinic acid and esters
e.g. $EtSO_2H$	ethanesulfinic acid
RS(=O)(=O)OH, RSO_2OR'	sulfonic acid and esters (sulfonates)
e.g. $PhSO_3H$	benzenesulfonic acid
e.g. $PhSO_2OEt$	ethyl benzenesulfonate
RS(=O)R′	sulfoxide
e.g. MeSOEt	ethyl methyl sulfoxide
RS(=O)(=O)R′	sulfone
e.g. $PhSO_2Et$	ethyl phenyl sulfone
RC(=S)R′	thioketone
e.g. MeC(=S)Et	ethyl methyl thioketone (butane-2-thione*)

Table 1 *(continued)*

(iii) Thiocarboxylic acids

(a) Monothioic acids

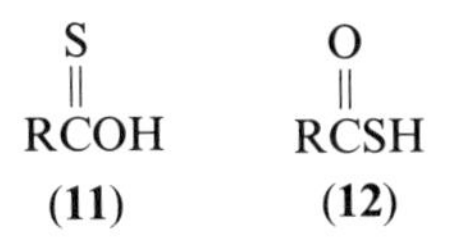

Normally (**11**) and (**12**) are indistinguishable; either form can be isolated by salt or ester formation. Compound (**11**) is termed the *O*-acid (thiono) and (**12**) the *S*-acid (thiolo). For simple acids, the prefix thio is used; for example, $MeCH_2CH_2COSH$ is thiobutyric acid. For more complex structures, the suffix thioic acid is used; for example,

$$\underset{6}{PhCH_2}(Me)\underset{5}{C}{=}\underset{4}{C}H(CH_2)_2\underset{1}{C}OSH$$

is 5-methyl-6-phenylhex-4-enethioic acid.

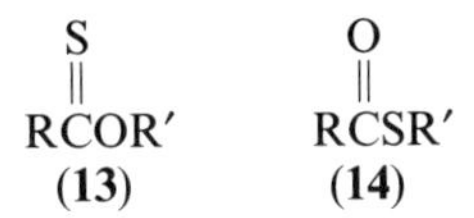

The corresponding esters are termed thioates. Thus, (**13**) is the *O*-alkyl thioate and (**14**) is the *S*-alkyl thioate; for example,

$$\underset{4}{Me}\underset{3}{CH_2}\underset{2}{CH_2}\underset{1}{C}(=S)OMe$$

is *O*-methyl butanethioate.

(b) Dithioic acids

$$RC(=S)SH$$

These are named as dithioic acids; for example, $Me(CH_2)_4CS_2H$ is hexanedithioic acid. The corresponding salts and esters are termed dithioates; thus, $\underset{3}{Me}\underset{2}{CH_2}\underset{1}{C}S_2Na$ is sodium propanedithioate.

(iv) Radicals

$—SO_3H$	sulfo
$—SO_2H$	sulfino
$—SOH$	sulfeno
$>NSO_2—$	sulfamoyl
$>NS—$	sulfenamoyl
$—SO_2Me$	mesyl
$—SO_2C_6H_4Me$-*p*	tosyl
$—SCN$	thiocyanato
$—NCS$	isothiocyanato
$—SCH$	thioformyl
$>C{=}S$	thiocarbonyl

(continued)

Table 1 *(continued)*

—SH	mercapto
—S—	thia

The radical names can be used as prefixes in compounds in which the sulfur moity is not the prime group; for example,

HO_3S—⟨benzene ring⟩—CO_2H

is p-sulfobenzoic acid.

*Denotes the IUPAC name.

Thioethers (RSR′) are known as sulfides and are named similarly to ethers; for instance,

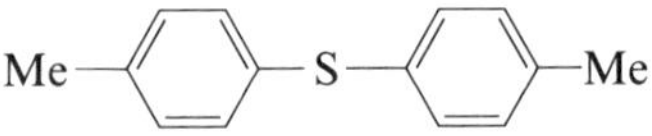

Me—⟨benzene ring⟩—S—⟨benzene ring⟩—Me

is di-*p*-tolyl sulfide and

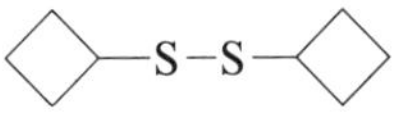

⟨cyclobutyl⟩—S—S—⟨cyclobutyl⟩

is named as dicyclobutyl disulfide. In the IUPAC system, the compounds are *p*-(tolylthio)toluene and cyclobutyldithiocyclobutane, respectively.

The various sulfur acids are named as derivatives of the parent hydrocarbons:

$MeCH_2CH_2CH_2SOH$	is	butanesulfenic acid,
$MeCH_2SO_2H$	is	ethanesulfinic acid and
$MeSO_3H$	is	methanesulfonic acid.

Esters and salts are named by replacing the terminal 'ic' by 'ate'; thus, $MeCH_2SOMe$ is methyl ethanesulfenate.

Unlike alcohols, sulfides can be oxidised to the hypervalent sulfur compounds, namely sulfoxides

$$\underset{\displaystyle \overset{\|}{O}}{RSR'}$$

and sulfones

$$\overset{\displaystyle O \atop \|}{\underset{\displaystyle \| \atop O}{RSR'}}$$

For example, $MeSOCH_2CH_2CH_2Me$ is butyl methyl sulfoxide.

Thioaldehydes are generally very unstable. Thioketones may sometimes be named by putting the prefix 'thio' in front of the name of a simple ketone, e.g. thiobenzophenone, but generally by replacing the terminal 'one' of ketones by 'thione' thus,

=S

is cyclohexanethione and

4 3 O 5 O1 2 $\underset{1}{C}\underset{2}{CH_2}\underset{3}{CH_2}\underset{4}{Me}$

is termed 1-(2-furyl)butane-1-thione.

Carbothioic acids can exist either as RC(=S)OH or RC(=O)SH, but either form may be isolated as salts or esters. The former type are *O*-acids (or thiono) and the latter are termed *S*-acids (or thiolo). For instance, $\underset{3}{Me}\underset{2}{CH_2}\underset{1}{C}(=S)OH$ is *O*-propanethionic acid. In esters and salts (thioates), the ending 'ic' is replaced by 'ate', so $\underset{4}{Me}\underset{3}{CH_2}\underset{2}{CH_2}\underset{1}{C}(=S)OCH_2Cl$ is *O*-chloromethyl butanethioate.

A sulfur atom in a chain or as part of a ring is termed 'thia' as a prefix when there is another group of higher priority present. For example, $\underset{6}{Me}\underset{5}{CH_2}\underset{4}{S}\underset{3}{CH_2}\underset{2}{CH_2}\underset{1}{C}ONH_2$ is 4- thiahexanamide and

4 3 Me 5 6 S1 2

is 3-methylthiapyran.

References

1. F. A. Cotton and G. Wilkinson, *Advanced Inorganic Chemistry*, 5th Edn, Wiley, New York, 1988, Chap. 13, p. 493.
2. A. G. Massey, *Main Group Elements*, Ellis Horwood, New York, 1990, p. 301.
3. *Sulfur in Organic and Inorganic Chemistry* (Ed. A. Senning), Dekker, New York: (a) H. Schumann, Vol. 3, 1972, Chap. 21; (b) H. H. Szmant, Vol. 1, 1971, Chap. 5; (c) K. L. Loening, Vol. 3, 1972, Chap. 28.
4. L. A. Damani, in *Sulfur-Containing Drugs and Related Organic Compounds—Chemistry, Biochemistry and Toxicology* (ed. L. A. Damani), Vol. 1, Part A, Ellis Horwood, Chichester, 1989, Chap. 1.
5. E. W. Godley, *Naming Organic Compounds: A Systematic Instruction Manual,* Ellis Horwood, Chichester, 1989.

2 SYNTHESIS OF ORGANOSULFUR COMPOUNDS

Organosulfur compounds can be obtained using elemental sulfur (S_8) and a variety of simple, readily available sulfur compounds, e.g. hydrogen sulfide, carbon disulfide, phosphorus pentasulfide, sodium sulfide, sodium sulfite, sulfur dioxide, sulfur trioxide, sulfuric acid, sulfur dichloride, thionyl chloride and sulfuryl chloride.[1,2]

Sulfur

The direct action of elemental sulfur on organic compounds in the presence of bases provides a general method of sulfuration that often results in a complex mixture of products.[3,4] Sulfur has been extensively used in the dehydrogenation of saturated compounds; thus, alkanes by heating with sulfur above 200 °C yield unsaturated products; for instance, in this way, cyclohexane is converted to benzene. Alkenes are generally much more reactive than alkanes and often react at 130 °C; for instance, methylcyclohexene (**1**) yields the dithiathiones (**2**) and (**3**) (Scheme 1).

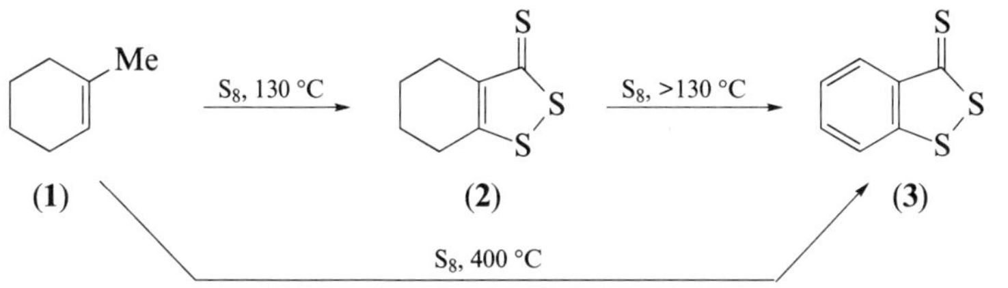

Me
S_8, 130 °C
S
S
S
S_8, >130 °C
S
S
S
(1)
(2)
(3)
S_8, 400 °C

Scheme 1

Activated unsaturated compounds containing highly acidic carbon—hydrogen bonds react selectively with sulfur under mild conditions; for example, terminal alkynes (**4**) give the alkynyl thiols (**5**) (Scheme 2).

Benzylic halides and diphenyl ether on heating with sulfur afford ether thiophenes (**6**), benzothiophenes (**7**) (the Veronkov reaction) or phenothiazole (**8**) (Scheme 3).

$$\underset{(\mathbf{4})}{RC{\equiv}CH} \xrightarrow{S_8,\ \text{heat}} \underset{(\mathbf{5})}{RC{\equiv}CSH}$$

Scheme 2

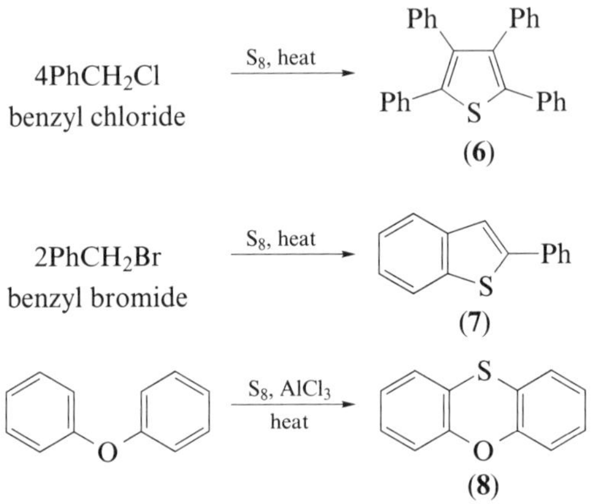

$$\underset{\text{benzyl chloride}}{4PhCH_2Cl} \xrightarrow{S_8,\ \text{heat}} \text{tetraphenylthiophene } (\mathbf{6})$$

$$\underset{\text{benzyl bromide}}{2PhCH_2Br} \xrightarrow{S_8,\ \text{heat}} \text{2-phenylbenzothiophene } (\mathbf{7})$$

$$\text{diphenyl ether} \xrightarrow[\text{heat}]{S_8,\ AlCl_3} \text{phenoxathiin } (\mathbf{8})$$

Scheme 3

Sulfur also reacts with Grignard reagents, and this reaction provides a route to thiols and sulfides (see Chapter 4, pp. 41 and 44) (Scheme 4).

$$RMgX \xrightarrow{S_8} RSMgX \xrightarrow{RX} R_2S$$

$$RSMgX \xrightarrow{H_3O^+} RSH$$

Scheme 4

Sulfur can be activated by the presence of bases, e.g. amines (R_2NH). Such proton-bearing amines cleave the S_8 ring and also will react with the thiolation products. This procedure is used in the synthesis of thioamides (**9**) from the methylarenes (**10**) (Scheme 5).

$$\underset{(\mathbf{10})}{ArMe} + S_8 + R_2NH \longrightarrow \underset{(\mathbf{9})}{Ar{-}C({=}S){-}NR_2}$$

Scheme 5

Another important route to thioamides is the Willgerodt–Kindler reaction (1923), in which an alkyl aryl ketone is heated with sulfur and an amine at 130 °C (see Chapter 8, p.137) (Scheme 6). In this reaction, the carbonyl group is reduced, while the terminal methyl group is oxidised.[3] The primary step in the reaction is probably the formation of an enamine (**11**) by reaction of the amine on the ketone; the enamine is subsequently isomerised and thiolated to yield the stable thioamide (**12**) (Scheme 7).

$$ArCO(CH_2)_nMe \xrightarrow[130\ °C]{S_8,\ R_2NH} Ar(CH_2)_{n+1}CSNR_2$$

Scheme 6

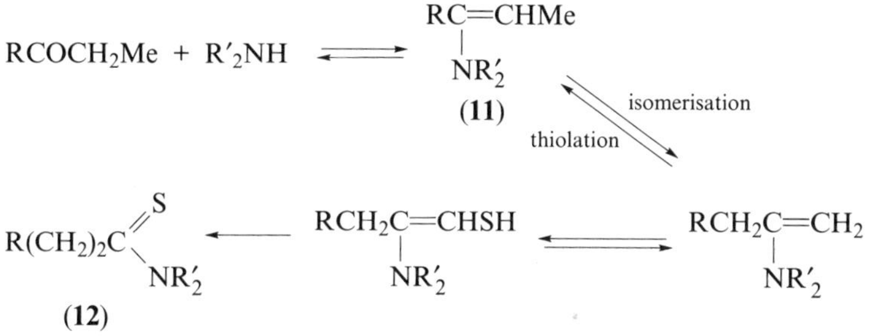

$$RCOCH_2Me + R'_2NH \rightleftharpoons RC(NR'_2){=}CHMe\ (\mathbf{11})$$

$$RC(NR'_2){=}CHMe\ (\mathbf{11}) \xrightarrow{isomerisation} RCH_2C(NR'_2){=}CH_2$$

$$RCH_2C(NR'_2){=}CH_2 \rightleftharpoons RCH_2C(NR'_2){=}CHSH \xrightarrow{\text{thiolation}} R(CH_2)_2C(=S)NR'_2\ (\mathbf{12})$$

Scheme 7

Enamines are sufficiently basic and nucleophilic to activate elemental sulfur and undergo thiolation at room temperature to give very high yields of the thioamides (Scheme 8). This step would appear to hold the key to the mechanism of the Willgerodt–Kindler reaction.

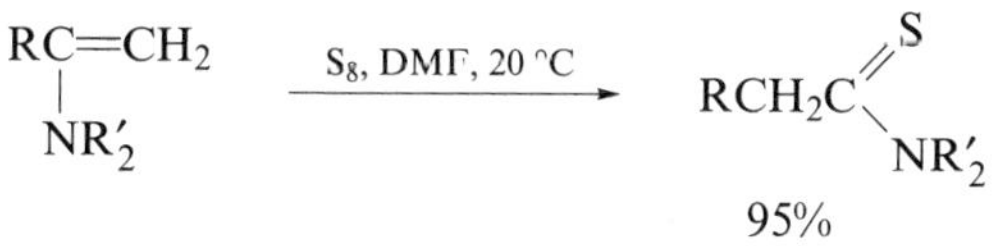

$$RC(NR'_2){=}CH_2 \xrightarrow{S_8,\ DMF,\ 20\ °C} RCH_2C(=S)NR'_2\quad 95\%$$

Scheme 8

Hydrogen sulfide

Hydrogen sulfide is a better nucleophile than water and will undergo nucleophilic addition to alkenes and alkynes in the presence of an acidic catalyst, e.g. sulfuric acid or aluminium chloride (Markownikoff addition). The reaction goes well with suitable alkenes and provides a preparative route to thiols (Scheme 9).

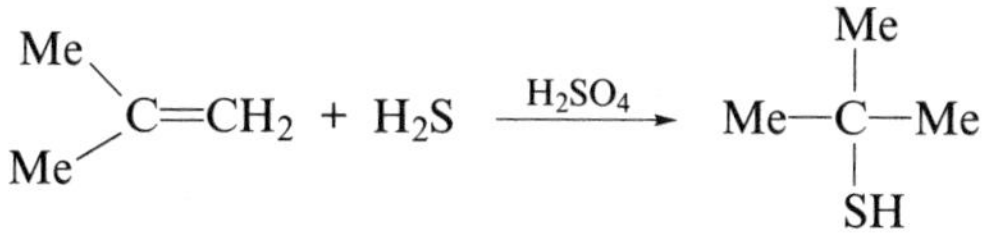

$$Me_2C{=}CH_2 + H_2S \xrightarrow{H_2SO_4} Me{-}C(Me)(SH){-}Me$$

Scheme 9

Hydrogen sulfide also reacts by nucleophilic substitution with alkyl halides and alcohols under appropriate conditions to form thiols (Scheme 10) (see Chapter 4, p. 42). The reaction with an alkyl halide is generally facilitated by using sodium hydrogen sulfide (NaSH) rather than hydrogen sulfide since NaSH is a much better nucleophile. Hydrogen sulfide also adds to ketones to give *gem*-dithiols (**13**) (Scheme 11).

$$\underset{(X = Cl, Br, I)}{RX} \xrightarrow[\text{catalyst}]{H_2S,\ \text{basic}} RSH \xleftarrow[\text{catalyst}]{H_2S,\ \text{acidic}} ROH$$

Scheme 10

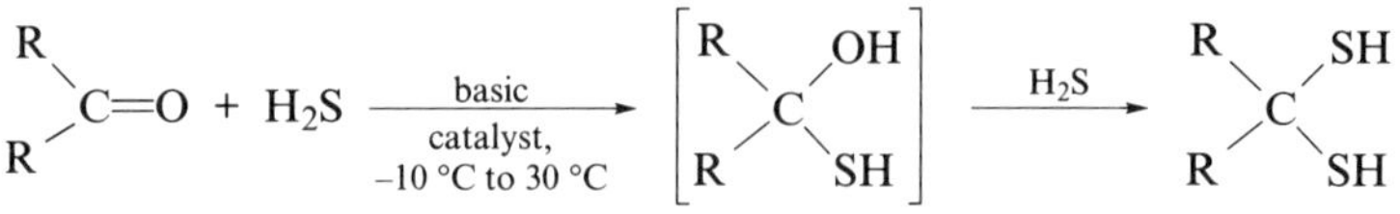

$$R_2C{=}O + H_2S \xrightarrow[\text{catalyst, } -10\,^\circ C \text{ to } 30\,^\circ C]{\text{basic}} \left[R_2C(OH)(SH) \right] \xrightarrow{H_2S} \underset{(\mathbf{13})}{R_2C(SH)_2}$$

Scheme 11

Carbon disulfide

Carbon disulfide (see Chapter 9, p. 149) can be used to prepare dithioacids (**14**) by insertion into reactive carbon–hydrogen bonds (Scheme 12).

$$R_3CH + CS_2 \longrightarrow \underset{(\mathbf{14})}{R_3CC({=}S)SH}$$

Scheme 12

Carbon disulfide reacts as an electrophile with amines in the presence of sodium hydroxide to give dithiocarbamates (**15**); the reaction involves nucleophilic addition of the amine to the electrophilic thiocarbonyl bond (Scheme 13). The reaction is industrially important for the manufacture of some widely used agricultural fungicides (see chapter 9, p. 148). Carbon disulfide undergoes a similar type of reaction with alcohols to give xanthates (**16**) (Scheme 14).

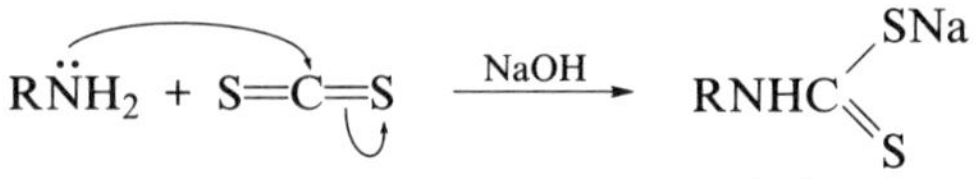

$$R\ddot{N}H_2 + S{=}C{=}S \xrightarrow{NaOH} \underset{(\mathbf{15})}{RNHC({=}S)SNa}$$

Scheme 13

$$ROH + CS_2 + NaOH \longrightarrow \underset{(\mathbf{16})}{ROCSSNa}$$

Scheme 14

Cellulose xanthate (see Chapter 8, p.135) is used in the manufacture of rayon and cellophane. Carbon disulfide, like carbon dioxide, reacts smoothly with Grignard reagents to give the corresponding dithioacids (**17**) (Scheme 15) (see Chapter 9, p.148).

$$RMgX + CS_2 \longrightarrow RCSSMgX \xrightarrow{H_3O^+} \underset{(\mathbf{17})}{RCSSH}$$

Scheme 15

Phosphorus pentasulfide

Phosphorus pentasulfide is a valuable reagent for conversion of carbonyl into thiocarbonyl groups; for instance, amides are converted to thioamides, ketones to thioketones (see Chapter 8, p.137), carboxylic acids to thioacids, and alcohols are converted to the corresponding phosphorodithioic acids (**18**) (Scheme 16).

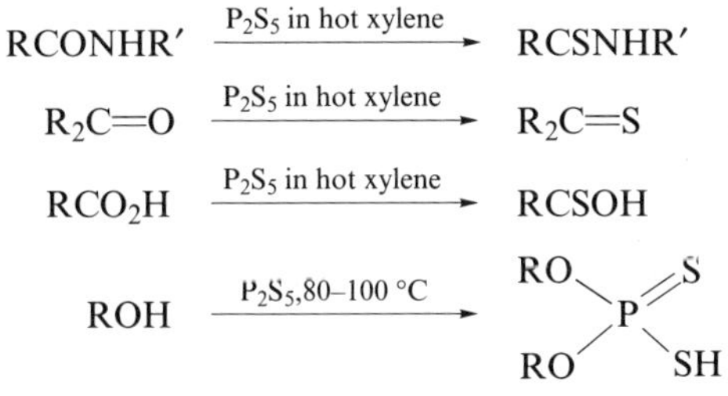

$$RCONHR' \xrightarrow{P_2S_5 \text{ in hot xylene}} RCSNHR'$$

$$R_2C{=}O \xrightarrow{P_2S_5 \text{ in hot xylene}} R_2C{=}S$$

$$RCO_2H \xrightarrow{P_2S_5 \text{ in hot xylene}} RCSOH$$

$$ROH \xrightarrow{P_2S_5, 80\text{–}100\ °C} \underset{(\mathbf{18})}{(RO)_2P({=}S)SH}$$

Scheme 16

Sodium hydrogen sulfide (NaSH) and sodium sulfide (Na_2S)

Sodium hydrogen sulfide can be used for the preparation of thiols (**19**) by reaction with alkyl halides. A large excess of sodium hydrogen sulfide must be employed to avoid formation of the dialkyl sulfide (Scheme 17) (see Chapter 4, p. 41).

$$RX + NaSH \xrightarrow{EtOH} \underset{(\mathbf{19})}{RSH} + NaX$$

Scheme 17

In the presence of sodium hydroxide, thiols react with alkyl halides to form the sulfides (**20**); the reaction occurs via the sodium thiolate and is analogous to the well-known Williamson synthesis of ethers and can also be applied to obtain unsymmetrical sulfides (Scheme 18). Symmetrical sulfides may be prepared directly by condensation of sodium sulfide with alkyl halides (Scheme 18). These reactions are of the S_N2 type, and consequently the optimum yields of sulfides are realised using primary alkyl halides.

$$RSH \xrightarrow{NaOH} RSNa \xrightarrow{R'X} \underset{\mathbf{(20)}}{RSR'} + NaX$$

$$2RX + Na_2S \xrightarrow{EtOH} R_2S + 2NaX$$

Scheme 18

Sodium hydrogen sulfite ($NaHSO_3$) and sodium sulfite (Na_2SO_3)

Both these reagents will react with alkyl halides in aqueous media to give the corresponding sulfonic acids (**21**). This procedure has been used extensively for the preparation of aliphatic sulfonic acids in good yields (see Chapter 7, p.100). Sodium hydrogen sulfite will also form sulfonic acids by addition to alkenes in the presence of peroxide catalysts (anti-Markownikoff reaction) (Scheme 19).

$$RX + NaHSO_3 \xrightarrow{EtOH,\ H_2O} \underset{\mathbf{(21)}}{RSO_3H} + NaX$$

$$RX + Na_2SO_3 \xrightarrow{EtOH,\ H_2O} RSO_3Na + NaX$$

$$RCH{=}CH_2 + NaHSO_3 \xrightarrow[catalyst]{peroxide} RCH_2CH_2SO_3Na$$

Scheme 19

Sodium hydrogen sulfite also adds to aldehydes and some ketones to form hydroxysulfonic acids (**22**) (Scheme 20). The nucleophilic addition generally goes well with aldehydes and methyl ketones of the type RCOMe, where R is a primary alkyl group. Since the water soluble products are readily converted back to the carbonyl compounds by treatment with dilute acid, the reaction provides

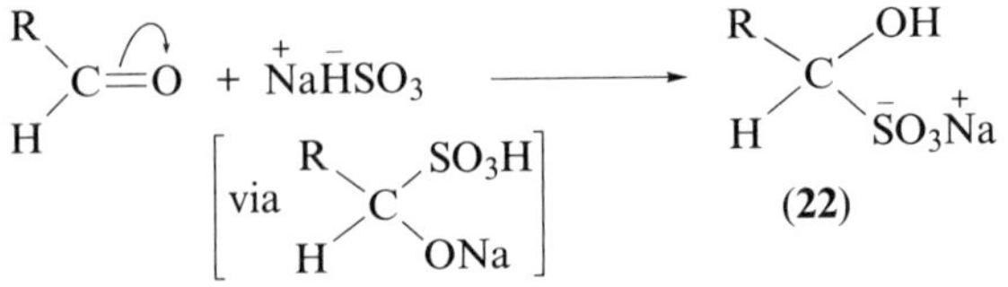

$$R(H)C{=}O + \overset{+}{Na}\overset{-}{H}SO_3 \longrightarrow \underset{\mathbf{(22)}}{R(H)C(OH)\overset{-}{S}O_3\overset{+}{Na}}$$

$$\left[\text{via } R(H)C(SO_3H)(ONa)\right]$$

Scheme 20

a useful method for purification of carbonyl compounds from non-carbonyl impurities.

Sulfur dioxide and sulfur trioxide

Sulfur dioxide and Sulfur trioxide both react with Grignard reagents to give sulfinic acids (**23**) or sulfonic acids (**24**) (Scheme 21).

$$\underset{\mathbf{(24)}}{RSO_3H} \xleftarrow{H_3O^+} RSO_3MgX \xleftarrow{SO_3} RMgX \xrightarrow{SO_2} RSO_2MgX \xrightarrow{H_3O^+} \underset{\mathbf{(23)}}{RSO_2H}$$

Scheme 21

The chlorosulfonation of organic molecules can be achieved by treatment with a mixture of sulfur dioxide and chlorine in the presence of ultraviolet light (the Reed reaction); this is a free radical process (Scheme 22). In this process, the liberated chlorine radical can then continue the radical chain reaction.

$$RH + SO_2 + Cl_2 \xrightarrow{h\nu} RSO_2Cl$$

$$Cl_2 \xrightarrow{h\nu} 2Cl\cdot \text{ (initiation step)}$$

$$RH + Cl\cdot \longrightarrow R\cdot + HCl$$

$$R\cdot + SO_2 \longrightarrow RSO_2\cdot$$

$$RSO_2\cdot + Cl_2 \longrightarrow RSO_2Cl + Cl\cdot$$

Scheme 22

Sulfur dioxide will participate in a concerted Diels–Alder-type reaction with 1,3-butadienes to yield cyclic sulfones (**25**) (Scheme 23). Sulfur dioxide also adds to reactive alkenes containing electron-withdrawing groups, e.g. ethyl acrylate (**26**), in the presence of formic acid and triethylamine (Scheme 23).

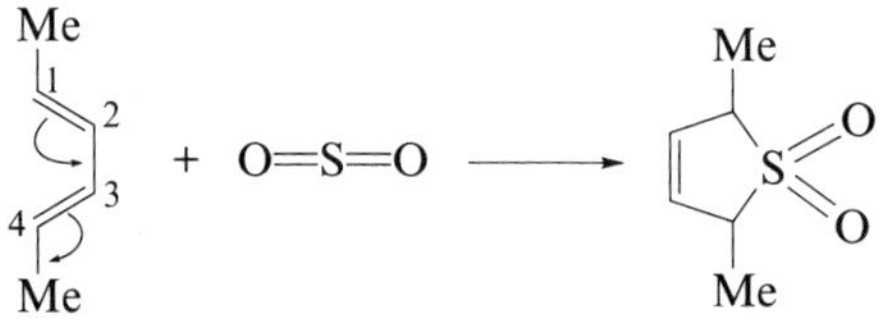

1,4-dimethyl-1,3-butadiene + O=S=O ⟶ (**25**)

$$\underset{\mathbf{(26)}}{H_2C{=}CHCO_2Et} + SO_2 \xrightarrow[\text{heat}]{HCO_2H,\ NMe_3} O_2S(CH_2CH_2CO_2Et)_2$$

Scheme 23

Sulfur trioxide is the active sulfonating agent in sulfuric acid and it has been used in the form of complexes, e.g. sulfur trioxide–pyridine, for the sulfonation of molecules that decompose in strongly acidic media.[5] The complex thus can be applied to obtain the sulfonic acids (**27**) and (**28**) from the parent furan and pyrrole heterocycles (**29**) and (**30**) (Scheme 24).

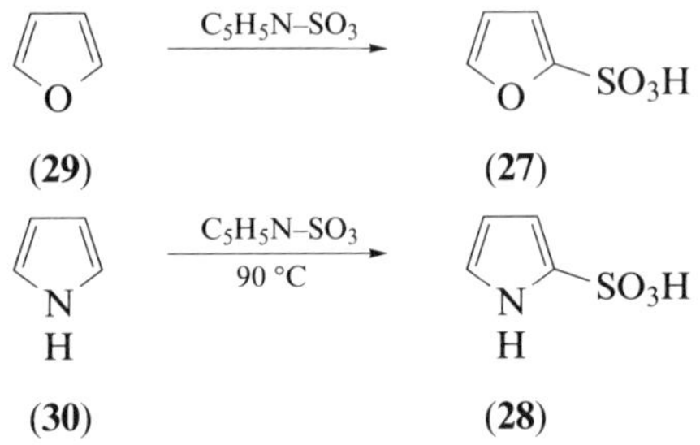

(**29**) $\xrightarrow{C_5H_5N–SO_3}$ (**27**) SO_3H

(**30**) $\xrightarrow[90\ °C]{C_5H_5N–SO_3}$ (**28**) SO_3H

Scheme 24

Sulfur trioxide as the dioxan complex will also sulfonate aldehydes and ketones containing reactive α-methylene hydrogen atoms (Scheme 25).

$$\underset{\alpha}{RCH_2COR'} \xrightarrow[20\ °C]{SO_3\text{–dioxan complex}} RCH(SO_3H)COR'$$

$$\text{e.g. } \underset{\alpha}{PhCOMe} \xrightarrow[20\ °C]{SO_3\text{–dioxan complex}} PhCOCH_2SO_3H$$

acetophenone

Scheme 25

Sulfuric acid

Concentrated or fuming sulfuric acid (oleum) is widely used for the direct sulfonation of aromatic compounds (see Chapter 7, p. 97).[5,6] The active sulfonating agent in sulfuric acid is the electrophile sulfur trioxide, and the sulfonating power of sulfuric acid is proportional to the concentration of SO_3. Consequently, fuming sulfuric acid, which contains excess sulfur trioxide, is a more powerful sulfonating agent than concentrated sulfuric acid. The sulfonation of an aromatic hydrocarbon is depicted in Scheme 26.

$$ArH + H_2SO_4 \rightleftharpoons ArSO_3H + H_2O$$

Scheme 26

To drive the reaction over to the sulfonic acid, an excess of sulfuric acid is generally employed which also assists the forward reaction by removing the water formed. Sulfonation is a bimolecular electrophilic substitution (S_E2) reaction and is therefore facilitated by the presence of electron-donating substituents on the aromatic nucleus and is made more difficult by the introduction of electron-withdrawing groups (see chapter 7, p. 98).

Sulfonation, unlike nitration, is a reversible reaction and may need to be driven to completion, for instance, by the use of a large excess of reagent. As a consequence of the reversibility of the reaction, the sulfonic acid group can be removed by heating with dilute acid, which allows it to be used as a protecting group in organic syntheses (see Chapter 7, p. 97).

Chlorosulfonic acid ($ClSO_3H$)

This is a powerful sulfonating agent approximately equivalent in strength to fuming sulfuric acid. Treatment of an aromatic compound with an equimolar amount of the reagent yields the corresponding sulfonic acid (**31**), but when an excess of

$$\underset{(\mathbf{32})}{RSO_2Cl} \xleftarrow[\text{(excess)}]{ClSO_3H} RH \xrightarrow[\text{(1 equiv.)}]{ClSO_3H} \underset{(\mathbf{31})}{RSO_3H} + HCl$$

Scheme 27

$$\underset{(\mathbf{34})}{RSO_2OR'} \xleftarrow[C_5H_5N]{R'OH} \underset{(\mathbf{32})}{RSO_2Cl} \xrightarrow[\text{base}]{R'NH_2} \underset{(\mathbf{33})}{RSO_2NHR'}$$

Scheme 28

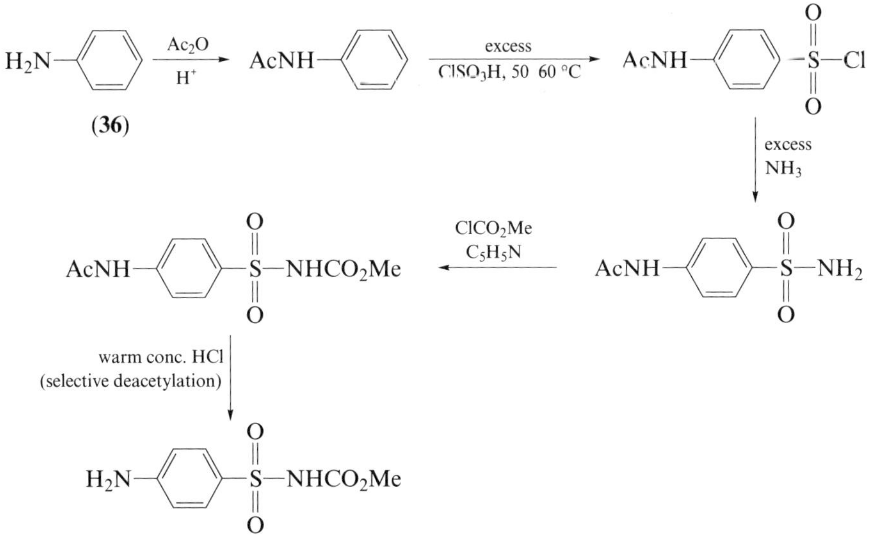

Scheme 29

chlorosulfonic acid is employed the sulfonyl chloride (**32**) is obtained (Scheme 27). The latter provides a very convenient direct synthetic route to sulfonyl chlorides,[7] which is valuable because sulfonyl chlorides (**32**) are important synthetic intermediates for the formation of other sulfonyl derivatives such as sulfonamides (**33**) and sulfonic acid esters (sulfonates) (**34**) (Scheme 28). The procedure has been extensively applied in the manufacture of a wide range of sulfonamide ('sulfa') drugs (see Chapter 11, p. 223) such as sulfanilamide (see Introduction, p. 4). The method is illustrated by the synthesis of the herbicide asulam (**35**) from aniline (**36**) (Scheme 29) (see Chapter 11, p. 234).

Sulfur monochloride, sulfur dichloride, thionyl chloride and sulfuryl chloride

All these reagents, (**37**)–(**40**) in Scheme 30, will react with aromatic compounds by Friedel–Crafts-type reactions to give organosulfur compounds, e.g. sulfides (**41**), sulfoxides (**42**) and symmetrical sulfones (**43**); the use of an alkane- or arenesulfonyl chloride (**32**) similarly affords a mixed sulfone (**44**) (Scheme 30).

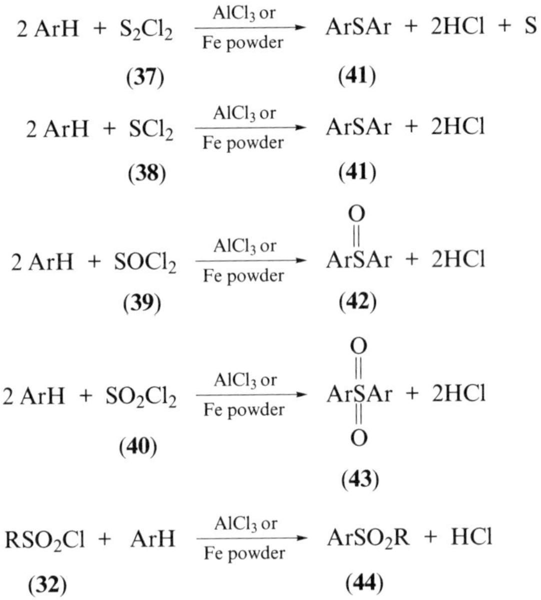

$$2\,ArH + \underset{(\mathbf{37})}{S_2Cl_2} \xrightarrow[\text{Fe powder}]{AlCl_3\text{ or}} \underset{(\mathbf{41})}{ArSAr} + 2HCl + S$$

$$2\,ArH + \underset{(\mathbf{38})}{SCl_2} \xrightarrow[\text{Fe powder}]{AlCl_3\text{ or}} \underset{(\mathbf{41})}{ArSAr} + 2HCl$$

$$2\,ArH + \underset{(\mathbf{39})}{SOCl_2} \xrightarrow[\text{Fe powder}]{AlCl_3\text{ or}} \underset{(\mathbf{42})}{ArS(=O)Ar} + 2HCl$$

$$2\,ArH + \underset{(\mathbf{40})}{SO_2Cl_2} \xrightarrow[\text{Fe powder}]{AlCl_3\text{ or}} \underset{(\mathbf{43})}{ArS(=O)_2Ar} + 2HCl$$

$$\underset{(\mathbf{32})}{RSO_2Cl} + ArH \xrightarrow[\text{Fe powder}]{AlCl_3\text{ or}} \underset{(\mathbf{44})}{ArSO_2R} + HCl$$

Scheme 30

In addition, sulfur monochloride (**37**) will react with *p*-chloroarylamines, like (**45**) in the presence of sodium hydroxide to give the sodium thiophenolate (**46**); this is the Herz reaction (Scheme 31) used in the synthesis of *o*-aminothiophenols, which are useful in the manufacture of thioindigo dyes (see Chapter 11, p. 221).

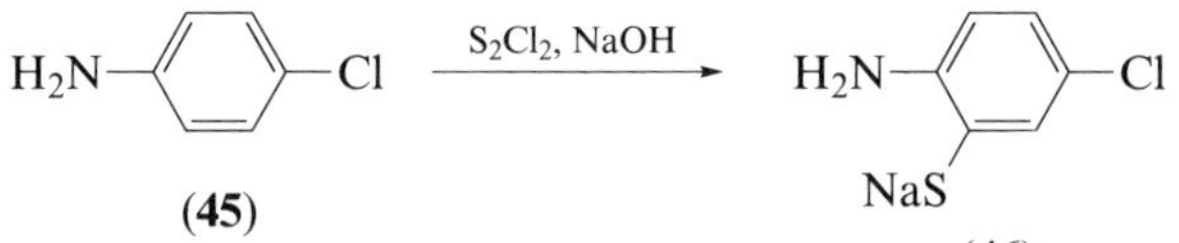

$$H_2N-C_6H_4-Cl \xrightarrow{S_2Cl_2,\ NaOH} H_2N-C_6H_3(SNa)-Cl$$

(45) (46)

Scheme 31

Thionyl chloride (**39**) in the presence of a few drops of DMF (catalyst) is a valuable reagent for the conversion of sulfonic acids (**31**) into the sulfonyl chlorides (**32**) (Scheme 32). This resembles the well-known reaction of thionyl chloride with carboxylic acids for the preparation of acyl chlorides, and has the advantage that both the by-products are gaseous.

$$\underset{(31)}{RSO_3H} + \underset{(39)}{SOCl_2} \xrightarrow{\text{DMF catalyst}} \underset{(32)}{RSO_2Cl} + HCl + SO_2$$

Scheme 32

Sulfuryl chloride (**40**) will chlorosulfonate organic molecules in the presence of a Lewis acid, but often ring chlorination also occurs leading to a mixture of products; however, improved yields of the sulfonyl chloride (**32**) often result from the use of the Grignard reagent derived from an alkyl or benzyl halide (Scheme 33).

$$RH + \underset{(40)}{SO_2Cl_2} \xrightarrow{AlCl_3} \underset{(32)}{RSO_2Cl} + [HCl]$$

$$RX \xrightarrow{Mg} RMgX \xrightarrow{SO_2Cl_2} \underset{(32)}{RSO_2Cl} + MgXCl$$

Scheme 33

Sulfuryl chloride (**40**) under the influence of heat or ultraviolet light chlorinates alkanes and alkylbenzenes to give a mixture of products; thus, with toluene (**47**), the free radical process yields compounds (**48**), (**49**) and (**50**), (Scheme 34). The radical process involves the formation of the intermediate stabilised benzyl radical leading to chlorination of the side chain.

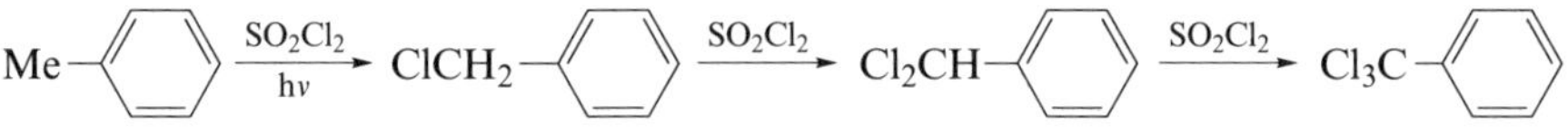

$$\underset{(47)}{Me-C_6H_5} \xrightarrow[h\nu]{SO_2Cl_2} \underset{(48)}{ClCH_2-C_6H_5} \xrightarrow{SO_2Cl_2} \underset{(49)}{Cl_2CH-C_6H_5} \xrightarrow{SO_2Cl_2} \underset{(50)}{Cl_3C-C_6H_5}$$

Scheme 34

This chapter illustrates how elemental sulfur and a number of cheap, readily available inorganic sulfur reagents can be utilised in the preparation of a range of organosulfur compounds. The reagents include hydrogen sulfide, carbon disulfide, phosphorus pentasulfide, sodium sulfide, sodium hydrogen sulfide, sodium sulfite, sodium hydrogen sulfite, sulfur dioxide, sulfur trioxide, sulfuric acid, chlorosulfonic acid, sulfur monochloride, sulfur dichloride, thionyl chloride and sulfuryl chloride. The reagents may be employed to synthesise thiols, sulfides, thioamides, thioketones, dithioacids, carbamates, xanthates, sulfinic acids, sulfonic acids, sulfonamides, some heterocyclic sulfur compounds, sulfoxides and sulfones, among others. In later chapters, further preparations and reactions of these and other classes of organosulfur compounds are described.

References

1. V. A. Usov and M. G. Veronkov, *Modern Principles of the Synthesis of Organosulfur Compounds*, in *The Chemistry of Organosulfur Compounds* (Ed. L. I. Belen'kii), Ellis Horwood, Chichester, 1990, Chap. 1, p. 13.
2. P. Page, *Organosulfur Chemistry-Synthetic Aspects*, Academic Press, London, 1995.
3. *Organic Chemistry of Sulfur* (Ed. S. Oae), Plenum Press, New York, 1977.
4. M. G. Veronkov, N. S. Vyazankin, E. N. Deryagina, A. S. Nakhmanovich and V. A. Usov, in *Reactions of Organic Compounds with Sulfur* (Ed. J. S. Pizey), Consultants Bureau, New York, 1987.
5. H. Cerfontain, *Mechanistic Aspects in Aromatic Sulfonation and Desulfonation*, Part 1, *Electrophilic Aromatic Sulfonation and Related Reactions*, Interscience, New York, 1967.
6. E. E. Gilbert, *Sulfonation and Related Reactions*, Wiley, New York, 1965.
7. P. Bassin, R. J. Cremlyn and F. J. Swinbourne, *Phosphorus, Sulfur and Silicon*, **56**, 254 (1991).

3 STRUCTURE–CHEMICAL RELATIONSHIPS IN ORGANOSULFUR COMPOUNDS

Organosulfur compounds may react either as electrophiles or nucleophiles depending on the nature of the atoms or groups attached to the sulfur atom. This dual functionality is possible because sulfur is capable of accommodating a positive or negative charge; this phenomenon is not shared by oxygen owing to its less expanded atomic structure. With electron-withdrawing moieties bonded to the sulfur atom like oxygen or halogen, the sulfur becomes an electrophilic centre and consequently such organosulfur molecules function as electrophiles. For example, sulfenyl (**1**), sulfinyl (**2**) and sulfonyl (**3**) halides (X = halogen) all undergo substitution reactions with nucleophiles (Nu^-) to yield the products (**4**)–(**6**) (Scheme 1).

Sulfenyl halides (**1**) will also undergo nucleophilic addition with alkenes (Scheme 1) to yield the episulfides (**7**), which can suffer a further nucleophilic addition to give the sulfides (**8**). The nucleophilic substitutions of sulfenyl (**1**) (see Chapter 4, p. 54) and sulfonyl (**3**) chlorides (see Chapter 7, p. 106) probably gen-

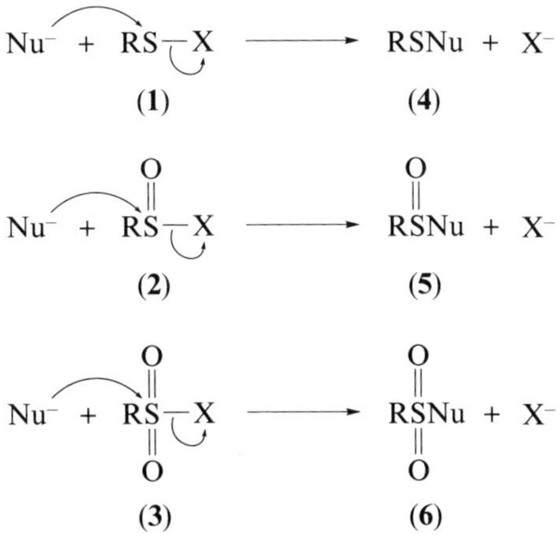

$$Nu^- + RS{-}X \longrightarrow RSNu + X^-$$

(**1**) → (**4**)

$$Nu^- + RS(=O){-}X \longrightarrow RS(=O)Nu + X^-$$

(**2**) → (**5**)

$$Nu^- + RS(=O)_2{-}X \longrightarrow RS(=O)_2Nu + X^-$$

(**3**) → (**6**)

Scheme 1 *(continued)*

Scheme 1 *(continued)*

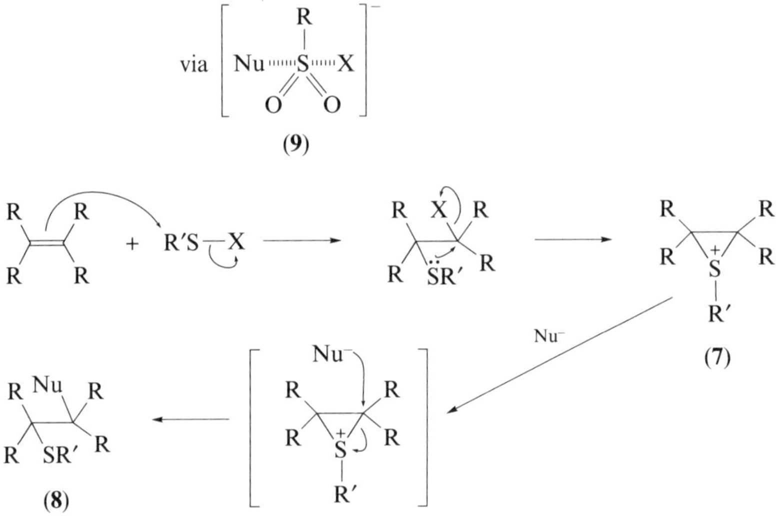

via [Nu····S(R)(=O)(=O)····X]⁻

(9)

R R + R'S—X → R X R / R SR' R → R R / R S⁺ R / R' **(7)**

Nu⁻

R Nu R / R SR' R **(8)** ← [Nu⁻ R R / R S⁺ R / R']

Scheme 1

erally proceed by the S_N2 reaction mechanism, which with the sulfonyl halide (**3**) involves the trigonal bipyramidal transition state (**9**) which subsequently collapses to give the sulfonyl derivative (**6**).[1a] The process essentially resembles the nucleophilic substitution reaction of acyl halides, except that sulfonyl chlorides are less reactive. The relative order of nucleophilicity of some nucleophiles towards the sulfonyl sulfur atom is reported to be as shown in Scheme 2. This relative order is similar to that obtained for substitution at the carbonyl carbon atom.[1a]

$$HO^- > RNH_2 > N_3^- > F^- > OAc^- > Cl^- > H_2O > I^-$$

Scheme 2

The reaction of sulfinyl chlorides (**2**) (X = Cl) with optically active alcohols is an important route to chiral sulfoxides (Andersen's method) (Scheme 3) (see Chapter 5, p. 63). The limitation in the procedure lies in the difficulty of separating the epimeric sulfinates (**10**) and (**11**). After their separation, the subsequent substitution reaction with the organolithium reagent generally occurs with complete inversion of configuration at the sulfur atom, yielding the chiral sulfoxides (**12**) and (**13**) (Scheme 3). Chiral sulfoxides are of considerable significance as intermediates in asymmetric synthesis (see Chapter 5, p. 65).[2]

An important property of the sulfur atom is its ability to act as an electrophilic centre when attached to methylene groups; thus, the latter are activated by an attached sulfur atom. The activation effect is enhanced when the sulfur is in the form of a sulfoxy (S=O) or sulfone (SO_2) group. All these sulfur moieties exert an

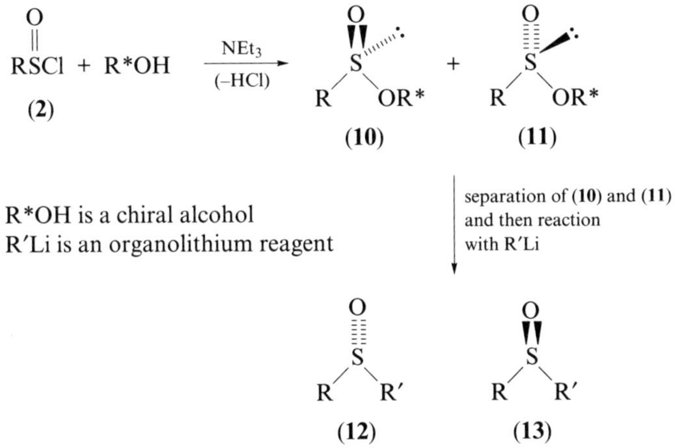

RSCl + R*OH (2) —NEt₃ (–HCl)→ (10) + (11)

R*OH is a chiral alcohol
R′Li is an organolithium reagent

separation of (10) and (11) and then reaction with R′Li

(12) (13)

Scheme 3

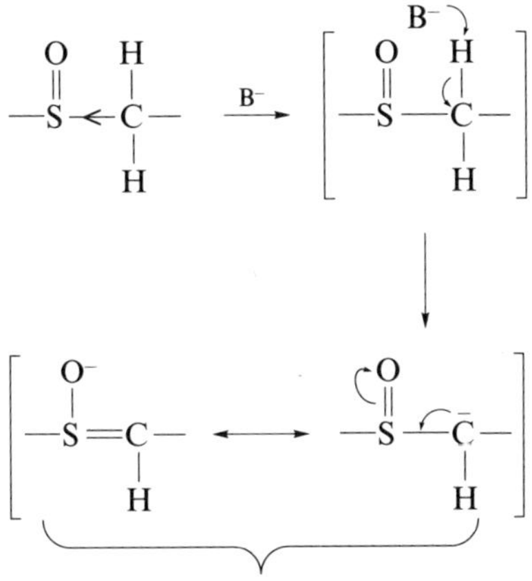

resonance-stabilised carbanion

Figure 1

electron-withdrawing (–I) effect which stabilises the derived carbanions as shown in Figure 1.

Such stabilised carbanion intermediates are of considerable importance in synthetic organic chemistry;[3] thus, they may be readily alkylated or acylated, e.g. in the conversion of the sulfone (**14**) into (**15**) and of the sulfoxide (**16**) into (**17**) (Scheme 4).

In addition 1,3-dithians can be converted into carbonyl compounds; thus, 1,3-dithiacyclohexane (**18**) (obtained by condensation of propane-1,3-dithiol and formaldehyde) can be alkylated with 3-chloroiodopropane (iodine is the more reactive halide and is selectively displaced) to yield eventually cyclobutanone (**19**) by the sequence shown in Scheme 4. This sequence shows the stabilising influence

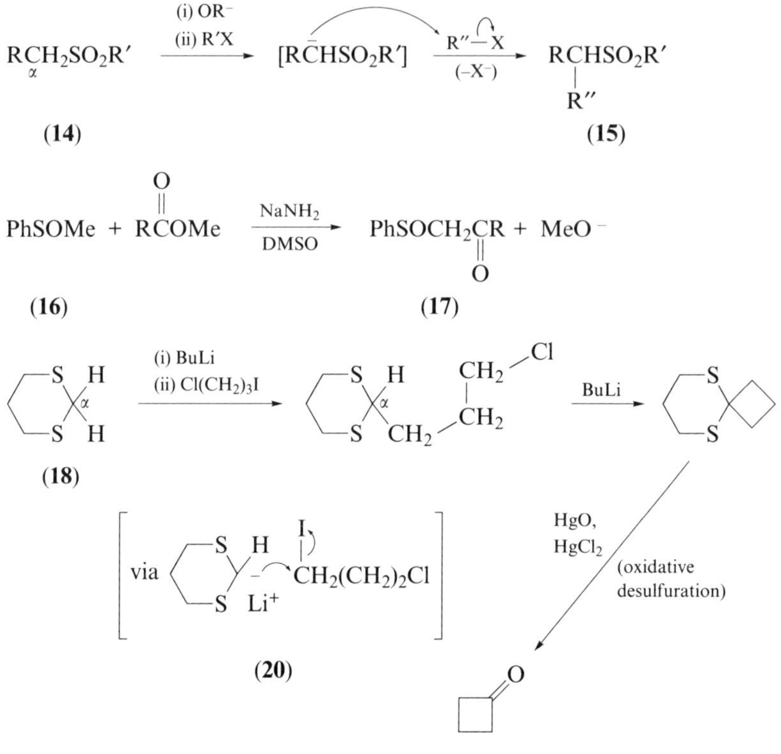

$$\underset{\alpha}{RCH_2SO_2R'}\ \textbf{(14)} \xrightarrow[\text{(ii) } R'X]{\text{(i) } OR^-} [R\bar{C}HSO_2R'] \xrightarrow[(-X^-)]{R''-X} RCH(R'')SO_2R'\ \textbf{(15)}$$

$$PhSOMe\ \textbf{(16)} + RC(=O)OMe \xrightarrow[DMSO]{NaNH_2} PhSOCH_2C(=O)R + MeO^-\ \textbf{(17)}$$

1,3-dithiane (18) $\xrightarrow[\text{(ii) } Cl(CH_2)_3I]{\text{(i) BuLi}}$ 2-(3-chloropropyl)-1,3-dithiane $\xrightarrow{BuLi}$ spiro-cyclobutane dithiane $\xrightarrow[\text{(oxidative desulfuration)}]{HgO,\ HgCl_2}$ cyclobutanone (19)

[via 2-lithio-1,3-dithiane Li^+ + $ICH_2(CH_2)_2Cl$] **(20)**

Scheme 4

of the two adjacent sulfur atoms on the intermediate carbanion (**20**) (see Chapter 6, p. 89). In the final step of these condensations, the sulfur atoms can, if necessary, be removed by reduction (hydrogenolysis) to give an alkane, by elimination to give an alkene, or by hydrolysis to yield a carbonyl compound as shown in Scheme 4 and Scheme 12 (see p. 34). In these reactions, the vacant *d*-orbitals on the sulfur atom assist in delocalising the electrons and hence the negative charge on adjacent atoms. For instance, in the reaction of methyl phenyl sulfide (**21**) and butyllithium, the resultant carbanion is stabilised by resonance structures which utilise the 3*d*-orbitals of sulfur (Scheme 5).

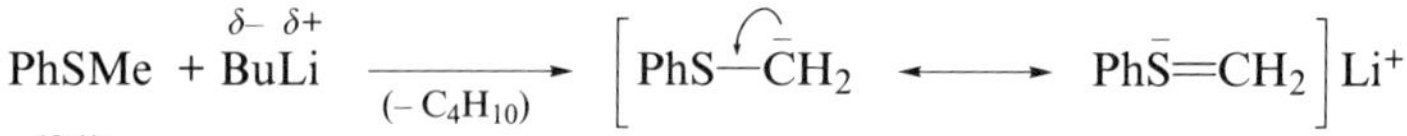

$$\underset{\textbf{(21)}}{PhSMe} + \overset{\delta-\ \delta+}{BuLi} \xrightarrow[(-C_4H_{10})]{} \left[PhS-\bar{C}H_2 \longleftrightarrow Ph\bar{S}{=}CH_2\right]Li^+$$

Scheme 5

The sulfoxy group of sulfoxides and sulfones is even more effective in delocalisation of a negative charge, e.g. in dimethyl sulfoxide (DMSO) (Scheme 6).

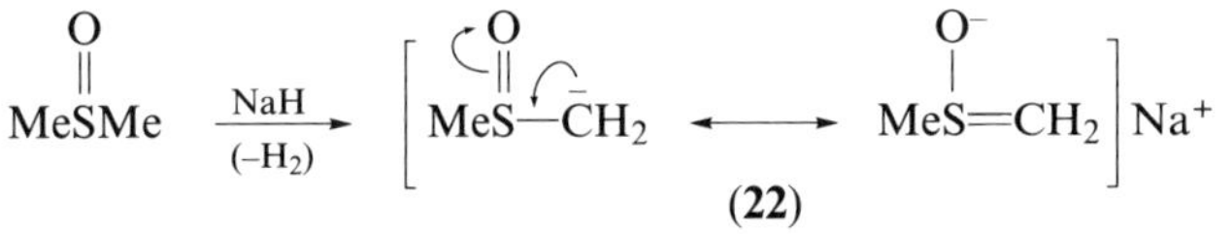

$$\mathrm{MeS(=O)Me} \xrightarrow[(-H_2)]{NaH} \left[\mathrm{MeS(=O)\text{—}\bar{C}H_2} \longleftrightarrow \mathrm{MeS(O^-)\text{=}CH_2} \right] Na^+$$

(22)

Scheme 6

Dimsylsodium (**22**) is an important reagent and it can be used for carbon–carbon bond formation (see Chapter 10, p. 187). DMSO may also be used in the Moffatt oxidation (see Chapter 5, p. 66) and by alkylation it can be converted into the corresponding dimethylsulfoxonium ylide which is valuable in organic synthesis (see Chapter 10, p. 187).[4]

Sulfur compounds may also act as nucleophiles; they are better nucleophiles than their oxygen analogues and generally react faster.[1b] For example, thiols are better nucleophiles than alcohols; this is illustrated by the reaction of 2-thioethanol with ethyl bromide to yield 2-ethylthioethanol (**23**) (Scheme 7).

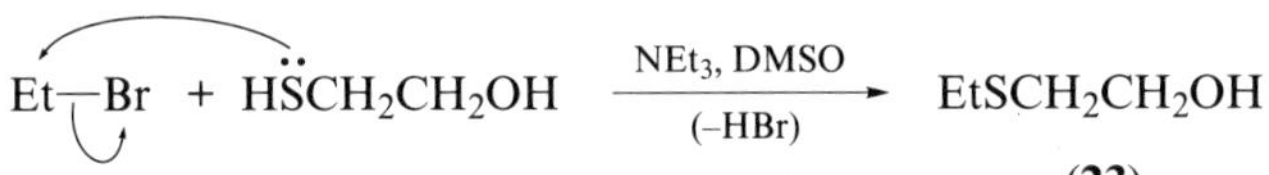

$$\mathrm{Et\text{—}Br} + \mathrm{H\ddot{S}CH_2CH_2OH} \xrightarrow[(-HBr)]{NEt_3,\ DMSO} \mathrm{EtSCH_2CH_2OH}$$

(23)

Scheme 7

The formation of 2-ethylthioethanol (**23**) involves selective nucleophilic attack by the mercapto (SH) moiety while the hydroxy group remains intact, showing that the former is a more powerful nucleophile. Thiols, as their sodio derivatives, will react with alkyl and even with aryl halides in polar aprotic solvents like DMSO and DMF to give good yields of dialkyl and diaryl sulfides (**24**) and (**25**) (Scheme 8).

$$\mathrm{RSNa} + \mathrm{R'X} \xrightarrow{DMSO} \mathrm{RSR'} + \mathrm{NaX}$$

(24)

$$\mathrm{ArSNa} + \mathrm{ArX} \xrightarrow{DMSO} \mathrm{ArSAr} + \mathrm{NaX}$$

(25)

Scheme 8

In these reactions, the sodium thiolates are much better nucleophiles than the free thiols, although the latter will readily condense with the more reactive acyl chlorides under mild conditions, providing a synthetic route to thiolocarboxylic acid esters (**26**) (Scheme 9).

$$\mathrm{RSH} + \mathrm{R'COCl} \longrightarrow \mathrm{R'COSR} + \mathrm{HCl}$$

(26)

Scheme 9

Thiolates will also displace nitrogen from diazonium salts (**27**) (Scheme 10).

$$\underset{\textbf{(27)}}{ArN_2X} \xrightarrow{RS^-} ArSR + N_2$$

Scheme 10

Thiols will also undergo nucleophilic addition reactions with aldehydes and ketones to give thioacetals (**28**) and dithioacetals (**29**); the former are unstable but the latter are stable (Scheme 11). There is a common method for protecting aldehydes and ketones which involves reaction with ethanedithiol (**30**) to give a cyclic dithioketal (**31**) (Scheme 12). The latter compound (**31**) can be disulfurised either by treatment with dilute acid to regenerate the carbonyl compound or with Raney nickel to form the hydrocarbon (**32**). This sequence provides a useful method for the conversion of a carbonyl group into a methylene group (Scheme 12).

$$R_2C{=}O + R'SH \xrightarrow{H^+} \underset{\textbf{(28)}}{R_2C(SR')(OH)} \xrightarrow[H^+]{R'SH} \underset{\textbf{(29)}}{R_2C(SR')_2}$$

Scheme 11

$$R_2C{=}O \xrightarrow[\textbf{(30)}]{HSCH_2CH_2SH,\ H^+} \underset{\textbf{(31)}}{R_2C(SCH_2CH_2S)} \xrightarrow[(-NiS)]{RaNi} \underset{\textbf{(32)}}{R_2CH_2}$$

(31) → R₂C=O: hydrolysis (dil. HCl) (oxidative desulfuration); (31) → (32): (reductive desulfuration)

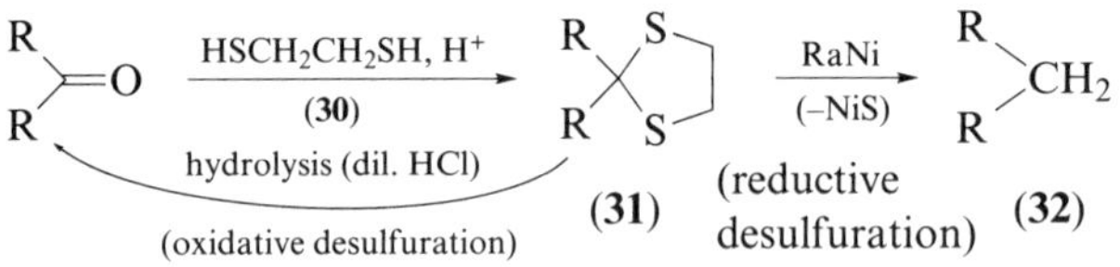

Scheme 12

Many organosulfur compounds can be resolved into optically active forms (enantiomers) owing to the presence of a chiral (asymmetric) sulfur atom;[5] important examples include sulfoxides and sulfonium salts. Chiral sulfoxides containing amino or carboxylic acid groups have been resolved by formation of the diastereoisomeric salts with *d*-camphor-10-sulfonic acid or *d*-brucine. The salts can then be separated by fractional crystallisation and the free optically isomeric sulfoxides liberated by acid hydrolysis. However, a more convenient synthetic procedure for the preparation of chiral sulfoxides of high optical purity is Andersen's method (see p. 30).

Sulfoxides possess a non-planar pyramidal configuration; consequently, chiral sulfoxides can exist as optical enantiomers (see Chapter 5, p. 63). This is illustrated by the structures (**33a**) and (**33b**) showing the optical isomers of ethyl methyl sulfoxide (Figure 2).

With cyclic sulfoxides, geometric isomerism is possible; for instance, 4-*t*-butylcyclohexanesulfoxide exists as the *cis*-isomer (**34a**) and *trans*-isomer (**34b**) (Figure 3).

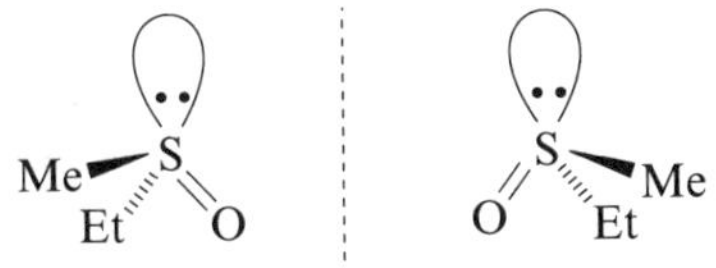

(**33a**) (**33b**)

Figure 2

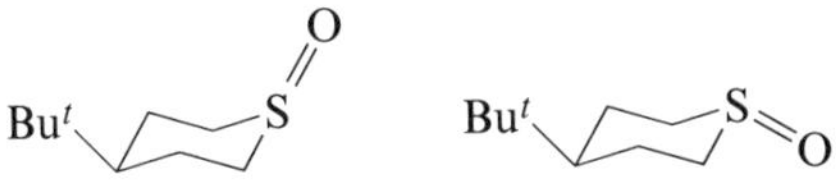

(**34a**) (e, a) (**34b**) (e, e)

Figure 3

The nucleophilic character of dialkyl sulfides is illustrated by their nucleophilic addition reaction with alkyl halides to form the corresponding sulfonium salts (**35**) (Scheme 13). Asymmetric sulfonium salts (**36**) have a tetrahedral configuration; therefore, like the analogous chiral saturated carbon compounds, they can be resolved into optical enantiomers (see Chapter 6, p. 81). They are, however, generally less optically stable than sulfoxides, but in sulfonium salts the unshared electron pair can hold its configuration at ordinary temperatures, unlike nitrogen in quaternary ammonium salts, enabling their resolution to be achieved.

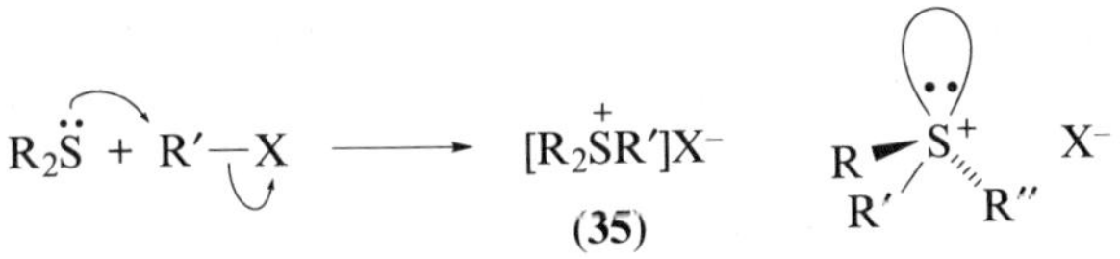

$R_2\ddot{S}$ + R′—X ⟶ $[R_2\overset{+}{S}R']X^-$ (**35**)

(**36**, R ≠ R′ ≠ R″)

Scheme 13

Sulfuranes

Sulfur, unlike oxygen, has the capacity to expand its valence shell beyond the normal octet of electrons to form hypervalent compounds such as sulfur tetrafluoride (SF_4) with 10 electrons in the outermost shell and sulfur hexafluoride (SF_6) containing 12 electrons in the valence shell.[6a,7] The chemistry of hypervalent sulfur started in 1873 with the discovery of the unstable compound sulfur tetrachloride (SCl_4). The existence of hypervalent sulfur compounds is an important feature of the chemistry of sulfur and the precise nature of the bonding in these molecules has remained a puzzling problem.

Pauling[8] (1939) considered that the bonding in hypervalent sulfur compounds involved *spd* hybridisation; however, if the 3*d*-orbitals are energetically too far above the energy levels of the 3*s*-orbital and 3*p*-orbitals, effective mixing to form

$d\pi$-type bonds will not occur. Coulson (1969) argued that the diffuse 3*d*-orbitals could be more effectively utilised for hybridisation when the electron affinity of the sulfur atom was increased by bonding to a highly electronegative element, e.g. fluorine or oxygen, since this would contract the orbitals and lower orbital energies.[6a,7] Polycoordinate bonding in hypervalent sulfur compounds like sulfur tetrafluoride and sulfur hexafluoride was considered to consist of either trigonal bipyramidal sp^3d or octahedral sp^3d^2 hybrids, respectively (Figure 4).

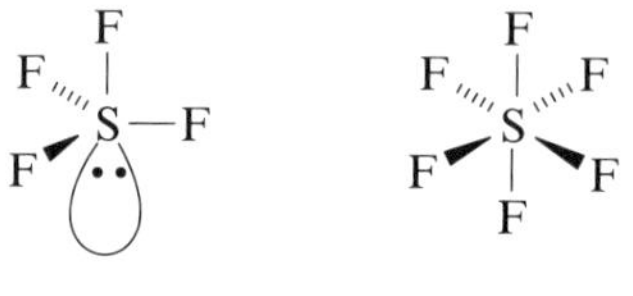

Figure 4

A large body of evidence has been collected to explain many chemical phenomena of organosulfur compounds in terms of the participation of the 3*d*-orbitals in the bonding; however, in the late 1960s the involvement of the 3*d*-orbitals was questioned on the basis of molecular orbital (MO) calculations, which suggested that it was relatively insignificant. More recently, the concept of a three-centre four-electron σ-bond, namely a hypervalent bond, was developed by Musher (1969) and Hoffmann (1972).[6a,6b] Currently, this view of the bonding in hypervalent sulfur molecules is favoured rather than the sp^3d-hybridised and sp^3d^2-hybridised models shown in Figure 4. The isolation of stable sulfuranes by Kapovits (1971) and Martin (1973)[6b] definitely established the structure of hypervalent sulfur compounds. An example of a stable sulfurane is afforded by the diphenyl diacyl sulfurane (**37**) (Figure 5); accordingly, the structure of sulfur tetrafluoride (**38**) is as depicted in Figure 5, containing two apical (a) and two equatorial (e) fluorine atoms with the lone electron pair occupying an equatorial position.

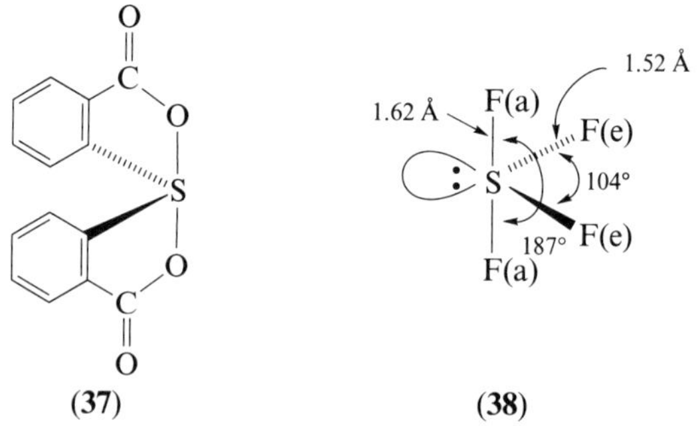

(**37**) (**38**)

Figure 5

The formation of the hypervalent bond in (**38**) may be visualised as the hypothetical combination of the sp^2-hybridised sulfur difluoride with the *p*-orbitals of two fluorine atoms. In hypervalent sulfur compounds like (**37**) and (**38**), the apical bonds formed by the *p*-orbitals are longer and more polar than the equatorial bonds. The special stability of many polycoordinated sulfur compounds is still concluded to be associated with the added orbital interaction with the energetically accessible 3*d*-orbitals. The discovery of sulfuranes has provided valuable insight into many nucleophilic substitution reactions on polycoordinate sulfur atoms and into ligand-coupling reactions which occur via σ-sulfurane intermediates.

A σ-sulfurane is a compound containing a sulfur atom singly bonded to four substituents by analogy with the hypothetical sulfurane molecule H_4S. In contrast, a sulfur ylide (see Chapter 10, p.182) in its fully covalent resonance structure (**39**) may be termed a π-sulfurane (Figure 6).

(**39**)

Figure 6

σ-Sulfuranes in which the sulfur atom is bonded to fluorine or oxygen atoms in addition to carbon have been discovered to be reasonably stable, e.g. compound (**37**). The mechanisms of nucleophilic substitutions at sulfinyl and sulfonyl sulfur (see p. 29) have been extensively studied to determine whether the reactions are two-step or one-step concerted processes (S_N2). If the reaction mechanisms are two-stage processes, they will involve sulfurane intermediates, and in support of this concept analogous compounds, e.g. (**40**), have been prepared (Figure 7).

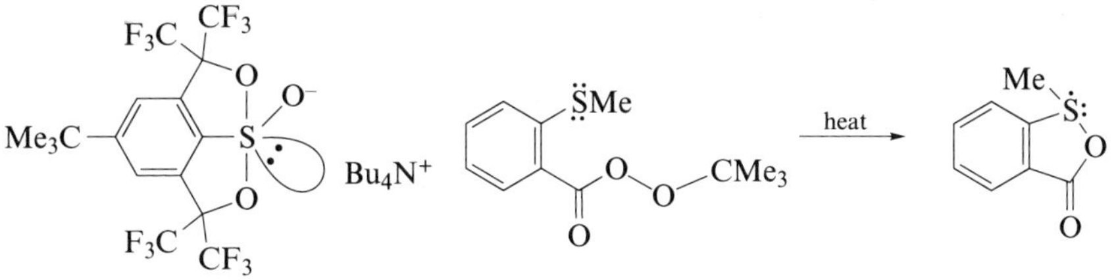

Bu_4N^+ heat

(**40**) (**41**) (**42**)

Figure 7

The substantial acceleration of radical ion formation observed in the homolysis of peroxide bonds in *ortho*-substituted sulfenyl perbenzoates, e.g. (**41**), is in agreement with the postulated formation of bridged radicals (**42**) (Figure 7).

Sulfuranes with hypervalent sulfur atoms possess expanded valence shells, and consequently the molecules are relatively unstable. The central sulfur atom can, however, attain the normal stable octet of electrons by extruding a ligand or a pair of ligands in an elimination process. The former results in ligand coupling

(Figure 8 and Schemes 14a and 14b), while the latter causes self-decomposition (Scheme 14c).

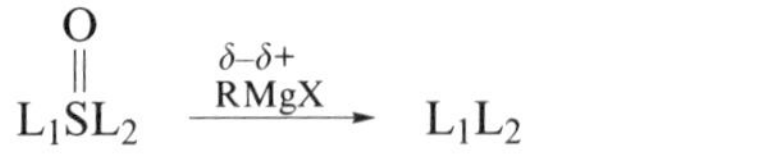

$$L_1S(=O)L_2 \xrightarrow[\delta-\delta+]{RMgX} L_1L_2$$

Where L_1 and L_2 are aromatic ligands

Figure 8

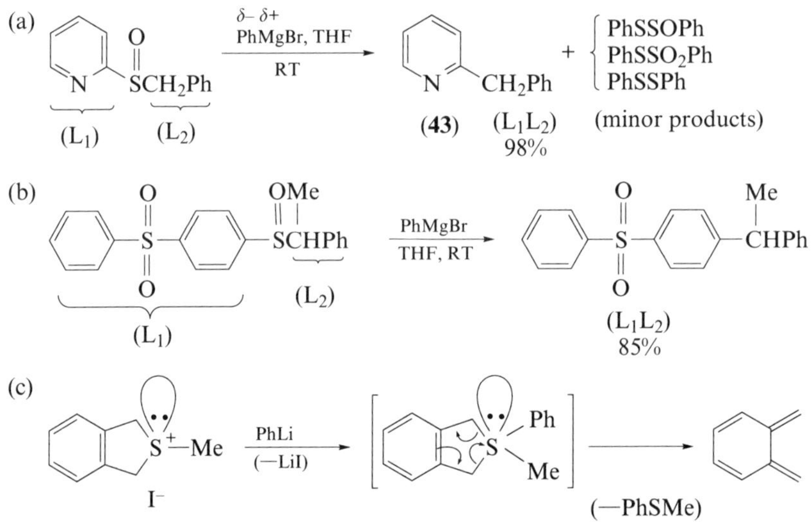

Scheme 14

Many ligand-coupling reactions occurring via σ-sulfurane intermediates involve treatment of aryl or heteroaryl sulfoxides with Grignard reagents (Scheme 14). The intramolecular nature of the ligand coupling reaction is demonstrated by the absence of cross-over products; thus, the formation of 2-benzylpyridine (**43**) (Scheme 14a) may proceed as shown in Scheme 15.

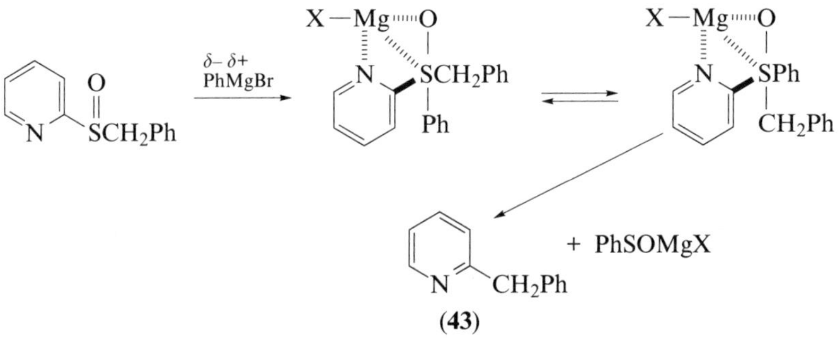

Scheme 15

References

1. J. March, *Advanced Organic Chemistry*, 4th edn, Wiley, New York, 1992: (a) p. 496; (b) p. 406.
2. R. K. Mackie, D. M. Smith and R. A. Aitken, *Guidebook to Organic Synthesis*, 2nd edn, Longman, Harlow, 1992.
3. A. Krief, in *Comprehensive Organic Synthesis* (Eds B. M. Trost and I. Fleming), Vol. 3, Pergamon Press, Oxford, 1991, p. 85.
4. B. M. Trost and L. S. Melvin, *Sulfur Ylides*, Academic Press, London, 1975.
5. A. Nudelman, *The Chemistry of Optically Active Sulfur Compounds*, Gordon and Breach, New York, 1984.
6. *Organic Sulfur Chemistry* (Eds F. Bernardi, I. G. M. Csizmadia and A. Mangini), Elsevier, Amsterdam, 1985: (a) S. Oae, Chap. 1; (b) R. A. Hayes and J. C. Martin, Chap. 8, p. 408.
7. E. Block, *Reactions of Organosulfur Compounds*, Academic Press, New York, 1978, Chap. 1.
8. L. Pauling, *The Nature of the Chemical Bond*, 1st Edn, Cornell University Press, Ithaca, NY, 1939.

4 THIOLS, SULFIDES AND SULFENIC ACIDS

Thiols, or thioalcohols (RSH),[1,2a] are sulfur analogues of alcohols and phenols in which the oxygen atom has been replaced by sulfur; they are derivatives of hydrogen sulfide in the same way that alcohols may be regarded as being derived from water. The volatile thiols have very unpleasant smells rather like the smell of hydrogen sulfide. In comparison with oxygen, sulfur has a larger atomic size, more diffuse electronic orbitals and appreciably lower electronegativity; consequently, sulfur–hydrogen bonds (bond energy $\simeq 80$ kcal mol^{-1}; 1 cal = 4.186 J) are weaker than oxygen–hydrogen bonds (100 kcal mol^{-1}). Thiols are therefore more acidic than the corresponding alcohols and form much weaker hydrogen bonds; this is shown by their relatively lower boiling points and aqueous solubility. For instance, methanethiol (MeSH) is a gas (cf. methanol, MeOH, a liquid, b.p. 65 °C) and ethanethiol (EtSH) is a liquid which boils at 37 °C (cf. ethanol, EtOH, b.p. 78 °C). The lack of hydrogen bonding in thiols is also indicated by the infrared spectra; the S—H bond is associated with a weak stretching absorption band (νSH 2600–2550 cm^{-1}) compared with νOH of 3350 cm^{-1} for alcohols. The S—H absorption band does not alter significantly with changes in the concentration, solvent or physical state of the substrate, indicative of little hydrogen bonding.

Several thiols occur naturally; for example, skunk secretion contains 3-methyl-1-butanethiol and cut onions evolve 1-propanethiol, and the thiol group of the natural amino acid cysteine plays a vital role in the biochemistry of proteins and enzymes (see Introduction, p. 2). Primary and secondary thiols may be prepared from alkyl halides (RX) by reaction with excess sodium thiolate (S_N2 nucleophilic substitution by HS^-) or via the Grignard reagent and reaction with sulfur. Tertiary thiols can be obtained in good yields by addition of hydrogen sulfide to a suitable alkene. Thiols can also be prepared by reduction of sulfonyl chlorides (Scheme 1).[1a,2a]

A problem with the first route is that even with excess of the thiolate, it is difficult to avoid some further reaction leading to formation of the sulfide (R_2S). The secondary reaction can be avoided by starting with thiourea (**1a**) and obtaining the thiol via the intermediate *S*-alkylisothiuronium salt (**2**) by subsequent hydrolysis (Scheme 2). The formation of the isothiuronium salt (**2**) depends on the relative instability of the thione (C=S) bond in thiourea (**1a**), which therefore exists mainly as the enethiol form isothiourea (**1b**) which then reacts with the halide. *S*-Benzylisothiuronium salts are used in organic qualitative analysis for

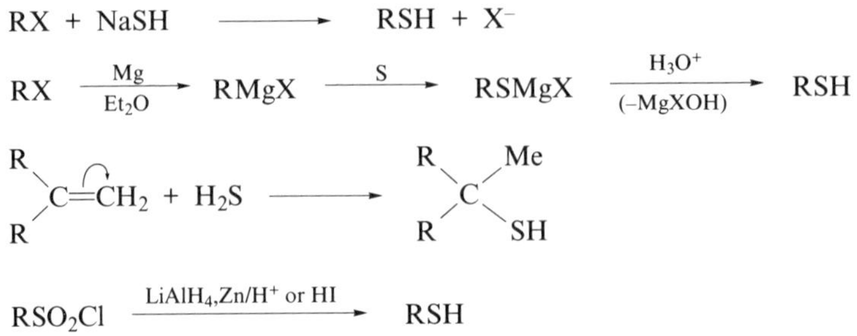

$$RX + NaSH \longrightarrow RSH + X^-$$

$$RX \xrightarrow[Et_2O]{Mg} RMgX \xrightarrow{S} RSMgX \xrightarrow[(-MgXOH)]{H_3O^+} RSH$$

$$R_2C{=}CH_2 + H_2S \longrightarrow R_2C(Me)SH$$

$$RSO_2Cl \xrightarrow{LiAlH_4, Zn/H^+ \text{ or } HI} RSH$$

Scheme 1

the characterisation of carboxylic acids with which they form nicely crystalline salts (**3**) by anion exchange (Scheme 2).

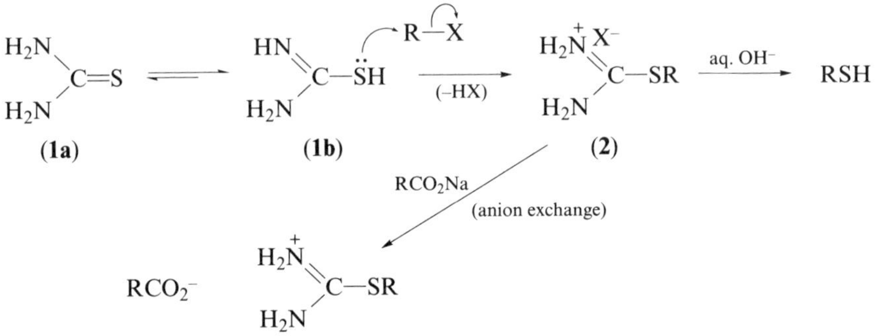

$$(H_2N)_2C{=}S \;(\mathbf{1a}) \rightleftharpoons HN{=}C(NH_2){-}\ddot{S}H \;(\mathbf{1b}) \xrightarrow[(-HX)]{R{-}X} H_2\overset{+}{N}{=}C(NH_2){-}SR\; X^- \;(\mathbf{2}) \xrightarrow{aq.\ OH^-} RSH$$

$$(\mathbf{2}) \xrightarrow[\text{(anion exchange)}]{RCO_2Na} RCO_2^-\; H_2\overset{+}{N}{=}C(NH_2){-}SR \;(\mathbf{3})$$

Scheme 2

Thiols, unlike alcohols, form insoluble salts (mercaptides) (**4**) and (**5**) by reaction with heavy metals like mercury and lead (Scheme 3); this is the origin of the former name 'mercaptan' for thiols, which comes from the Latin *mercurium captans*, meaning mercury seizing. In modern nomenclature the name 'thiol' is preferred to 'mercaptan', although the prefix 'mercapto' is still allowed for the unsubstituted SH radical.

$$2\,RSH + HgO \longrightarrow \underset{(\mathbf{4})}{Hg(SR)_2} + H_2O$$

$$2\,RSH + Pb(OCOMe)_2 \longrightarrow \underset{(\mathbf{5})}{Pb(SR)_2} + 2\,MeCO_2H$$

Scheme 3

The reaction accounts for the high toxicity of lead and mercury to living organisms because they react with vital cellular thiol enzymes, thereby poisoning them.

Thiols will undergo nucleophilic addition to aldehydes and ketones, whereas alcohols only react with aldehydes (Scheme 4). Thiols also react with acid chlorides to yield esters.

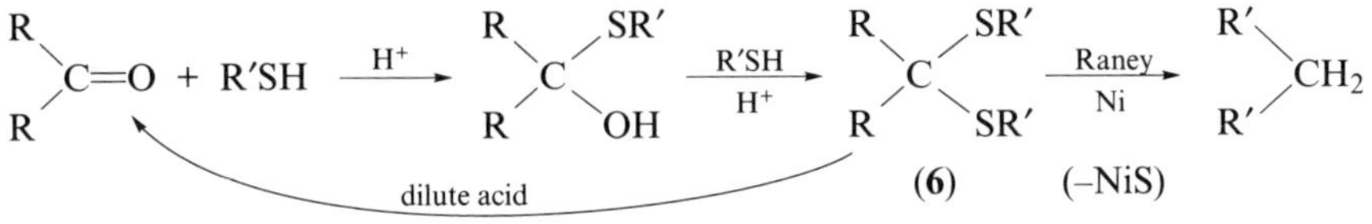

$$R_2C{=}O + R'SH \xrightarrow{H^+} R_2C(SR')(OH) \xrightarrow[H^+]{R'SH} R_2C(SR')_2\ \mathbf{(6)} \xrightarrow[Ni]{Raney} R'_2CH_2\ (-NiS)$$

(6) $\xrightarrow{\text{dilute acid}}$ $R_2C{=}O$

Scheme 4

The thioacetals (R = H) and thioketals (**6**) can be used for the protection of aldehydes and ketones since on treatment with dilute acid they are converted back to the original carbonyl substrates.[1b] They may also be applied in the conversion of a carbonyl to a methylene group by reaction with Raney nickel, an important general procedure of desulfurisation (Scheme 4).[3]

The thiol sulfur atom is a powerful nucleophile and can participate in nucleophilic substitution at a saturated carbon atom when this is attached to a good leaving group L (Scheme 5). An illustrative example is provided by the condensation of a thiol with an acyl halide to yield the corresponding thiolocarboxylic acid ester (Scheme 5).

$$R\ddot{S}H + R'{-}L \longrightarrow RSR' + HL$$

$$R\ddot{S}H + R'CO{-}X \longrightarrow RSCOR' + HX$$

thiolocarboxylic acid ester

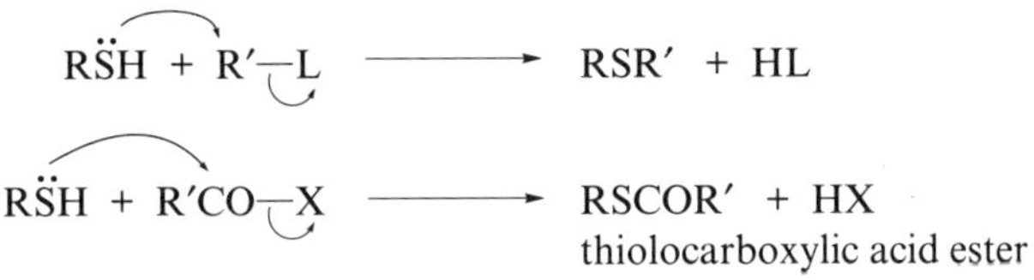

Scheme 5

The oxidation of thiols follows a completely different course as compared with the oxidation of alcohols, because the capacity of the sulfur atom to form hypervalent compounds allows it to become the site of oxidation. Thiols are readily oxidised to disulfides by mild oxidants such as atmospheric oxygen, halogens or iron(III) salts (Scheme 6). This type of reaction is unique to thiols and is not undergone by alcohols, it is a consequence of the lower bond strength of the S—H as compared with the O—H bond, so that thiols are oxidised at the weaker S—H bonds, whereas alcohols are preferentially oxidised at the weaker C—H bonds (Scheme 7). The mechanism of oxidation of thiols may be either radical or polar or both (Scheme 6). The polar mechanism probably involves transient sulfenic acid intermediates like (**7**) and (**8**). In contrast, thiols react with more powerful oxidants, like potassium permanganate, concentrated nitric acid or hydrogen peroxide, to yield the corresponding sulfonic acids (**10**). This oxidation probably proceeds via the relatively unstable sulfenic (**7**) and sulfinic acids (**9**), which are too susceptible to further oxidation to be isolated (Scheme 8).

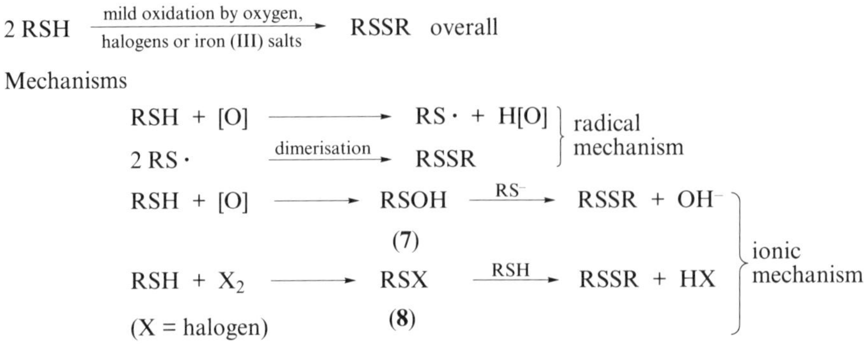

$$2\,RSH \xrightarrow[\text{halogens or iron (III) salts}]{\text{mild oxidation by oxygen,}} RSSR \quad \text{overall}$$

Mechanisms

$$\left.\begin{array}{l} RSH + [O] \longrightarrow RS\cdot + H[O] \\ 2\,RS\cdot \xrightarrow{\text{dimerisation}} RSSR \end{array}\right\} \text{radical mechanism}$$

$$\left.\begin{array}{l} RSH + [O] \longrightarrow \underset{(\mathbf{7})}{RSOH} \xrightarrow{RS^-} RSSR + OH^- \\ RSH + X_2 \longrightarrow \underset{(\mathbf{8})}{RSX} \xrightarrow{RSH} RSSR + HX \\ (X = \text{halogen}) \end{array}\right\} \text{ionic mechanism}$$

Scheme 6

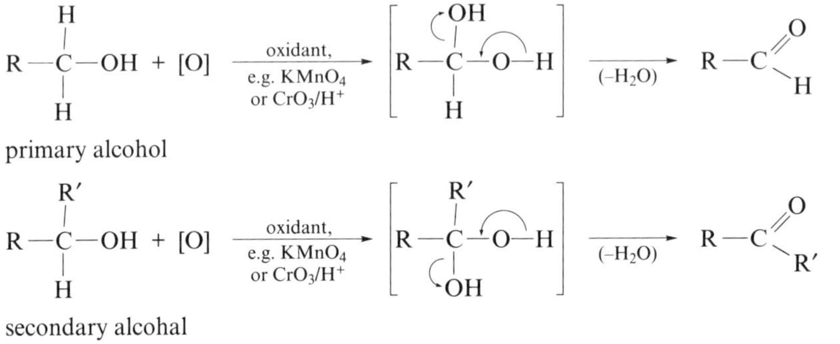

primary alcohol: $R{-}CH_2{-}OH + [O] \xrightarrow[\text{e.g. } KMnO_4 \text{ or } CrO_3/H^+]{\text{oxidant,}} [R{-}CH(OH){-}O{-}H] \xrightarrow{(-H_2O)} R{-}C(=O){-}H$

secondary alcohal: $R{-}CHR'{-}OH + [O] \xrightarrow[\text{e.g. } KMnO_4 \text{ or } CrO_3/H^+]{\text{oxidant,}} [R{-}CR'(OH){-}O{-}H] \xrightarrow{(-H_2O)} R{-}C(=O){-}R'$

Scheme 7

$$RSH \xrightarrow{[O]} \underset{(7)}{RSOH} \xrightarrow{[O]} \underset{(\mathbf{9})}{RS(=O)OH} \xrightarrow{[O]} \underset{(\mathbf{10})}{RS(=O)_2OH}$$

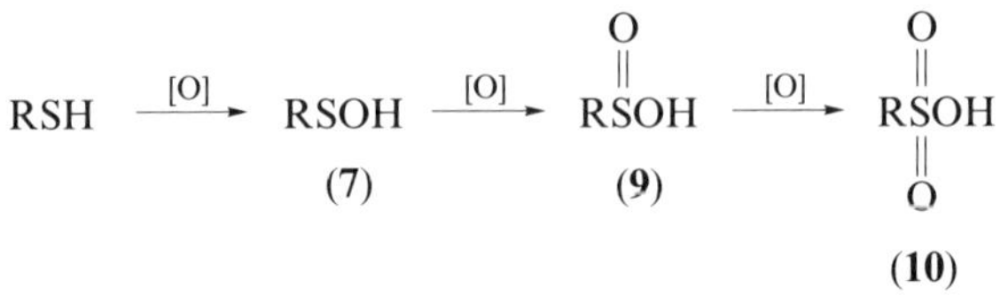

Scheme 8

Sulfides, or thioethers, are sulfur analogues of ethers, and like ethers they can be either symmetrical (R_2S) or unsymmetrical (RSR′, where R and R′ are different). Sulfides can be prepared from alkyl halides by a Williamson-type synthesis with sodium hydrogen sulfide, sodium thiolate or sodium sulfide; from alkyl or aryl halides via the Grignard reagent (**11**); from alkenes by radical-catalysed addition of thiols; or by reduction of sulfoxides (Scheme 9).[2b]

The addition of thiols to alkenes is a free radical reaction that is catalysed by formation of phenyl radicals from the dibenzoyl peroxide catalyst (**12**), and in which the R′ S· radical can continue the radical chain reactions as shown in Scheme 10. In this reaction, the initial cleavage of dibenzoyl peroxide (**12**) is facilitated by the relatively weak peroxide (O—O) bond; the subsequently formed

$$RX + R'SNa \longrightarrow RSR' + NaX$$

$$2\,RX + Na_2S \longrightarrow R_2S + 2\,NaX$$

$$RX \xrightarrow{Mg} \underset{(\mathbf{11})}{RMgX} \xrightarrow{S} RSMgX \xrightarrow{R'X} RSR' + MgX_2$$

$$RCH{=}CH_2 + \overset{\delta-\ \delta+}{R'SH} \xrightarrow[(\mathbf{12})]{PhC(O)OOC(O)Ph} \underset{(\mathbf{13})}{RSCH_2CH_2SR'}$$

(anti-Markownikoff addition)

$$RS(O)R' \xrightarrow[\text{(reduction)}]{LiAlH_4,\ Pd/H_2 \text{ or } Ph_3P} RSR'$$

Scheme 9

$$\underset{(\mathbf{12})}{PhC(O)OOC(O)Ph} \xrightarrow{h\nu} 2PhC(O)\cdot \longrightarrow 2Ph\cdot + 2\,CO_2$$

$$Ph\cdot + R'SH \longrightarrow C_6H_6 + R'S\cdot$$

$$RCH{=}CH_2 + R'S\cdot \longrightarrow R\dot{C}HCH_2SR' \xrightarrow{R'SH} \underset{(\mathbf{13})}{RCH_2CH_2SR'} + R'S\cdot$$

Scheme 10

alkylthio (R′S·) free radical is an electrophilic species which preferentially attacks the most electron-rich carbon atom leading to the formation of the thiol (**13**).

Cyclic sulfides (episulfides) are known and may be prepared by reaction of an epoxide (**14**) with potassium thiocyanate to give the episulfide (**15**), or by reaction of sodium sulfide with 1,4- or 1,5-dihalides; for example, (**16**) yields the episulfide (**17**) (Scheme 11). The mechanism of the first reaction in Scheme 11 involves initial nucleophilic addition of the thiocyanate anion to the epoxide (**14**) followed by nucleophilic substitution to yield the episulfide (**15**) (Scheme 12). The second reaction is an intermolecular followed by an intramolecular nucleophilic substi-

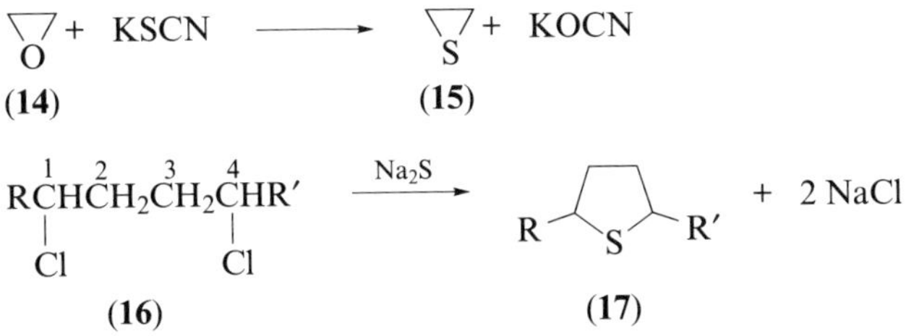

$$\underset{(\mathbf{14})}{\text{epoxide}} + KSCN \longrightarrow \underset{(\mathbf{15})}{\text{episulfide}} + KOCN$$

$$\underset{(\mathbf{16})}{\overset{1}{R}\overset{2}{C}H(Cl)\overset{3}{C}H_2\overset{4}{C}H_2CH(Cl)R'} \xrightarrow{Na_2S} \underset{(\mathbf{17})}{\text{2-R-5-R'-thiolane}} + 2\,NaCl$$

Scheme 11

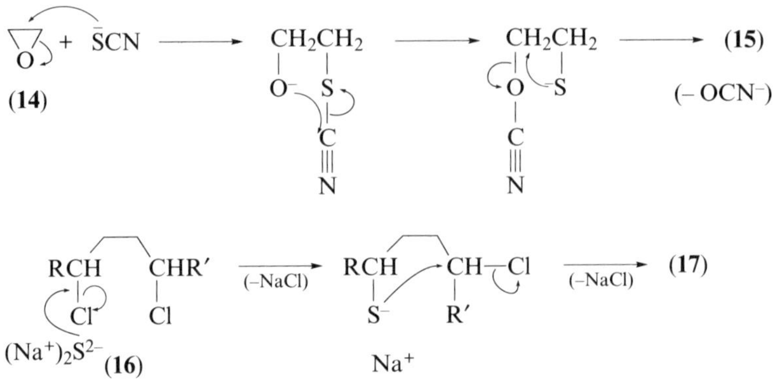

$$\text{(14)} + \bar{S}CN \longrightarrow \underset{O^- \quad S-C\equiv N}{CH_2CH_2} \longrightarrow \underset{O-C\equiv N \quad S}{CH_2CH_2} \longrightarrow \textbf{(15)} \quad (-OCN^-)$$

$$\underset{Cl}{RCH}\text{—}\underset{Cl}{CHR'} \;(16),\; (Na^+)_2S^{2-} \xrightarrow{(-NaCl)} \underset{S^-}{RCH}\text{—}\underset{R'}{CH}\text{—}Cl \;\; Na^+ \xrightarrow{(-NaCl)} \textbf{(17)}$$

Scheme 12

tution by the sulfide anion on the 1,4-dihalide (**16**) to give the episulfide (**17**), as shown in Scheme 12.

Another synthetic route to episulfides is by reaction of a diazoalkane (**18**) with either sulfur or a thioketone (**19**) (Scheme 13).

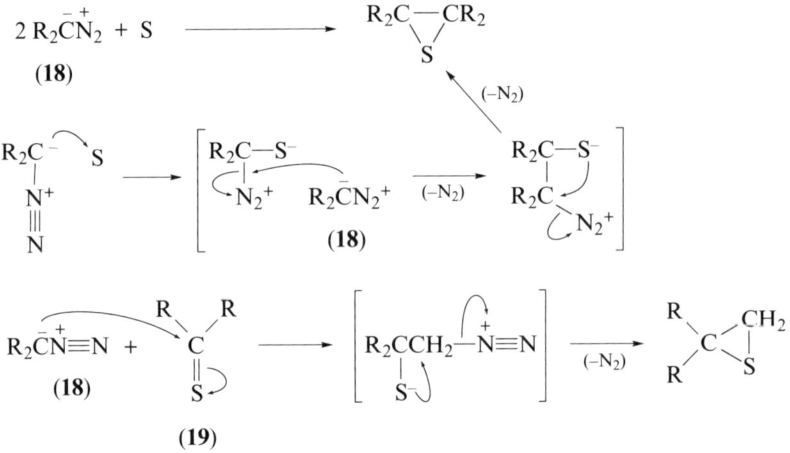

$$2\,R_2\overset{-}{C}\overset{+}{N}_2 \;(18) + S \longrightarrow R_2C\text{—}CR_2 \text{ (bridged by S)}$$

$$R_2C^-\text{—}N^+\equiv N + S \longrightarrow \left[R_2C(\text{—}S^-)\text{—}N_2^+ \quad R_2\overset{-}{C}N_2^+ \; (18) \right] \xrightarrow{(-N_2)} \left[R_2C(\text{—}S^-)\text{—}R_2C\text{—}N_2^+ \right] \xrightarrow{(-N_2)} R_2C\text{—}CR_2 \text{ (bridged by S)}$$

$$R_2\overset{-}{C}\overset{+}{N}\equiv N \;(18) + R_2C{=}S \;(19) \longrightarrow \left[R_2C(S^-)CH_2\text{—}\overset{+}{N}\equiv N \right] \xrightarrow{(-N_2)} R_2C\text{—}CH_2 \text{ (bridged by S)}$$

Scheme 13

Cyclic sulfides like (**15**) have similar reactivity to the analogous oxides (**14**); thus, they are attacked by nucleophilic reagents, e.g. amines, water and alkoxides, involving scission of the episulfide ring (Scheme 14).

Episulfides (**15**) can also be converted to alkenes by treatment with phosphorus(III) compounds, like triphenylphosphine (Scheme 15), the reaction occurring via a four-centre transition state similar to that involved in the Wittig reaction.

Sulfides contain a nucleophilic sulfur atom and consequently undergo important reactions involving the lone electron pairs on sulfur; thus, they react with

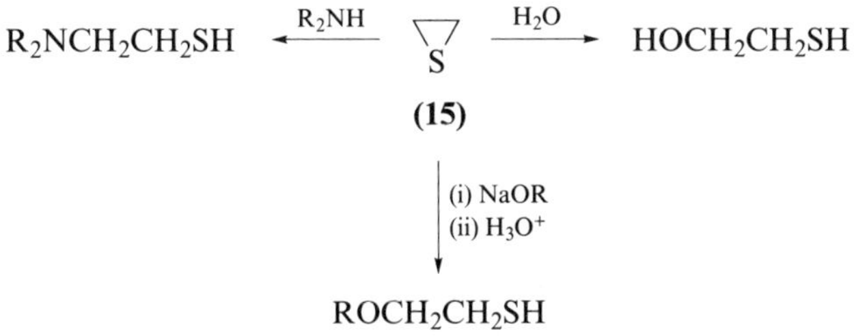

$$R_2NCH_2CH_2SH \xleftarrow{R_2NH} \underset{\textbf{(15)}}{\text{thiirane}} \xrightarrow{H_2O} HOCH_2CH_2SH$$

(i) NaOR
(ii) H_3O^+

$ROCH_2CH_2SH$

Scheme 14

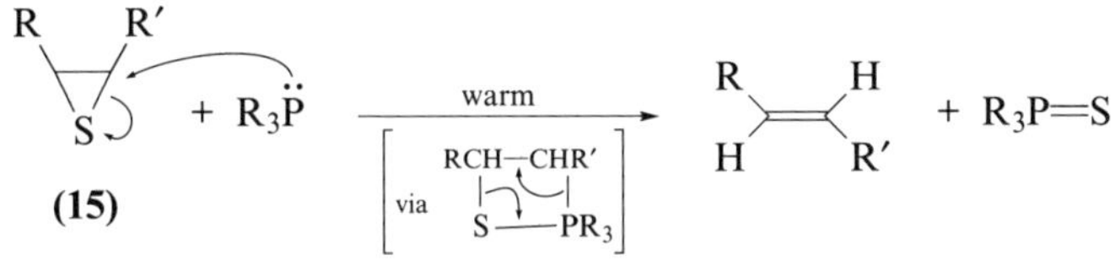

$$\text{(15)} + R_3\ddot{P} \xrightarrow[\text{via } [RCH-CHR' / S-PR_3]]{\text{warm}} RCH{=}CHR' + R_3P{=}S$$

Scheme 15

$$RR'\ddot{S} + R''-X \longrightarrow RR'R''S^+X^- \quad \textbf{(20)}$$

$$RR'\ddot{S} + X-X \longrightarrow RR'S^+X\,X^- \quad \textbf{(21)}$$

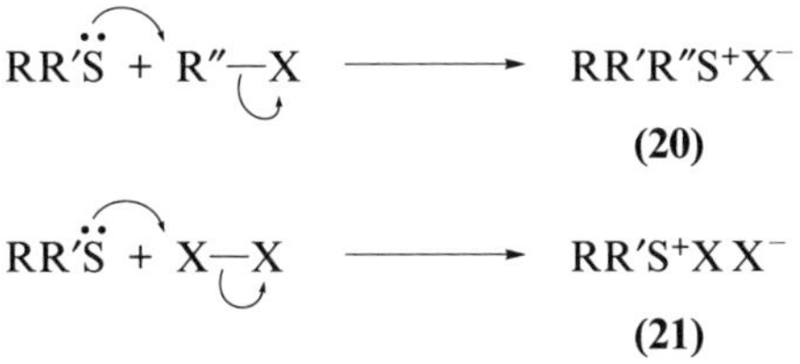

Scheme 16

alkyl halides to form sulfonium salts (**20**) and with halogens to yield dihalides (**21**) (Scheme 16).

Sulfonium salts (**20**) have a non-planar tetrahedral configuration, and consequently the structure is chiral when all the attached groups, namely R, R′ and R″, are different. Such sulfonium salts can generally be resolved into quite stable optical isomers (see Chapter 6, p. 83). Sulfonium salts are analogous to quaternary ammonium salts and the hydroxides are strong bases which behave similarly on heating (cf. the Hofmann elimination of quaternary ammonium hydroxides).

Triethylsulfonium hydroxide (**22**) is obtained by treatment of the corresponding iodide with moist silver oxide (Scheme 17). The hydroxide (**22**) by heating

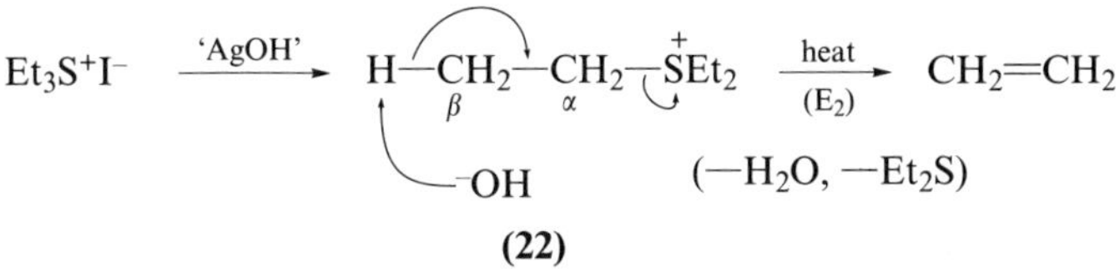

$$Et_3S^+I^- \xrightarrow{\text{'AgOH'}} H-\underset{\beta}{CH_2}-\underset{\alpha}{CH_2}-\overset{+}{S}Et_2 \;\; {}^-OH \;\; \textbf{(22)} \xrightarrow[(E_2)]{\text{heat}} CH_2{=}CH_2 \quad (-H_2O, -Et_2S)$$

Scheme 17

affords ethene by a bimolecular (E2) elimination reaction involving the β-hydrogen atom, as depicted in Scheme 17.

Sulfides are easily oxidised at the sulfur atom by sources of electrophilic oxygen; in the oxidation process, sulfur accepts electrons in its *d*-orbitals, leading to hypervalent sulfur compounds. The first oxidation product is the sulfoxide (**23**), and this is further oxidised to the stable sulfone (**24**) (Scheme 18); sulfones have many applications in synthetic organic chemistry (see Chapter 10, p.195).

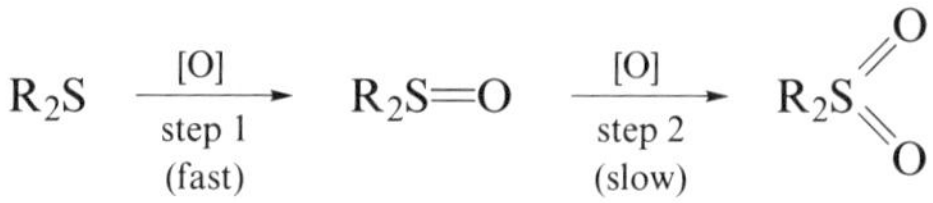

$$R_2S \xrightarrow[\text{step 1 (fast)}]{[O]} R_2S{=}O \, (\mathbf{23}) \xrightarrow[\text{step 2 (slow)}]{[O]} R_2S({=}O)_2 \, (\mathbf{24})$$

Scheme 18

Various oxidants have been used in the oxidation of sulfides such as hydrogen peroxide and a number of different peroxycarboxylic acids (RCO_3H), the latter being the more powerful oxidising agents. In the oxidation (Scheme 18), step 1 is much faster than step 2, and consequently many sulfoxides can be prepared by this route; for instance, benzyl methyl sulfoxide (**26**) can be obtained from benzyl methyl sulfide (**25**) (Scheme 19). More powerful oxidants or more vigorous conditions effect the direct conversion of sulfides into the corresponding sulfones; thus, (**27**) and (**28**) are transformed into (**29**) and (**30**), respectively (Scheme 20).

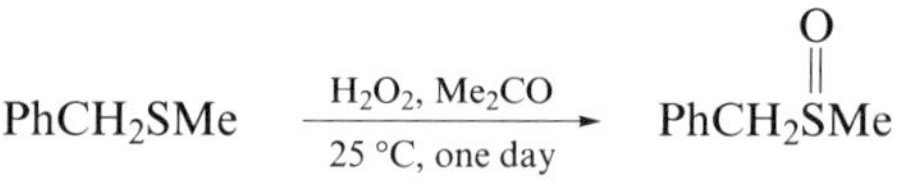

$$PhCH_2SMe \, (\mathbf{25}) \xrightarrow[25\,°C,\ \text{one day}]{H_2O_2,\ Me_2CO} PhCH_2S({=}O)Me \, (\mathbf{26}) \ (77\%)$$

Scheme 19

$$[Me(CH_2)_{15}]_2S \, (\mathbf{27}) \xrightarrow[90\,°C]{H_2O_2,\ HOAc} [Me(CH_2)_{15}]_2\ SO_2 \, (\mathbf{29}) \ >90\%$$

$$Ph_2S \, (\mathbf{28}) \xrightarrow{KMnO_4,\ H^+,\ \text{heat}} Ph_2SO_2 \, (\mathbf{30}) \ 70\%$$

Scheme 20

Sulfides are easily desulfurised by treatment with Raney nickel (Scheme 21).[3] Other methods of desulfuration are given in Chapter 10, p. 214. Desulfuration is important in many synthetic transformations in which a sulfur atom is introduced to provide the required reactivity but then has to be removed since it is not required in the final product. As has been previously noted (see Chapter 3, p.30), the attachment of electronegative sulfur atoms activates suitably placed methylene hydrogen atoms (reactive α-hydrogens) towards nucleophilic attack, allowing the formation of stabilised carbanions which can undergo further reactions. The effect is especially pronounced with 1,3-dithia compounds like (**31**), which can be transformed into carbonyl compounds, e.g. (**32**) (Scheme 22). A similar example is the conversion of 1,3-dithiacyclohexane into cyclobutanone (see Chapter 3, p. 32).

$$RSR' \xrightarrow{\text{Raney Ni}} RH + R'H + NiS$$

Scheme 21

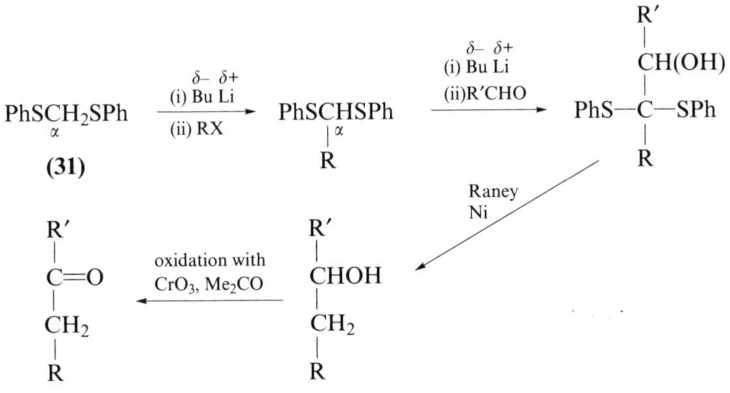

$$PhSCH_2SPh\ (\mathbf{31}) \xrightarrow[\text{(ii) RX}]{\text{(i) Bu Li}} PhSCH(R)SPh \xrightarrow[\text{(ii) R'CHO}]{\text{(i) Bu Li}} PhS{-}C(R)(CH(OH)R'){-}SPh \xrightarrow{\text{Raney Ni}} R'CHOHCH_2R \xrightarrow{\text{oxidation with } CrO_3, Me_2CO} R'C(=O)CH_2R\ (\mathbf{32})$$

Scheme 22

The ability of thiol groups to form disulfide bridges reversibly is essential in the biological reduction of sulfate and in the biosynthesis of cysteine (**33**) and methionine (**34**) from cystine (**35**) (Scheme 23).[4]

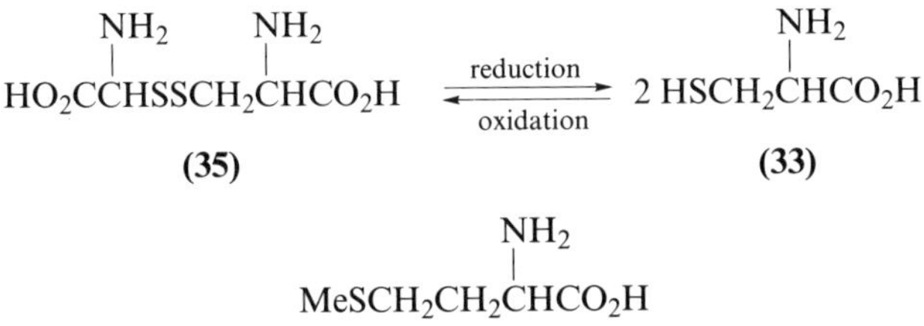

$$HO_2CCH(NH_2)SSCH_2CH(NH_2)CO_2H\ (\mathbf{35}) \underset{\text{oxidation}}{\overset{\text{reduction}}{\rightleftharpoons}} 2\ HSCH_2CH(NH_2)CO_2H\ (\mathbf{33})$$

$$MeSCH_2CH_2CH(NH_2)CO_2H\ (\mathbf{34})$$

Scheme 23

The oxidative properties of lipoic acid (see p. 51) also depend on the ease of interconversion of thiol and disulfide bonds. Methionine (**34**) is one of the eight essential amino acids for human nutrition and is deficient in the diet of vegans; luckily, this amino acid can be synthesised for use as a nutritive food additive. One synthetic route uses 2-chloroethyl methyl sulfide (**36**) and phthalimide (**37**), as shown in Scheme 24.

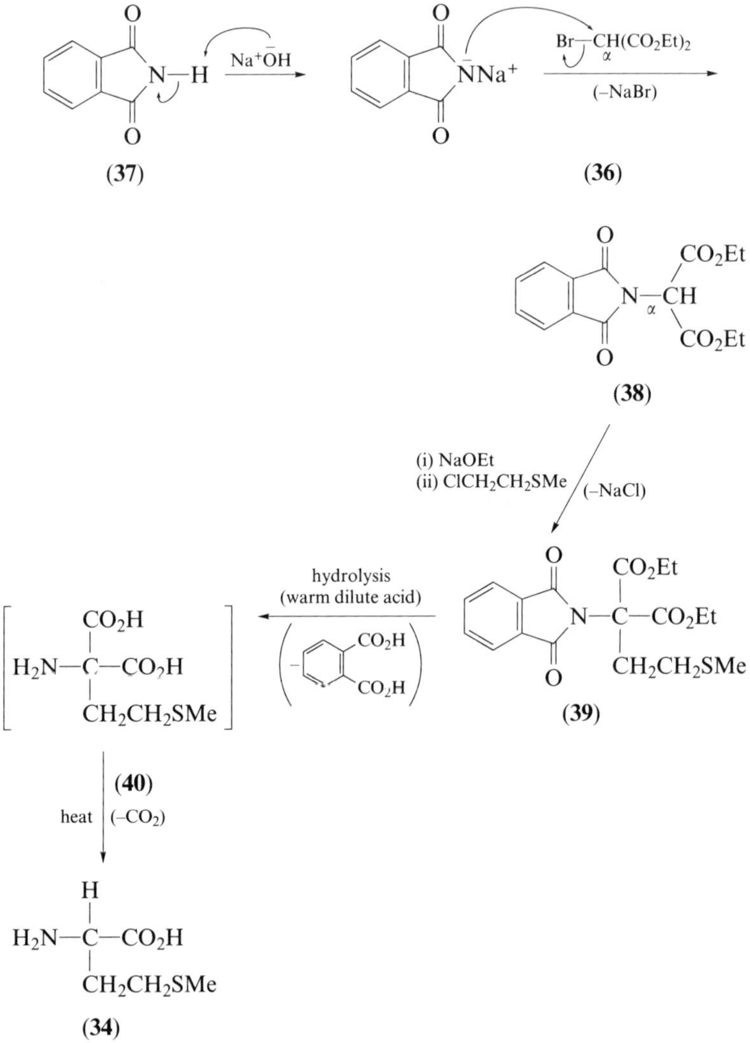

O O N—H $Na^+\bar{O}H$ (**37**) → $\bar{N}Na^+$ $Br—CH(CO_2Et)_2$ α (–NaBr) (**36**) →

N—CH α CO_2Et CO_2Et (**38**)

(i) NaOEt
(ii) $ClCH_2CH_2SMe$ (–NaCl)

N—C—CO_2Et, CO_2Et, CH_2CH_2SMe (**39**)

hydrolysis (warm dilute acid) (– $C_6H_4(CO_2H)_2$)

[$H_2N—C(CO_2H)_2—CH_2CH_2SMe$] (**40**)

heat (–CO_2)

$H_2N—CH(CO_2H)—CH_2CH_2SMe$ (**34**)

Scheme 24

The synthetic sequence (Scheme 24) utilises the acidity of the N—H bond in phthalimide (**37**) to form the sodio derivative which undergoes a nucleophilic substitution reaction with diethyl bromomalonate to yield the phthalimido derivative (**38**). This retains one reactive α-hydrogen atom and consequently forms a sodio derivative with sodium ethoxide, and this is condensed with the sulfide (**36**) to give (**39**). Heating (**39**) with dilute acid causes hydrolysis to form the malonic acid derivative (**40**); the latter by heating suffers loss of one mole of carbon dioxide to give methionine (**34**). Such selective monodecarboxylation is characteristic of the effect of heating a malonic acid derivative.

Disulfide bridges play a vital role in maintaining the essential three-dimensional structure of proteins, in which the polypeptide chains are held together by disulfide bridges between two cysteine units (Scheme 25).[4] The first step in determination of the primary structure of a polypeptide involves breaking the disulfide bridges by oxidation with performic acid and separating the resultant sulfonic acids (Scheme 25). Two important examples of proteins containing disulfide bridges are the hormones insulin and vasopressin. Insulin is used in the treatment of diabetes owing to its ability to control glucose metabolism. Vasopressin is an antidiuretic hormone which controls water excretion in the body by causing a contraction in the blood vessels and hence an increase in blood pressure.

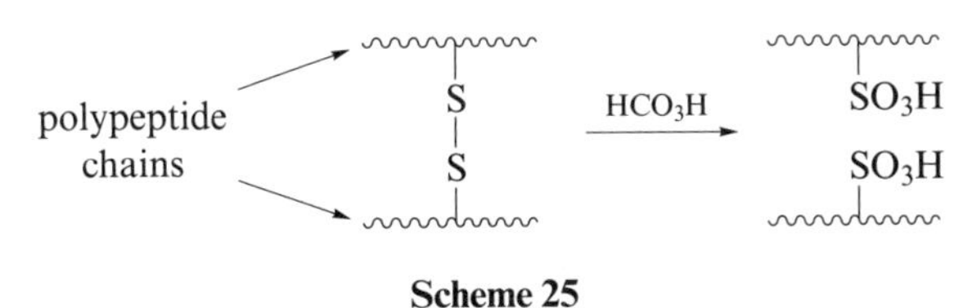

Scheme 25

Reversible acylation is important in biochemical reactions, and sulfur functions as an activator in many such processes. Coenzyme A (**41**) is an acyl group carrier which is involved in lipid oxidation and the biosynthesis of lipids and steroids. The active form of coenzyme A is the thiol ester (acyl coenzyme A) (**42**), which is more stable than coenzyme A (**41**) and hence functions as an efficient acyl group donor for a substrate RH (Scheme 26).

$$\underset{\textbf{(42)}}{\mathrm{Me\overset{\overset{\displaystyle O}{\|}}{C}SCoA}} + \mathrm{RH} \longrightarrow \mathrm{R\overset{\overset{\displaystyle O}{\|}}{C}Me} + \underset{\textbf{(41)}}{\mathrm{HSCoA}}$$

Scheme 26

Lipoic acid (**43**) is another naturally occurring disulfide; it is a growth factor and is the cofactor required for the enzymic oxidation of pyruvic acid in microorganisms. Lipoic acid (**43**) is an oxidising agent which is reduced to the thiol (**44**), and the latter can be subsequently reoxidised (Scheme 27).

The synthesis of disulfides is discussed later in this chapter (see p. 55).

Coenzyme A, lipoic acid and the tripeptide glutathione (γ-glutamylcysteinylglycine) frequently feature in biological systems as coenzymes. These coenzymes

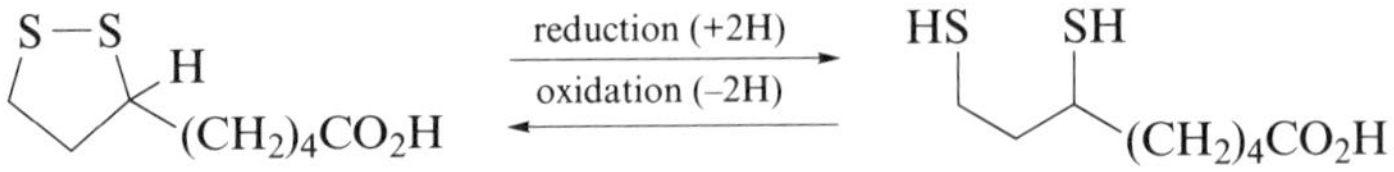

reduction (+2H)
oxidation (–2H)

(43) (44)

Scheme 27

catalyse a wide variety of biological reactions by acting as powerful nucleophiles, good leaving groups and mediators of oxido-reductive processes. Their activity is often enhanced by solvent effects, traces of metal ions and conformational factors.[4]

Sulfenic acids and derivatives[2c]

Sulfenic acids (**45**) are generally quite unstable; they easily dimerise and eliminate water to form thiol sulfinates (**46**) (Scheme 28). Several sulfenic acids have, however, been isolated and many of these are stabilised by hydrogen bonding to a carbonyl or amino group. The first sulfenic acid to be isolated was the anthraquinone derivative (**47**) in 1912. Sulfenic acids have been postulated as transient intermediates in many chemical and biochemical processes, e.g. the oxidation of thiol groups in proteins and the thermolysis of sulfoxides, including the acid-catalysed rearrangement of penicillin sulfoxides (**48**) to cephalosporins (**49**) (Scheme 29)

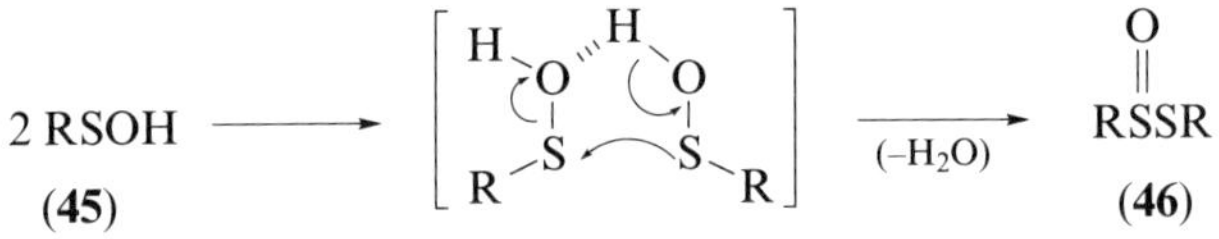

2 RSOH (45) → [transition state] →(–H_2O) RS(O)SR (46)

Scheme 28

(47) (48) (50) (49)

EtOAc, reflux

$MeSO_3H$, $AcNMe_2$

(R = phthalimido, R′ = *p*-$NO_2C_6H_4CH_2$)

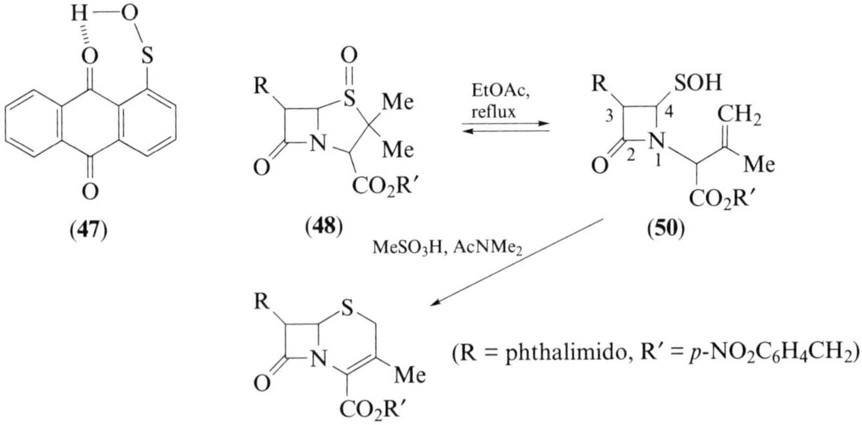

Scheme 29

(see Chapter 5, p. 70 and Chapter 11, p. 226). Much of the chemistry of penicillin sulfoxides (**48**) is related to the relatively high stability of the intermediate 2-oxazetidine-4-sulfenic acid (**50**), which has been isolated as a stable crystalline solid.

In contrast to the free acids, sulfenate esters, amides and halides are more stable. Disulfides (**52**) can be obtained from thiols by mild oxidation (see p. 57), and sulfenyl chlorides (**51**) can in turn be prepared from disulfides (**52**) by treatment with chlorine (Scheme 30). Sulfenyl chlorides (**51**) react with alcohols to give esters, e.g. the methyl sulfenate (**53**) which on alkaline hydrolysis yields the sulfenic acid (**45**) (Scheme 30).

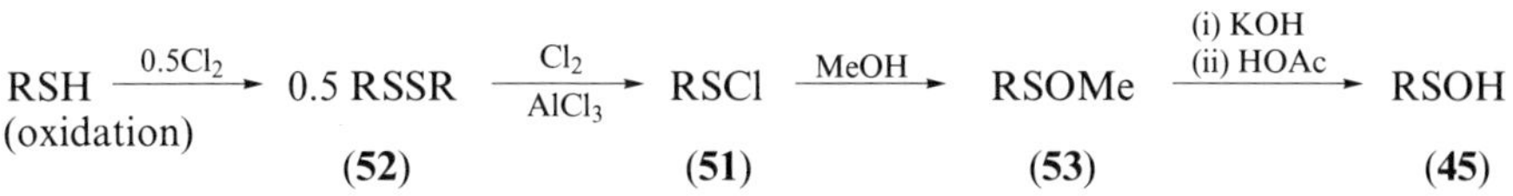

$$\underset{\text{(oxidation)}}{\text{RSH}} \xrightarrow{0.5\text{Cl}_2} \underset{(\mathbf{52})}{0.5\ \text{RSSR}} \xrightarrow[\text{AlCl}_3]{\text{Cl}_2} \underset{(\mathbf{51})}{\text{RSCl}} \xrightarrow{\text{MeOH}} \underset{(\mathbf{53})}{\text{RSOMe}} \xrightarrow{\text{(i) KOH, (ii) HOAc}} \underset{(\mathbf{45})}{\text{RSOH}}$$

Scheme 30

Some sulfenic acids have been generated by flash vacuum pyrolysis of alkyl sulfoxides; thus, *t*-butanesulfenic acid (**54**) was detected as an intermediate in the thermolysis of the *t*-butyl sulfoxide (**55**) in various solvents (Scheme 31). Sulfenic acids (**45**) may also be obtained by thermolysis of thiol sulfinates (**56**) (Scheme 31). The intermediate sulfenic acids formed in these reactions can be characterised by IR and NMR spectroscopy and may be trapped by addition to ethyl acrylate (**57**) (Scheme 32).

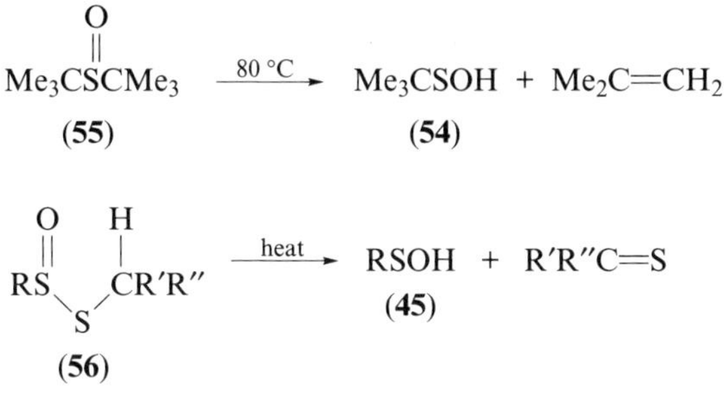

$$\underset{(\mathbf{55})}{\text{Me}_3\text{CS(=O)CMe}_3} \xrightarrow{80\ ^\circ\text{C}} \underset{(\mathbf{54})}{\text{Me}_3\text{CSOH}} + \text{Me}_2\text{C=CH}_2$$

$$\underset{(\mathbf{56})}{\text{RS(=O)–S–CHR'R''}} \xrightarrow{\text{heat}} \underset{(\mathbf{45})}{\text{RSOH}} + \text{R'R''C=S}$$

Scheme 31

$$\underset{(\mathbf{45})}{\overset{\delta-\ \delta+}{\text{RSOH}}} + \underset{(\mathbf{57})}{\text{CH}_2\text{=CHCO}_2\text{Et}} \longrightarrow \left[\text{cyclic transition state: RS}\cdots\text{O}\cdots\text{H},\ \text{H}_2\text{C–CH(CO}_2\text{Et)}\right] \longrightarrow \underset{(\mathbf{58})}{\text{RS(=O)CH}_2\text{CH}_2\text{CO}_2\text{Et}}$$

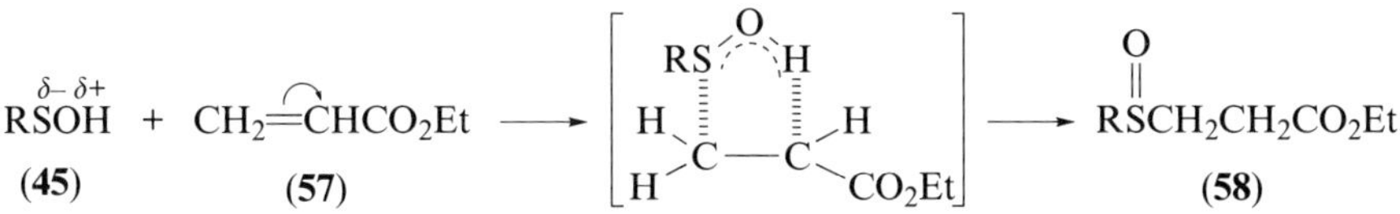

Scheme 32

REACTIONS

Sulfenic acids, esters and halides are reduced to thiols by many reagents, e.g. hydrazine, lithium aluminium hydride and metallic sodium. These addition reactions are extensively employed for trapping sulfenic acids, e.g. with ethyl acrylate (**57**) to give the adduct (**58**) (Scheme 32).

Sulfenic acid derivatives (RSX) such as esters, amides and halides are reactive owing to the relative weakness of the S—X bond, and they consequently react with nucleophilic reagents (Nu^-) (Scheme 33). In such nucleophilic substitutions at the sulfenyl sulfur atom, the reaction mechanism may be S_N1 (Scheme 34).

$$RSX + Nu^- \longrightarrow RSNu + X^-$$

$$(X = OH, Cl, OR', NR'_2, SCN)$$

Scheme 33

$$RSX \xrightarrow[\text{slow}]{H^+ \text{ or Lewis acid}} RS^+ + X^- \quad \text{step 1}$$

$$RS^+ + Nu^- \xrightarrow{\text{fast}} RSNu \quad \text{step 2}$$

Scheme 34

The well-known reaction of elemental sulfur with benzene may follow the S_N1 mechanistic pathway via the *in situ* formation of the benzenesulfenyl cation which immediately reacts with another aromatic nucleus to give thianthrene (**59**) and diphenyl sulfide (Scheme 35).

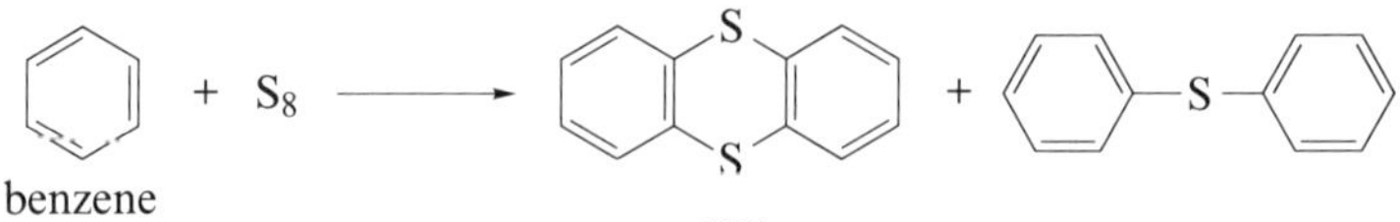

benzene + S_8 ⟶ (**59**) + Ph–S–Ph

Scheme 35

However, for the majority of substitution reactions at the sulfenyl sulfur atom, most of the available data favour the operation of the bimolecular S_N2-type reaction mechanism involving the σ-sulfurane intermediate (**60**) (see chapter 3, p. 35), as depicted in Scheme 36.

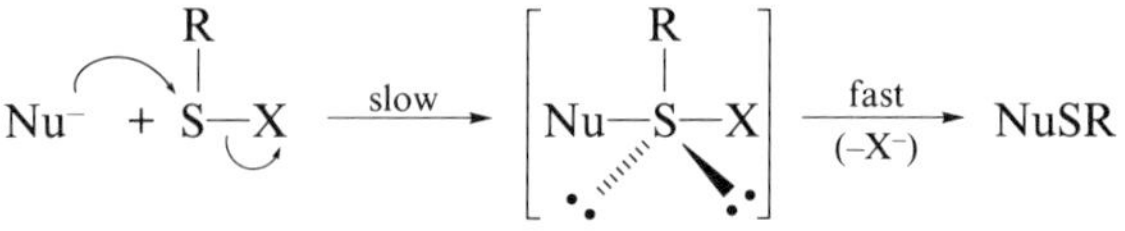

$$Nu^- + R\text{–}S\text{–}X \xrightarrow{\text{slow}} [Nu\text{–}S(R)\text{–}X] \xrightarrow[(-X^-)]{\text{fast}} NuSR$$

(**60**)

Scheme 36

Alkane- and arenesulfenyl chlorides add to alkenes and alkynes; thus, the addition of 2,4-dinitrobenzenesulfenyl chloride (**61**) may be used to prepare solid derivatives of alkenes, e.g. the adduct (**62**) (Scheme 37). The Markownikoff addition of a sulfenyl chloride to an alkene is stereospecifically *trans*; the adduct may be sequentially treated with a Lewis acid and a nucleophilic reagent to achieve the introduction of a new carbon–carbon bond (**64**) via the episulfenium ion intermediate (**63**) (Scheme 38). The procedure can also be applied to the synthesis of unsaturated sulfides, e.g. (**65**) (Scheme 39).

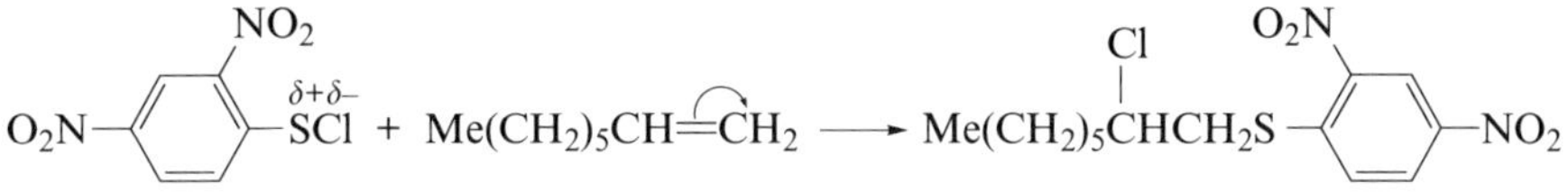

O_2N–$C_6H_3(NO_2)$–$\overset{\delta+\delta-}{SCl}$ + $Me(CH_2)_5CH{=}CH_2$ ⟶ $Me(CH_2)_5CH(Cl)CH_2S$–$C_6H_3(NO_2)_2$

(**61**) (**62**)

Scheme 37

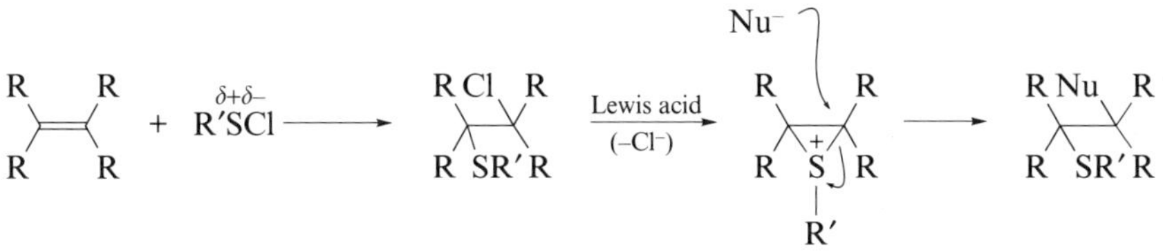

$R_2C{=}CR_2$ + $\overset{\delta+\delta-}{R'SCl}$ ⟶ R Cl R / R SR′ R —Lewis acid (–Cl⁻)⟶ (**63**) [Nu⁻, episulfonium S^+–R′] ⟶ R Nu R / R SR′ R (**64**)

Scheme 38

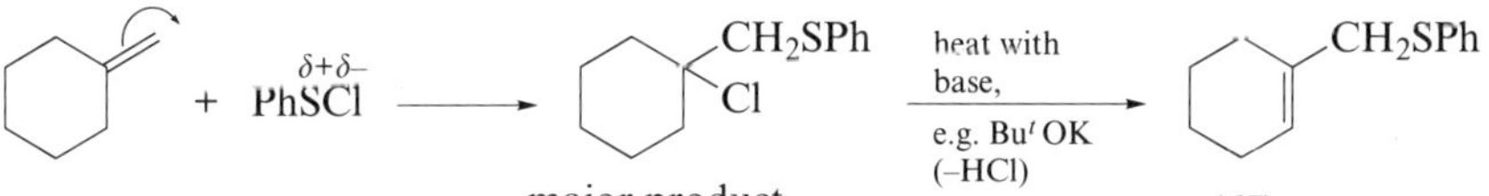

methylenecyclohexane + $\overset{\delta+\delta-}{PhSCl}$ ⟶ 1-chloro-1-(CH_2SPh)cyclohexane (major product) —heat with base, e.g. Bu^tOK (–HCl)⟶ 1-(CH_2SPh)cyclohexene (**65**)

Scheme 39

The reaction of sulfenyl chlorides with imides, e.g. maleimide, succinimide or phthalimide (**66**), provides a useful route for the preparation of disulfides (**67**) (Scheme 40) (see p. 57). The intermediate sulfenylimide (**68**) is quite stable but reacts with a thiol with expulsion of the imide moiety, which is a good leaving

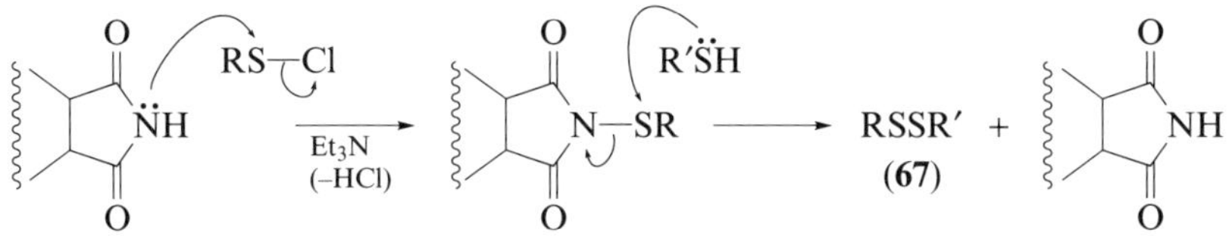

imide NH (**66**) + RS–Cl —Et_3N (–HCl)⟶ imide N–SR (**68**) + R′SH ⟶ RSSR′ (**67**) + imide NH

Scheme 40

group, to yield the disulfide (**67**). The method is also useful to obtain unsymmetrical disulfides.

Sulfenyl chlorides undergo the Friedel–Crafts reaction with suitable aromatic compounds to produce the corresponding sulfides; thus, benzenesulfenyl chloride condenses with anisole to give *p*-methoxyphenyl phenylsulfide (Scheme 41).

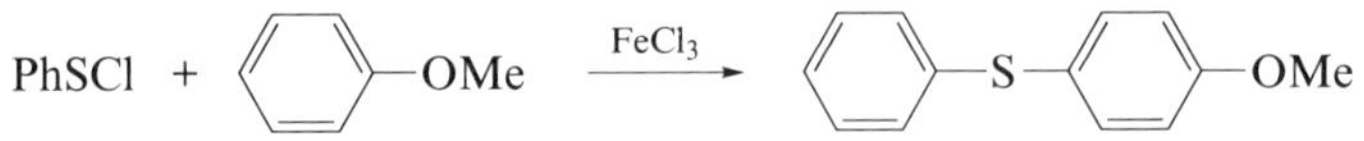

$$PhSCl + C_6H_5OMe \xrightarrow{FeCl_3} C_6H_5{-}S{-}C_6H_4{-}OMe$$

Scheme 41

Trichloromethanesulfenyl chloride (**69**), a useful synthetic intermediate is manufactured by reaction of carbon disulfide and chlorine in the presence of iodine catalyst (Scheme 42). The sulfenyl chloride (**69**) is employed in the preparation of several commercial agricultural fungicides such as captan (see Chapter 9, p. 151).

$$2\,CS_2 + 5\,Cl_2 \xrightarrow{I_2} \underset{\mathbf{(69)}}{2\,Cl_3CSCl} + S_2Cl_2$$

Scheme 42

Disulfides and polysulfides

Organic disulfides in general have considerable biological significance (see Introduction, p. 2). Important compounds of this type include the amino acid cystine (see p. 49), several proteins (see p. 51) and the growth factor lipoic acid (see p. 51). The biological oxidation–reduction catalyst glutathione (**70**) is the most common cellular thiol; it is the tripeptide γ-glutamylcysteinylglycine (Figure 1).

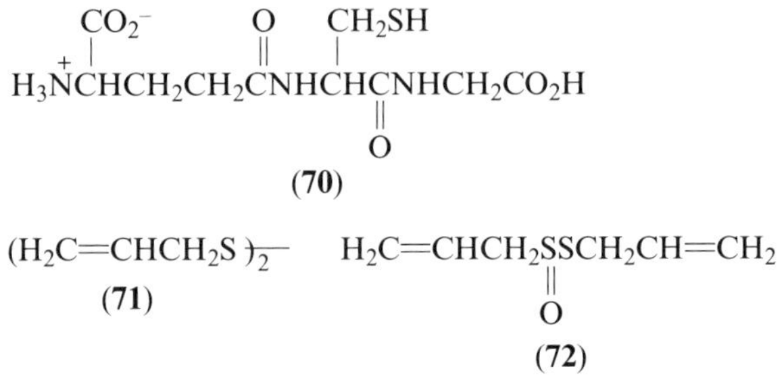

$$H_3\overset{+}{N}CH(CO_2^-)CH_2CH_2C(=O)NHCH(CH_2SH)C(=O)NHCH_2CO_2H$$

(70)

$$(H_2C{=}CHCH_2S)_2 \qquad H_2C{=}CHCH_2S(=O)SCH_2CH{=}CH_2$$

(71) **(72)**

Figure 1

Glutathione functions as a reducing agent because of the ease of its oxidation to the corresponding disulfide (glutathione disulfide). Organic disulfides are found in plants, especially in the genus *Allium*—onion, garlic and leek. Garlic contains diallyl disulfide (**71**) and this may be a factor in its antibacterial properties while the antibacterial agent allicin (**72**) has been isolated from onions and the structure contains a disulfide bond (Figure 1).

Symmetrical disulfides (**73**) may be prepared by reaction of alkyl halides with disodium disulfide (Scheme 43). The product is contaminated with tri- and polysulfides owing to the presence of impurities in the disodium disulfide; however, lower members of the series of dialkyl disulfides may be purified by fractional distillation. Disulfides can also be obtained from thiols by mild oxidation, e.g. by treatment with iodine or dimethyl sulfoxide (DMSO) (Scheme 44). In the reaction with iodine, the hydriodic acid formed must be removed, otherwise the disulfide is largely reduced back to the thiol by hydriodic acid which is a powerful reducing agent. Pure unsymmetrical disulfides are more difficult to prepare owing to their tendency to undergo disproportionation; they can, however, be synthesised from thiols by treatment with imides (see p. 59) or sulfenyl halides (**51**) (Scheme 45).

$$2\,RX + Na_2S_2 \longrightarrow \underset{(\mathbf{73})}{RSSR} + 2\,NaX$$

Scheme 43

$$2\,RSH + I_2 \rightleftharpoons \underset{(\mathbf{73})}{RSSR} + 2\,HI$$

$$2\,RSH + Me_2SO \longrightarrow RSSR + Me_2S + H_2O$$

Scheme 44

$$R\overset{-\,+}{S}Na + \underset{(\mathbf{51})}{R'S{-}X} \longrightarrow \underset{(\mathbf{73})}{RSSR'} + NaX$$

Scheme 45

Unsymmetrical trisulfides have biological significance; several have been isolated from plants of the onion family, e.g. allyl propyl trisulfide. Interest in this class of compound was stimulated by the discovery of a biologically active compound shown to be the trisulfide (**74**) (Figure 2) isolated from the tropical plant *Petiveria alliacea* found in Africa and other countries.[5] The trisulfide (**74**) shows insecticidal and antimicrobial activity.

$$\underset{(\mathbf{74})}{PhCH_2SSSCH_2CH_2OH}$$

Figure 2

Tri- and tetrasulfides are more unstable than disulfides and the unsymmetrical members are particularly liable to suffer disproportionation to give mixtures of products. Several methods are used for the synthesis of symmetrical dialkyl trisulfides (**75**); thus, they may be obtained by reaction of sulfur dichloride with thiols

(Scheme 46). However, this route does not afford pure trisulfides and the products are contaminated with both di- and tetrasulfides; the optimum preparative procedure involves the reaction of the appropriate *S*-alkyl- or *S*-arylthioisothiuronium chloride (**76**) with dimethylamine (Scheme 47).

$$2\,RSH + SCl_2 \longrightarrow 2\,RSSSR\;(\mathbf{75}) + 2HCl$$

Scheme 46

$$RSSC(=\overset{+}{N}H_2Cl^-)NH_2\;(\mathbf{76}) \xrightarrow{Me_2NH} RSSSR\;(\mathbf{75})$$

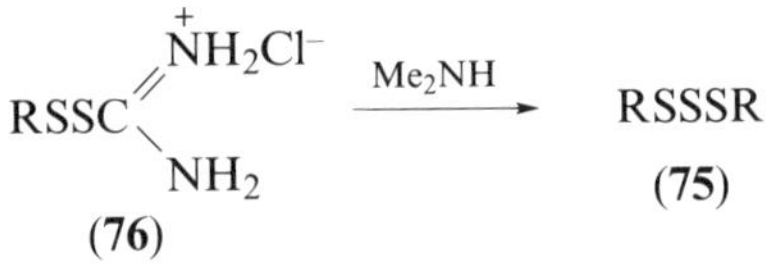

Scheme 47

The thioisothiuronium chloride (**76**) is prepared by reaction of thiourea (**1a**) with the appropriate thiol in the presence of hydrogen peroxide (Scheme 48). The mechanism of the reaction probably involves initial oxidation of the thiol to the sulfenic acid (**45**) and the subsequent reaction with thiourea (**1a**) (Scheme 49).

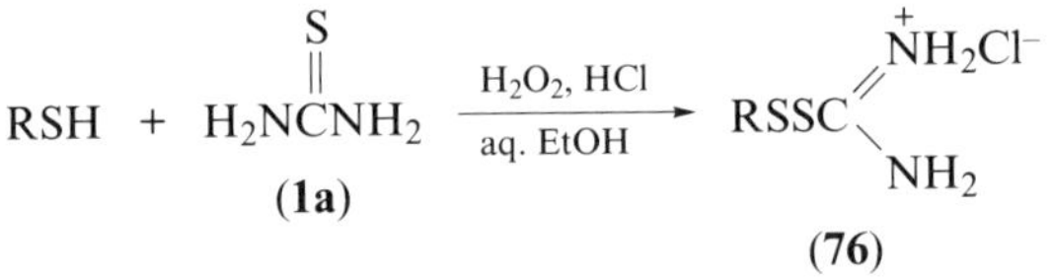

$$RSH + H_2NC(=S)NH_2\;(\mathbf{1a}) \xrightarrow[\text{aq. EtOH}]{H_2O_2,\ HCl} RSSC(=\overset{+}{N}H_2Cl^-)NH_2\;(\mathbf{76})$$

Scheme 48

$$RSH \xrightarrow{H_2O_2} RSOH\;(\mathbf{45})$$

$$(H_2N)_2C{=}S\;(\mathbf{1a}) \rightleftharpoons HN{=}C(NH_2){-}\ddot{S}H\;(\mathbf{1b}) + RS{-}\overset{+}{O}H_2 \xrightarrow[(-H_2O)]{HCl} RSSC(=\overset{+}{N}H_2Cl^-)NH_2\;(\mathbf{76})$$

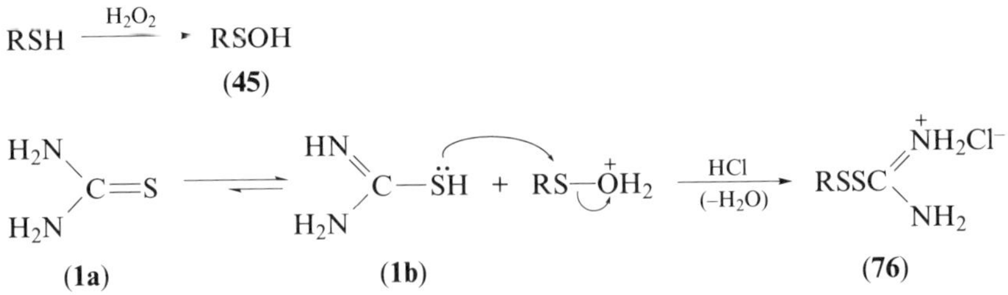

Scheme 49

The synthesis of unsymmetrical trisulfides is more difficult than that of the symmetrical analogues owing to their facile disproportionation; they are of more interest because of their occurrence in nature and their biological activity, e.g. the natural product (**74**). One route to this compound is by condensation of benzyl thiol (**77**) with sulfur dichloride and 2-mercaptoethanol (**78**) (Scheme 50).

Another route involves condensation of phthalimide (**79**) with sulfur monochloride to give *N,N′*-thiobis(phthalimide) (**80**) which is reacted with benzyl thiol

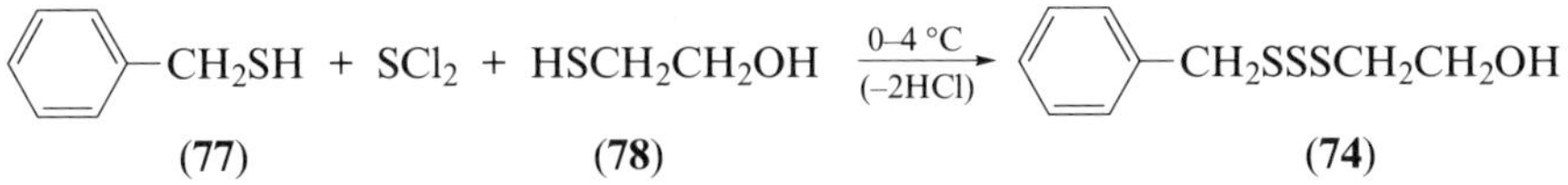

$$C_6H_5CH_2SH + SCl_2 + HSCH_2CH_2OH \xrightarrow[(-2HCl)]{0\text{–}4\ °C} C_6H_5CH_2SSSCH_2CH_2OH$$

(77) (78) (74)

Scheme 50

to give the disulfide (**81**) by displacement of the phthalimido moiety. In the final stage, the disulfide (**81**) is treated with 2-mercaptoethanol (**78**) to give the required unsymmetrical trisulfide (**74**) (Scheme 51). This reaction illustrates the capacity of the phthalimido moiety to act as a leaving group and demonstrates that the thiol group is a better nucleophile than the hydroxyl group, because it is the thiol group that reacts preferentially with the disulfide (**81**) in the final stage of Scheme 51. Scheme 50 can be modified to provide a useful synthesis of unsymmetrical tetrasulfides like (**82**); in this case, sulfur monochloride is condensed with the appropriate thiol and 2-mercaptoethanol (**78**) (Scheme 52)

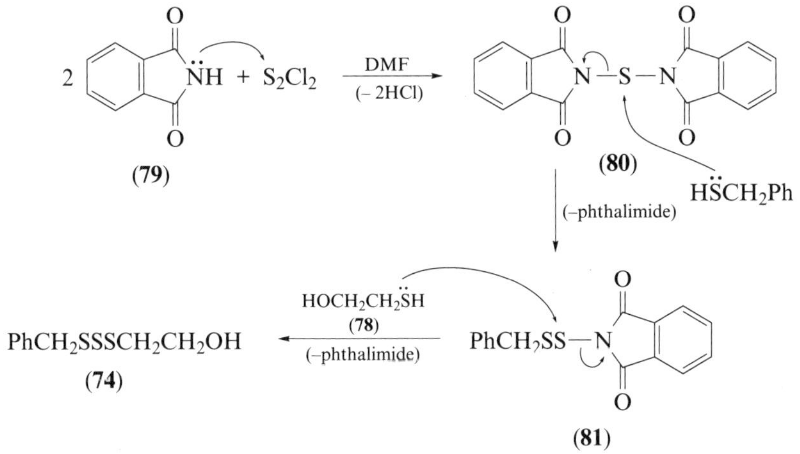

2 (79) NH + S_2Cl_2 —DMF (− 2HCl)→ N—S—N (80)

(80) + $HSCH_2Ph$ —(−phthalimide)→ $PhCH_2SS$—N (81)

(81) + $HOCH_2CH_2SH$ (78) —(−phthalimide)→ $PhCH_2SSSCH_2CH_2OH$ (74)

Scheme 51

$$RSH + S_2Cl_2 + HSCH_2CH_2OH \xrightarrow[(-2HCl)]{0\text{–}4\ °C} RSSSSCH_2CH_2OH$$

(78) (82)

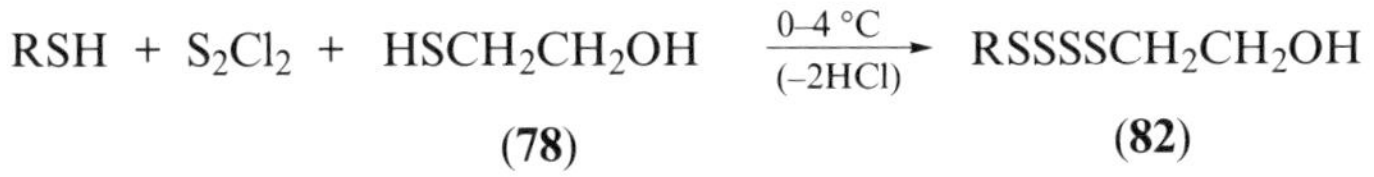

Scheme 52

Examination of a series of sulfides of type $ArS_nCH_2CH_2OH$ ($n = 1\text{–}4$) showed that the monosulfides ($n = 1$) and disulfides ($n = 2$) were stable at room temperature for one year, but the trisulfides ($n = 3$) and tetrasulfides ($n = 4$) were not stable under these conditions and decomposed to mixtures of products.

In their chemical reactions, disulfides (**73**) are sensitive to heating, reduction and oxidation as a consequence of the relative weakness of the disulfide bond; thus, on warming, dibenzyl disulfide (**83**) yields stilbene (**84**) (Scheme 53).

$$\underset{\textbf{(83)}}{PhCH_2SSCH_2Ph} \xrightarrow{\text{heat}} \underset{\textbf{(84)}}{PhCH{=}CHPh} + H_2S + S$$

Scheme 53

Disulfides are reduced to thiols by the majority of reducing agents, e.g. Na_2S, NaSH, $LiAlH_4$ and Zn/H^+ (Scheme 54). In this reduction, the diaryl disulfides are more easily reduced than the dialkyl analogues; thus, sodium borohydride ($NaBH_4$) reduces the former but not the latter, since ArS^- is a better leaving group than $AlkS^-$.

$$\underset{\textbf{(73)}}{4\,RSSR} + 2\,Na_2S + 6\,NaOH \longrightarrow 8\,RSNa + Na_2S_2O_3 + 3\,H_2O$$

Scheme 54

Disulfides (**73**) are oxidised by hydrogen peroxide to the corresponding sulfenic acids (**45**); dialkyl disulfides may also be converted to thiol sulfinates by treatment with perbenzoic acid; thus, diallyl disulfide (**71**) yields allicin (**72**), an essential component of garlic (Scheme 55). The latter reaction involves a nucleophilic attack of the sulfur atom on the oxygen of the peroxide group, and therefore the more electron-rich sulfur atom is preferentially oxidised. For instance, methyl phenyl disulfide (**85**) yields the sulfoxide (**86**) (Scheme 56).

$$\underset{\textbf{(73)}}{RSSR} + H_2O_2 \longrightarrow \underset{\textbf{(45)}}{2\,RSOH}$$

$$\underset{\textbf{(71)}}{(H_2C{=}CHCH_2S)_2} \xrightarrow{PhCO_3H} \underset{\textbf{(72)}}{H_2C{=}CHCH_2S(O)SCH_2CH{=}CH_2}$$

Scheme 55

$$\underset{\textbf{(85)}}{PhSSMe} \xrightarrow{MeCO_3H} \underset{\textbf{(86)}}{PhS(O)SMe}$$

Scheme 56

The lability of the disulfide bridge is also shown in the metabolism of disulfides, in which the disulfide bond is reduced to the corresponding thiol.[4] For example, disulfiram (**87**), a drug used in the treatment of alcoholism, is reduced *in vivo* to *N*,*N*-diethyldithiocarbamic acid (**88**), the biologically active metabolite (Scheme 57).

$$Et_2NC(=S)SSC(=S)NEt_2 \xrightarrow[\textit{in vivo}]{2[H]} 2Et_2N{-}C(=S)SH$$

(87) **(88)**

Scheme 57

References

1. *The Chemistry of the Thiol Group* (Ed. S. Patai), Wiley, New York, 1974: (a) J. L. Wardell, Part 1, p. 193; (b) R. K. Olsen and J. O. Currie, Part 2, p. 519.
2. *Comprehensive Organic Chemistry* (Eds D. H. R. Barton and W. D. Ollis), Vol. 3, Part II, Pergamon Press, Oxford, 1979: (a) G. C. Barrett, p. 3; (b) G. C. Barrett, p. 33; (c) D. R. Hogg, p. 261.
3. G. R. Pettit and E. E. van Tamelin, *Org. React.*, **12**, 356 (1962).
4. *Organic Sulfur Chemistry: Biochemical Aspects* (Eds S. Oae and T. Okuyama), CRC Press, Boca Raton, FL, 1992.
5. E. T. Ayodele, *Studies in the Synthesis and Fungicidal Activity of some Benzyl 2-Hydroxyethyl Oligosulfides and Related Compounds*, PhD Thesis, University of North London, 1994.

5 SULFOXIDES AND SULFONES

Sulfoxides (**1**)[1,2] are generally prepared by controlled oxidation of sulfides (**2**) (see Chapter 4, p. 48) (Scheme 1). The choice of oxidant and the reaction conditions are critical to avoid further oxidation to the sulfone (see Chapter 10, p. 195). On a small scale, the preferred reagents are (i) sodium metaperiodate in aqueous methanol (0 °C), (ii) *m*-chloroperbenzoic acid (MCPBA) in dichloromethane or ethyl acetate, the latter being more useful since it can be used at lower temperatures (–40 °C), and (iii) *t*-butyl hypochlorite in methanol (Scheme 1).

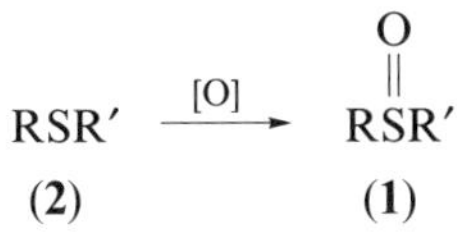

$$\underset{(\mathbf{2})}{RSR'} \xrightarrow{[O]} \underset{(\mathbf{1})}{R\overset{O}{\overset{\|}{S}}R'}$$

Scheme 1

The asymmetric oxidation of an alkyl aryl sulfide (**2**) (R = alkyl, R′ = aryl) with a chiral oxidant, e.g. a peroxotitanium complex (a Sharpless reagent), can provide a route to chiral sulfoxides; thus, methyl phenyl sulfide (**2**) (R = Me, R′ = Ph) yields the corresponding sulfoxide (**1**) (R = Me, R′ = Ph) in 80% yield and with 89% e.e. This asymmetric oxidation may also be achieved microbiologically with *Aspergillus niger*, often with high stereoselectivity. Sulfoxides may also be obtained by hydrolysis of halosulfonium salts (**3**), and this route has been applied to prepare ^{18}O-labelled sulfoxides (**1**). The conversion may also be achieved by treating a solution of the sulfide (R_2S) in acetic acid, excess water and pyridine (one equivalent) with bromine.

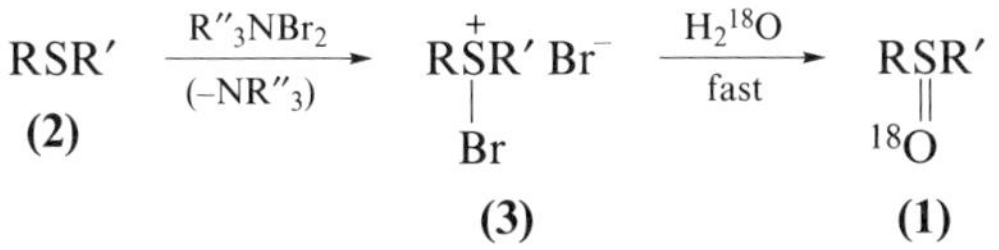

$$\underset{(\mathbf{2})}{RSR'} \xrightarrow[(-NR''_3)]{R''_3NBr_2} \underset{(\mathbf{3})}{R\overset{+}{\underset{\underset{Br}{|}}{S}}R'\ Br^-} \xrightarrow[fast]{H_2{}^{18}O} \underset{(\mathbf{1})}{R\underset{\underset{^{18}O}{\|}}{S}R'}$$

Scheme 2

Chiral sulfoxides are conveniently synthesised by Andersen's method (1962), in which a chiral sulfinate (**4**) is treated with a Grignard reagent. The reaction involves a sulfinyl group transfer and occurs with complete stereochemical inversion at the sulfur atom (see Chapter 3, p. 30) (Scheme 3). In this reaction, the yields are sometimes improved by replacing the Grignard reagent by other organometallics, e.g. lithium dialkylcuprates (R_2CuLi). Cholesterol may be used instead of (–)-menthol as the chiral alcohol, this modification being valuable for the synthesis of chiral dialkyl sulfoxides (Scheme 4).

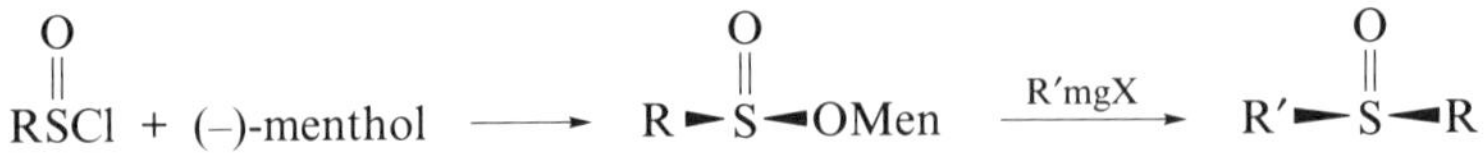

$$\mathrm{RS(=O)Cl} + (-)\text{-menthol} \longrightarrow \underset{(\mathbf{4})}{\mathrm{R{-}S(=O){-}OMen}} \xrightarrow{\mathrm{R'mgX}} \underset{(\mathbf{1})}{\mathrm{R'{-}S(=O){-}R}}$$

Scheme 3

$$\mathrm{MeS^*(=O)OChol} \xrightarrow{\mathrm{RMgX}} \mathrm{MeS^*(=O)R}$$

Scheme 4

Optically active sulfinates (**5**) may be obtained by reaction of sulfinyl chlorides (**6**) with a chiral alcohol in the presence of an optically active tertiary amine (Scheme 5).

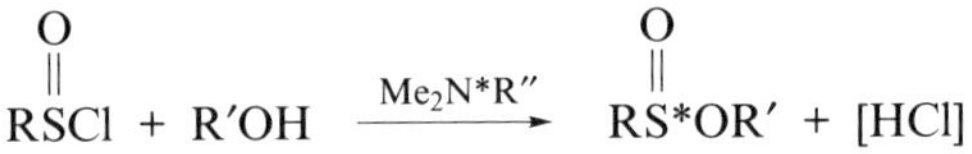

$$\underset{(\mathbf{6})}{\mathrm{RS(=O)Cl}} + \mathrm{R'OH} \xrightarrow{\mathrm{Me_2N^*R''}} \underset{(\mathbf{5})}{\mathrm{RS^*(=O)OR'}} + [\mathrm{HCl}]$$

Scheme 5

Certain sulfoxides can be prepared by rearrangement of the sulfenate esters; thus, allyl arenesulfenate esters (**7**), obtained by condensation of the sulfenyl chloride with allyl alcohol (**8**), spontaneously rearrange to the allyl sulfoxides (**9**) (Scheme 6). The rearrangement also occurs with alkynic alcohols; for instance, trichloromethanesulfenyl chloride (**10**) reacts with propargyl alcohol (**11**) to form the allenic sulfoxide (**12**) (Scheme 7).

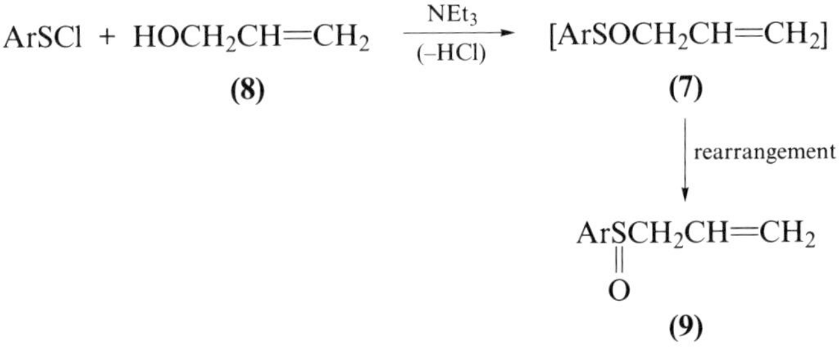

$$\mathrm{ArSCl} + \underset{(\mathbf{8})}{\mathrm{HOCH_2CH{=}CH_2}} \xrightarrow[(-\mathrm{HCl})]{\mathrm{NEt_3}} \underset{(\mathbf{7})}{[\mathrm{ArSOCH_2CH{=}CH_2}]} \xrightarrow{\text{rearrangement}} \underset{(\mathbf{9})}{\mathrm{ArS(=O)CH_2CH{=}CH_2}}$$

Scheme 6

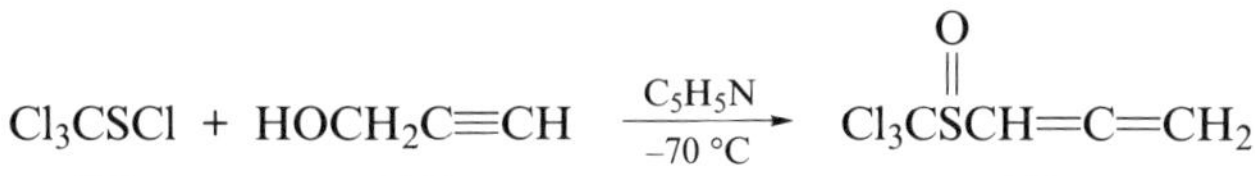

$$Cl_3CSCl + HOCH_2C{\equiv}CH \xrightarrow[-70\,°C]{C_5H_5N} Cl_3CS(=O)CH{=}C{=}CH_2$$

(10) **(11)** **(12)**

Scheme 7

Structure of sulfoxides

X-ray studies of dimethyl sulfoxide (DMSO) show that the molecule is a rather distorted pyramid with the sulfur atom at the apex (Figure 1). If the lone electron pair on the sulfur atom is taken into account, the structure may be described as tetrahedral with the sulfur atom at the centre of the tetrahedron. Sulfoxides are therefore structural analogues of amines and phosphines, and hence they should be capable of being resolved into optical enantiomers like (**12a**) and (**12b**) (Figure 1) (see Chapter 3, p. 34). The resolution was first achieved by Phillips (1926) with *m*-methylcarboxyphenyl methyl sulfoxide (**1**) (R = *m*-$C_6H_4CO_2Me$, R′ = Me). Chiral sulfoxides also occur naturally; for instance, the *d*-isomer of *S*-allyl-L-cystine sulfoxide (**1**) (R = $CH_2CH{=}CH_2$, R′ = $CH_2CH(CO_2^-)NH_3^+$) has been isolated from garlic.

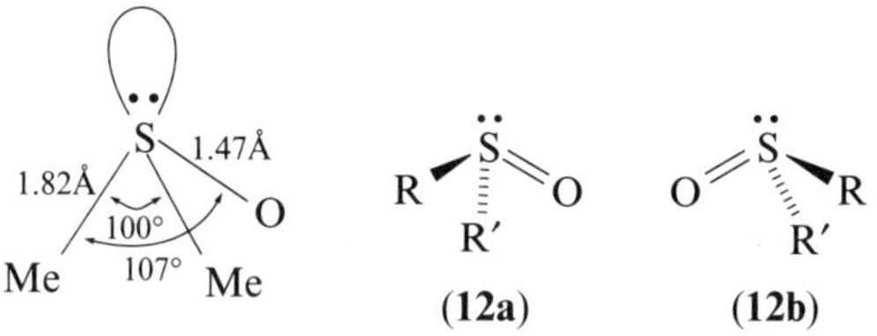

1.82Å 1.47Å 100° 107° S O Me Me

R S O R′ (**12a**) O S R R′ (**12b**)

Figure 1

In sulfoxides (**1**), the nature of the sulfur–oxygen bond remains somewhat doubtful. It is sometimes depicted as a single bond (S→O or $S^+{—}O^-$), which emphasises the polar character, or as a double bond (S=O). In this book, the double-bonded structure is preferred on the basis of bond length measurements (see Chapter 1, p. 10).

Sulfoxides (**1**) may be oxidised to sulfones by treatment with hydrogen peroxide (in acetic acid) or peroxycarboxylic acids (see Chapter 10, p. 195). Sulfoxides can also be reduced to sulfides by powerful reducing agents, e.g. lithium aluminium hydride. The reduction can also be achieved by reaction with dichloroborane (**13**) in THF at 0 °C without affecting other functional groups like CO_2R, COCl, CN or NO_2, and even side reactions with aldehyde and ketone groups are relatively slow (Scheme 8).

Chiral sulfoxides will act as directing groups in asymmetric reactions. An example is in the conjugate addition of a Grignard reagent to an enone (**14**) to yield the chiral 3-methylcyclopentanone (**15**) (Scheme 9). Chiral sulfoxides may also be applied to the asymmetric synthesis of biologically active compounds.[3]

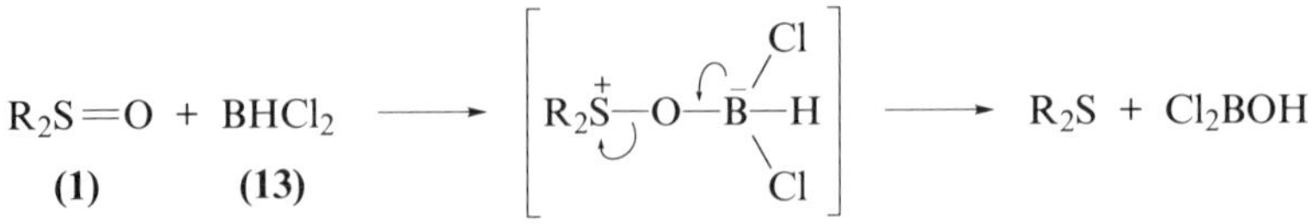

$$\underset{(\mathbf{1})}{R_2S{=}O} + \underset{(\mathbf{13})}{BHCl_2} \longrightarrow \left[R_2\overset{+}{S}{-}O{-}\overset{-}{B}HCl_2\right] \longrightarrow R_2S + Cl_2BOH$$

Scheme 8

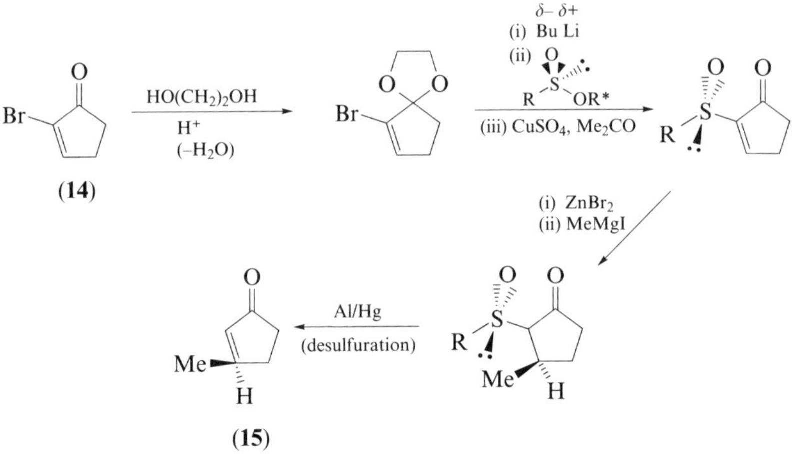

(14) → $HO(CH_2)_2OH$, H^+ $(-H_2O)$ → (i) Bu Li ($\delta-$ $\delta+$); (ii) R–S(O)–OR*; (iii) $CuSO_4$, Me_2CO → (i) $ZnBr_2$; (ii) MeMgI → Al/Hg (desulfuration) → (15)

Scheme 9

Sulfoxides (**1**) can react as nucleophiles, this being demonstrated by their ability to form sulfoxonium salts (**16**) with strong acids (Scheme 10). The nucle-

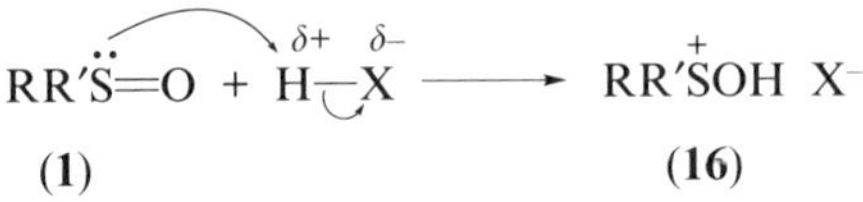

$$\underset{(\mathbf{1})}{RR'\ddot{S}{=}O} + \overset{\delta+}{H}{-}\overset{\delta-}{X} \longrightarrow \underset{(\mathbf{16})}{RR'\overset{+}{S}OH\ X^-}$$

Scheme 10

ophilicity of sulfoxides is exploited in the Moffatt oxidation of phenacyl halides[4] (**17**) to phenyl glyoxals (**18**) on dissolution in DMSO at room temperature in the presence of sodium hydrogen carbonate. Later work showed that with heating, the reaction can be extended to benzylic halides (**19**) and tosylates (**20**) (Scheme 11).

The mechanism involves a bimolecular nucleophilic substitution reaction (S_N2) of the hydrogen atom or tosylate group by DMSO to give an alkoxysulfonium intermediate (**21**), which is subsequently converted to the carbonyl compound by attack of the base (B^-) by two possible routes (Scheme 12).

In the Pfitzner–Moffatt oxidation,[4] an alcohol is oxidised by treatment with a mixture of dicyclohexylcarbodiimide (DCC) (**22**), DMSO and an acid source, e.g. polyphosphoric acid, sulfur trioxide–pyridine, pyridine trifluoroacetate or acetic

$$ArCOCH_2X \xrightarrow[9\ h]{DMSO,\ NaHCO_3,\ 25\ °C} ArCOCHO$$

(**17**) (**18**)

$$ArCH_2X \xrightarrow[5\ min]{DMSO,\ NaHCO_3,\ 100\ °C} ArCHO$$

(**19**)

$$RCH_2OTs \xrightarrow[3\ min]{DMSO,\ NaHCO_3,\ 150\ °C} RCHO$$

(**20**)

(X = halogen)

Scheme 11

(i) $RCH_2-X + \ddot{O}=SMe_2 \xrightarrow{(-X^-)} RCH(H)-O-\overset{+}{S}Me_2$ (**21**) (with B^-) $\xrightarrow{(-BH)} RCHO + Me_2S$

or

(ii) $RCH_2O\overset{+}{S}(Me)CH_2-H$ (with B^-) $\xrightarrow{(-BH)}$ [$RCH(H)-O-\overset{+}{S}Me-\overset{-}{C}H_2$] $\rightarrow RCHO + Me_2S$

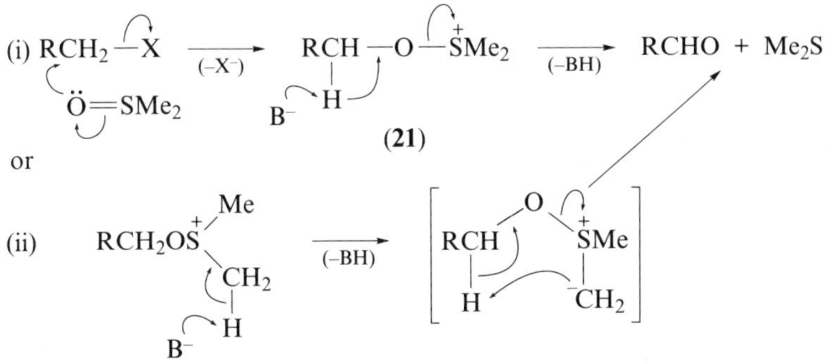

Scheme 12

anhydride. The acidic materials serve to activate the DMSO and the oxidation has three steps: (i) activation of DMSO, (ii) formation of the alkoxysulfonium salt (**21**) and (iii) decomposition of the salt (**21**) (Scheme 13).

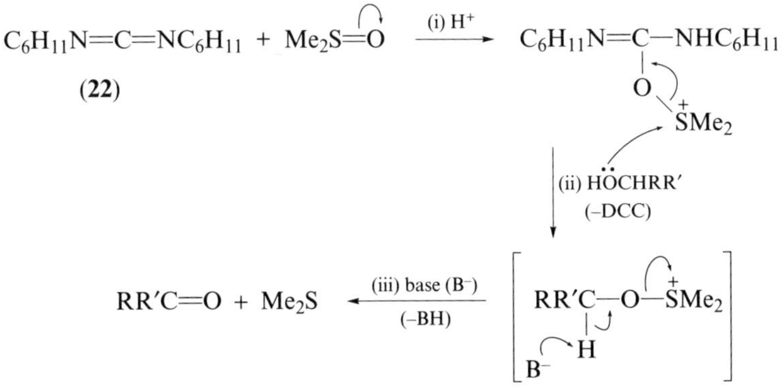

$C_6H_{11}N=C=NC_6H_{11}$ (**22**) $+ Me_2S=O \xrightarrow{(i)\ H^+} C_6H_{11}N=C(-O-\overset{+}{S}Me_2)-NHC_6H_{11}$

$\xrightarrow[(-DCC)]{(ii)\ H\ddot{O}CHRR'}$ [$RR'C(H)-O-\overset{+}{S}Me_2$, B^-] (**21**)

$\xrightarrow[(-BH)]{(iii)\ base\ (B^-)} RR'C=O + Me_2S$

Scheme 13

The Moffatt oxidation is especially valuable for the oxidation of a primary alcohol, because the conditions required are quite mild and avoid any possibility of further oxidation to the carboxylic acid.

In the widely used Swern oxidation[4] of primary and secondary alcohols to carbonyl compounds via alkoxysulfonium ylides, the DMSO is activated by oxalyl chloride (**23**) (Scheme 14).

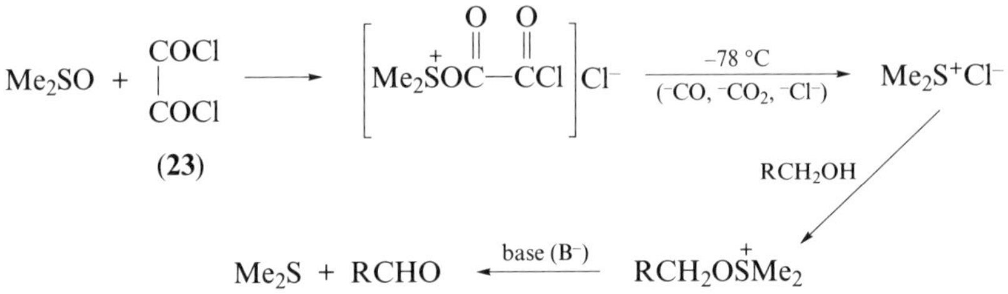

$$Me_2SO + \underset{(\mathbf{23})}{(COCl)_2} \longrightarrow \left[Me_2\overset{+}{S}OC(=O)-C(=O)Cl\right]Cl^- \xrightarrow[(^-CO,\ ^-CO_2,\ ^-Cl^-)]{-78\ ^\circ C} Me_2S^+Cl^-$$

$$Me_2S^+Cl^- \xrightarrow{RCH_2OH} RCH_2O\overset{+}{S}Me_2 \xrightarrow{\text{base }(B^-)} Me_2S + RCHO$$

Scheme 14

The Pummerer rearrangement[5]

A sulfoxide containing at least one α-hydrogen atom is reduced to a sulfide with concomitant oxidation at the α-carbon atom. The Pummerer rearrangement is generally effected by treatment of the sulfoxide (**24**) with acetic anhydride to yield the corresponding α-acetoxy sulfide (**25**) (Scheme 15).

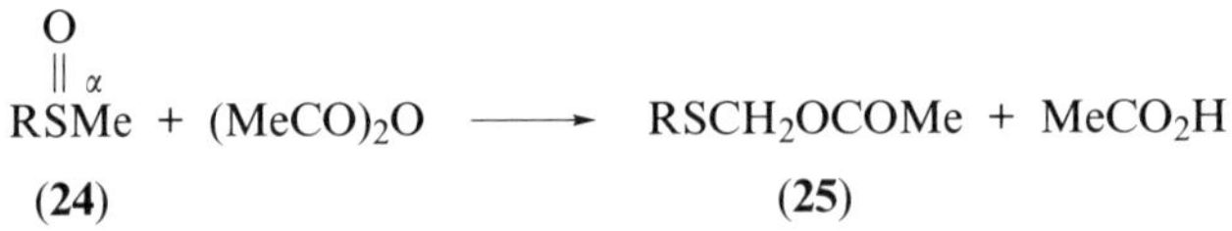

$$\underset{(\mathbf{24})}{RS(=O)Me^{\alpha}} + (MeCO)_2O \longrightarrow \underset{(\mathbf{25})}{RSCH_2OCOMe} + MeCO_2H$$

Scheme 15

The mechanism of the Pummerer rearrangement is probably as shown in Scheme 16. The rearrangement involves four stages. First is the formation of an

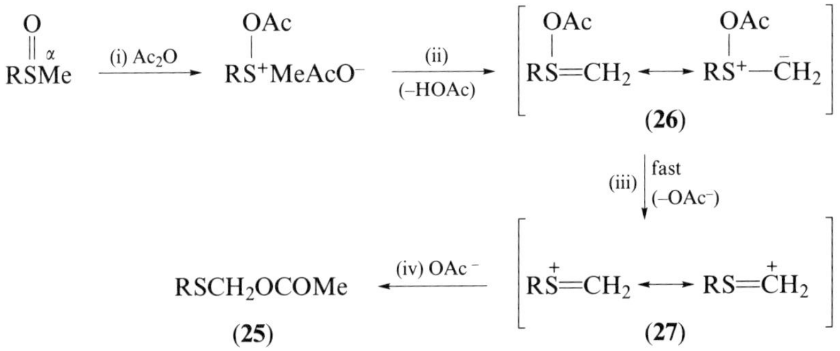

$$RS(=O)Me^{\alpha} \xrightarrow{\text{(i) } Ac_2O} RS^+(OAc)Me\ AcO^- \xrightarrow[(-HOAc)]{\text{(ii)}} \left[RS(OAc){=}CH_2 \longleftrightarrow RS^+(OAc){-}\bar{C}H_2\right] (\mathbf{26})$$

$$(\mathbf{26}) \xrightarrow[\text{fast } (-OAc^-)]{\text{(iii)}} \left[R\overset{+}{S}{=}CH_2 \longleftrightarrow RS{=}\overset{+}{C}H_2\right] (\mathbf{27}) \xrightarrow{\text{(iv) } OAc^-} \underset{(\mathbf{25})}{RSCH_2OCOMe}$$

Scheme 16

intermediate acylsulfonium ylide (**26**), which rapidly decomposes to a sulfur-stabilised carbocation (**27**). The latter can be trapped by a suitably placed electron-rich centre to result in cyclisation, as shown in the Pummerer rearrangement of the β-keto sulfoxide (**28**) to give the bicyclic sulfide (**29**) (Scheme 17).

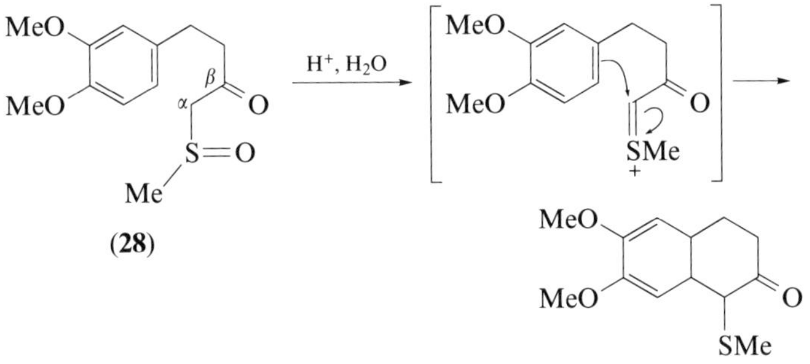

H^+, H_2O

(**28**)

(**29**)

Scheme 17

The Pummerer rearrangement can also be effected by other acid anhydrides, acid chlorides or hydrochloric acid.

Sulfoxides (**30**) on warming can undergo a [1,3]-sigmatropic rearrangement to the corresponding sulfenates (**31**) (Scheme 18)

$$\mathrm{Ar\overset{O}{\overset{\|}{S}}CH_2X} \xrightleftharpoons{\text{heat}} \mathrm{ArSOCH_2X}$$

(**30**) (**31**)

(X = OR or NRR′)

Scheme 18

Thermal and base-catalysed elimination

Sulfoxides on prolonged heating in the presence of strong bases eliminate sulfenic acids which can be trapped by the addition of activated alkenic compounds, e.g. ethyl acrylate (see Chapter 4, p. 53). Some examples are shown in Scheme 19. The thermolysis involves a *syn*-elimination as shown for the reaction of the diastereoisomeric 1,2-diphenylpropyl phenyl sulfoxide (**32**) to give *cis*-1,2-diphenylpropene (**33**) (Scheme 19).

The most important application of sulfoxide thermolysis is probably in the field of β-lactam antibiotics.[6] Penicillin sulfoxide (**34**) by heating is converted into the more stable isomer (**35**). The reaction occurs via the intermediate sulfenic acid (**36**). This sulfenic acid, by treatment with trimethyl phosphite in glacial acetic acid, can be reduced to the thiol acetate (**37**), which can be isolated (Scheme 20).

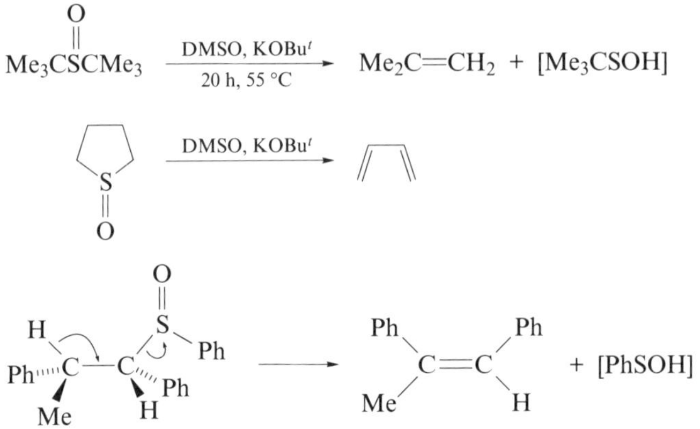

$$\mathrm{Me_3CS(O)CMe_3} \xrightarrow[\text{20 h, 55 °C}]{\text{DMSO, KOBu}^t} \mathrm{Me_2C{=}CH_2} + [\mathrm{Me_3CSOH}]$$

DMSO, KOBu^t

(32) (33) + [PhSOH]

Scheme 19

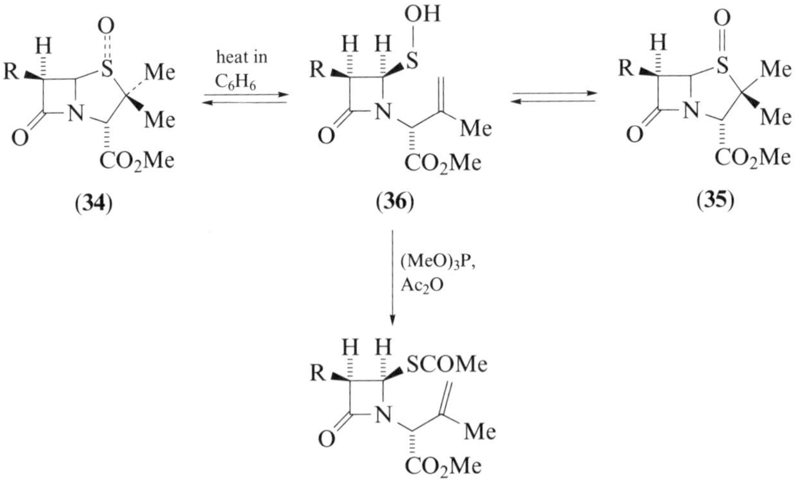

heat in C_6H_6

(34) (36) (35)

$(MeO)_3P$, Ac_2O

(37)

Scheme 20

Dimethyl sulfoxide (DMSO) can be used for carbon–carbon bond formation and as a precursor for the synthesis of stabilised sulfur ylides (see Chapter 3, p. 33). The sulfoxy (S=O) group, like the sulfone (SO_2) group, has a stabilising effect on an adjacent carbanion; hence, sulfinyl carbanions, like sulfonyl carbanions (see Chapter 10, p. 200), are useful reagents in organic synthesis. The carbanions (**39**) derived from alkylthioalkyl sulfoxides (**38**) are particularly valuable intermediates in syntheses (Scheme 21).

Carbanions such as (**39**) can be readily alkylated, and by subsequent hydrolytic desulfuration the products are converted into aldehydes (**40**) or ketones (**41**)

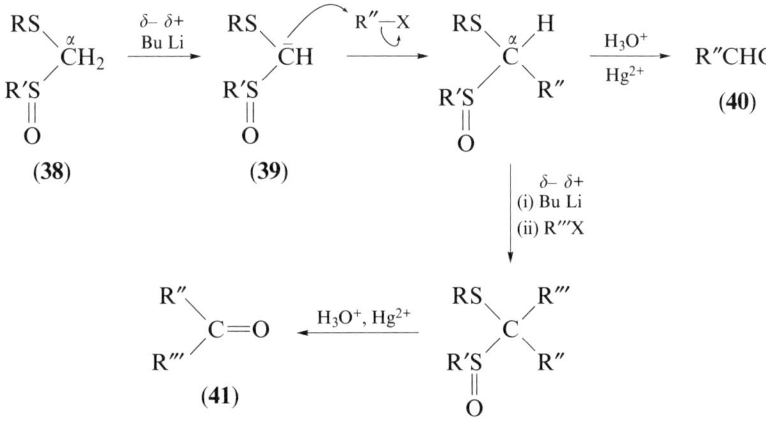

Scheme 21

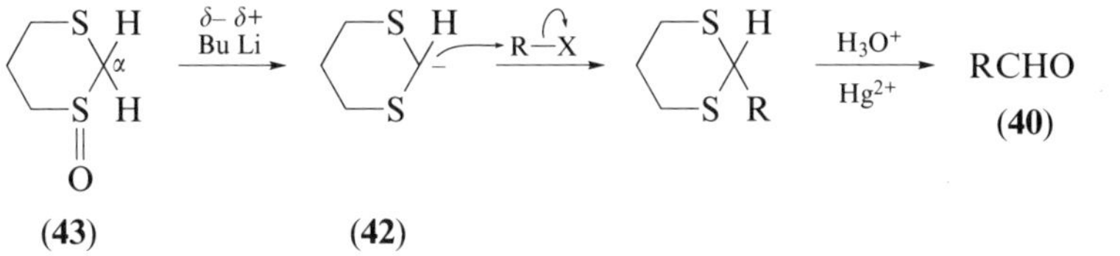

Scheme 22

(Scheme 21). Some important α-sulfinyl carbanions (**42**) are derived from dithioacetal *S*-oxides (**43**), and on subsequent alkylation and hydrolysis they yield the aldehydes (**40**) (Scheme 22).

Unlike the carbanions from 1,3-dithians (see Chapter 3, p. 31 and Chapter 6, p. 90), the sulfinyl carbanions have the ability to undergo Michael additions (conjugate or 1,4-additions) with α,β-unsaturated carbonyl compounds. For instance, the secondary carbanion (**44**) from the ethyl ethylthiomethyl sulfoxide (**45**) may be sucessively reacted with ethyl iodide and 3-butene-2-one to give heptan-2,5-dione (**46**) via the tertiary carbanion (**47**), as shown in Scheme 23. The carbanion (**47**) may also be condensed with propyl bromide, and hydrolysis of the product yields ethyl propyl ketone (**48**) (Scheme 23).

Sulfinyl carbanions easily add to aldehydes and ketones to yield β-ketosulfoxides, thus providing another route to substituted ketones. With α,β-unsaturated ketones (**49**), sulfinyl carbanions undergo 1,4-addition to give γ-ketosulfoxides (**50**) (Scheme 24).

Vinyl sulfoxides have been extensively employed as dienophiles in Diels–Alder [2 + 4] cycloadditions. For instance, the vinyl sulfoxide (**51**) adds to cyclopentadiene (**52**) to give the adducts (**53**) and (**54**) (Scheme 25). In the cycloaddition, the *endo*, *syn*-form (**54**) (80% yield) is favoured in comparison with the *endo*, *anti*-isomer (**53**).

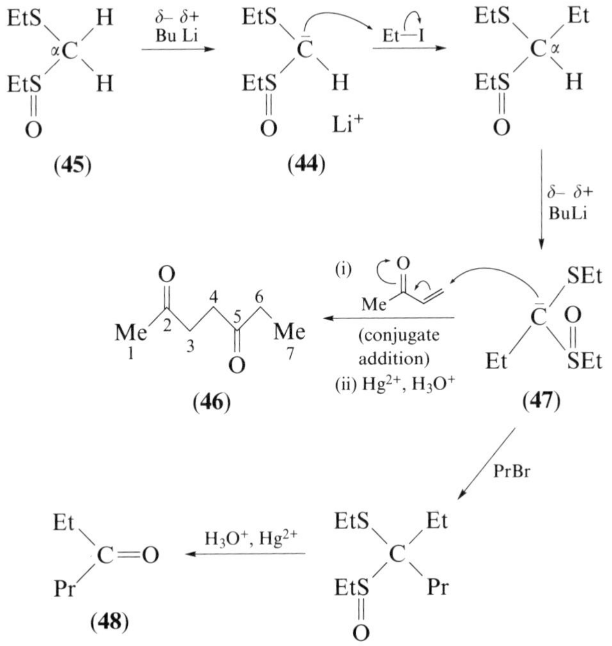

EtS, H, αC, EtS(=O), H (45) —δ− δ+ Bu Li→ EtS, C̄, EtS(=O), H, Li+ (44) —Et—I→ EtS, Et, Cα, EtS(=O), H —δ− δ+ BuLi→ SEt, C̄, O=SEt, Et (47)

(47) —(i) Me(C=O)CH=CH₂ (conjugate addition) (ii) Hg^{2+}, H_3O^+→ Me(1)–C(2)(=O)–(3)–(4)–C(5)(=O)–(6)–Me(7) (46)

(47) —PrBr→ EtS, Et, C, EtS(=O), Pr —H_3O^+, Hg^{2+}→ Et(Pr)C=O (48)

Scheme 23

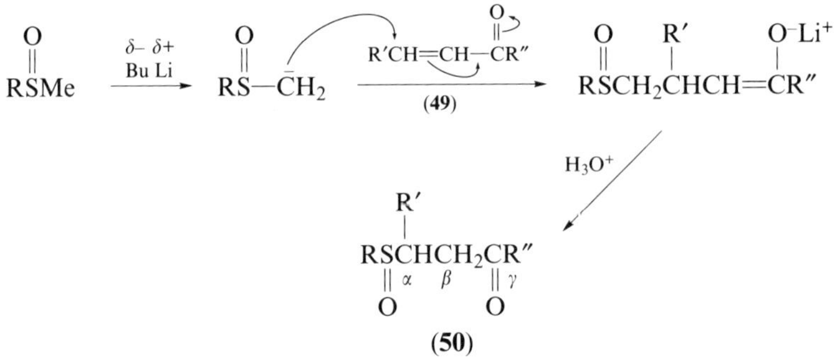

RS(=O)Me —δ− δ+ Bu Li→ RS(=O)—C̄H₂ —R′CH=CH—C(=O)R″ (49)→ RS(=O)CH₂CH(R′)CH=C(O⁻Li⁺)R″ —H_3O^+→ RS(=O)CH(R′)CH₂C(=O)R″ (α, β, γ) (50)

Scheme 24

PhS(=O)CH=CHCO₂H (51) + (52) ⟶ (53) [CO₂H, S(=O)Ph] + (54) [CO₂H, S(=O)Ph]

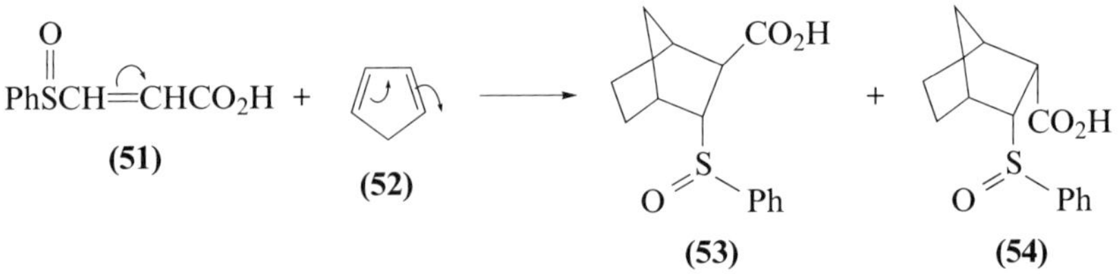

Scheme 25

Sulfones

Sulfones (**55**) are generally prepared by the oxidation of the appropriate sulfides (**2**) (Scheme 26).[7] The oxidation is generally performed by treatment with peroxycarboxylic acids, but other oxidants may be used (see Chapter 10, p. 195). Diarylsulfones are often obtained by the Friedel–Crafts reaction, and special methods are available for the synthesis of substituted sulfones such as vinyl, hydroxy and halosulfones (see Chapter 10, p.197). Sulfonyl ethers (**56**) can be prepared by reaction of a suitable chloroether (**57**) with the appropriate - sodium sulfinate (**58**) (Scheme 27).

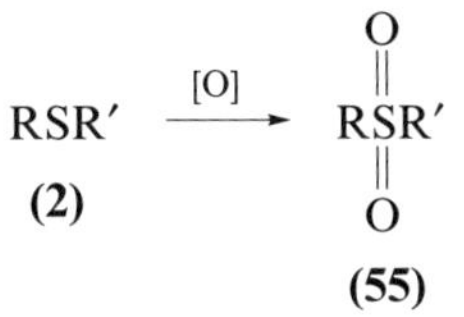

$$\underset{(\mathbf{2})}{RSR'} \xrightarrow{[O]} \underset{(\mathbf{55})}{RS(=O)_2R'}$$

Scheme 26

$$\underset{(\mathbf{57})}{ArOCH_2Cl} + \underset{(\mathbf{58})}{Ar'SO_2Na} \longrightarrow \underset{(\mathbf{56})}{ArOCH_2SO_2Ar'} + NaCl$$

Scheme 27

Epoxysulfones (**59**) are obtained by epoxidation of alkenyl sulfones (**60**) (Scheme 28).

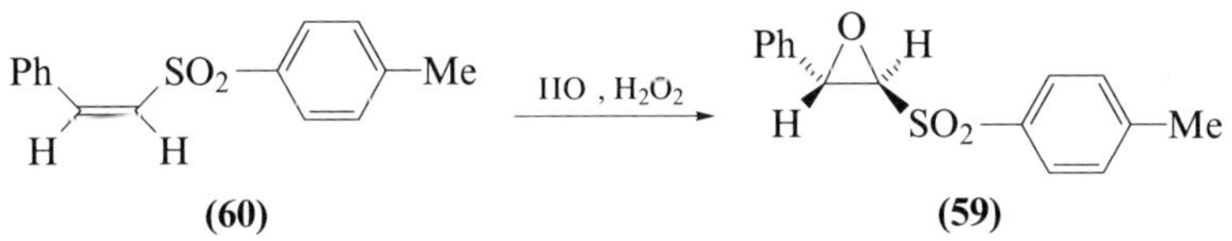

$$\underset{(\mathbf{60})}{PhCH{=}CH{-}SO_2{-}C_6H_4{-}Me} \xrightarrow{HO^-,\ H_2O_2} \underset{(\mathbf{59})}{Ph{-}CH\overset{O}{—}CH{-}SO_2{-}C_6H_4{-}Me}$$

Scheme 28

α-Ketosulfones (**61**) are prepared by reaction of a thiol (**62**) with an acyl chloride followed by oxidation of the intermediate acylthiolate (**63**) (Scheme 29).

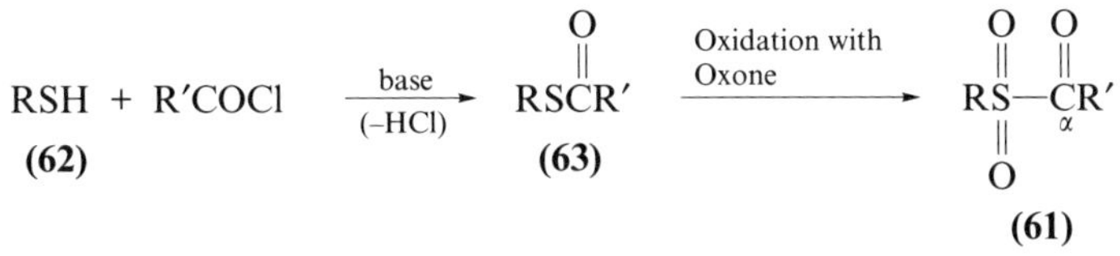

$$\underset{(\mathbf{62})}{RSH} + R'COCl \xrightarrow[(-HCl)]{base} \underset{(\mathbf{63})}{RSC(=O)R'} \xrightarrow{\text{Oxidation with Oxone}} \underset{(\mathbf{61})}{RS(=O)_2{-}\underset{\alpha}{C}(=O)R'}$$

Scheme 29

β-Ketosulfones (**64**) and γ-ketosulfones (**65**) may be obtained from enamines (**66**) or suitable α,β-unsaturated carbonyl compounds (**67**), as shown in Scheme 30.

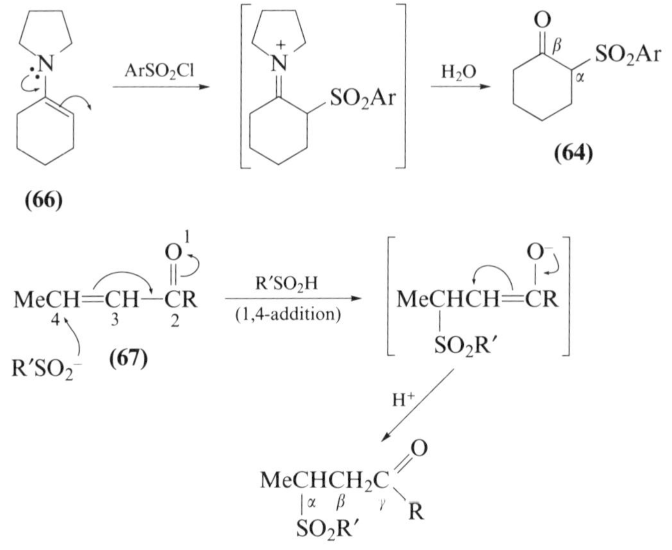

(66) $\xrightarrow{ArSO_2Cl}$ [SO_2Ar] $\xrightarrow{H_2O}$ (64)

$MeCH{=}CH{-}CR$ (67) $\xrightarrow[(1,4\text{-addition})]{R'SO_2H}$ [$MeCHCH{=}CR$, SO_2R'] $\xrightarrow{H^+}$ $MeCHCH_2C(O)R$, SO_2R' (65)

Scheme 30

Ketosulfones are valuable as synthetic intermediates because of their facile alkylation and acylation (see Chapter 10, p. 205).[7]

Aminosulfones (**68**) may be prepared by conjugate 1,4-addition of amines to α,β-unsaturated sulfones or by a Mannich-type reaction of a sulfinic acid with an aldehyde in the presence of an amine (Scheme 31).

$$ArSO_2Na + CH_2{=}O + PhNH_2{\cdot}HCl \longrightarrow ArSO_2CH_2NHPh \quad \textbf{(68)}$$

Scheme 31

Hydroxy and halosulfones are obtained as described in Chapter 10 (pp. 197 and 199, respectively). Sulfur-substituted sulfones, e.g. (**69**) and (**70**), can be synthesised by sulfenylation of sulfonyl carbanions by treatment with a sulfinate ester (**71**) or a thiophthalimide (**72**) (Scheme 32).

Sulfonyl carbanions are even more stable than sulfinyl carbanions and are consequently of greater significance in synthesis. They can be alkylated and acylated at the α-carbon atom using organolithium bases, and cyclic sulfones can be formed by intramolecular alkylation, (see Chapter 10, p. 202). Sulfonyl carbanions (**73**) also react with terminal epoxides, and this reaction is applicable for the synthesis of unsaturated alcohols (**74**) (Scheme 33).

Sulfonyl carbanions undergo aldol-type reactions with aldehydes and ketones to give β-hydroxy sulfones which can be converted into alkenes (the Julia reaction) (see Chapter 10, p. 197). With allyl methyl sulfones (**75**) and α,β-unsaturat-

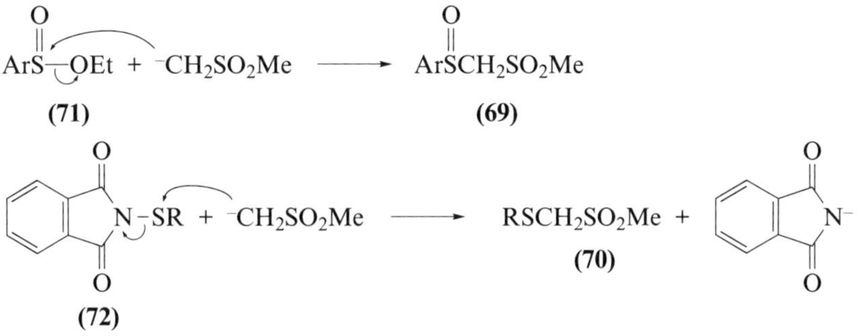

$$ArS(=O){-}OEt\ (\mathbf{71}) + {}^{-}CH_2SO_2Me \longrightarrow ArS(=O)CH_2SO_2Me\ (\mathbf{69})$$

$$(\mathbf{72}) + {}^{-}CH_2SO_2Me \longrightarrow RSCH_2SO_2Me\ (\mathbf{70}) + \text{phthalimide anion}$$

Scheme 32

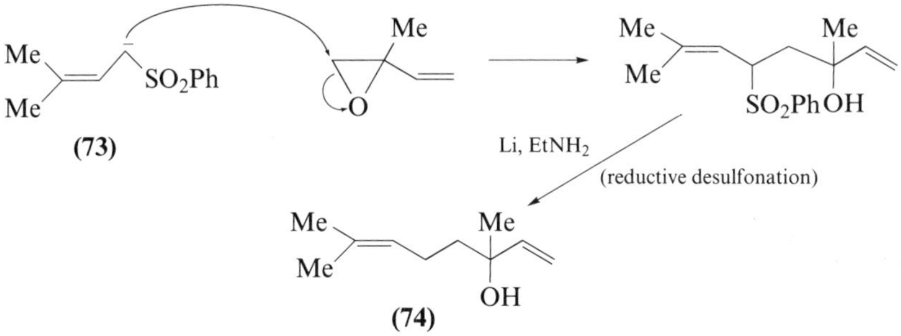

$$(\mathbf{73}) \longrightarrow \xrightarrow{\text{Li, }EtNH_2\text{ (reductive desulfonation)}} (\mathbf{74})$$

Scheme 33

ed carbonyl compounds (**63**), the initial conjugate addition is followed by proton transfer and intramolecular acylation of the sulfone to give the cyclic β-keto-sulfone (**76**) (Scheme 34).

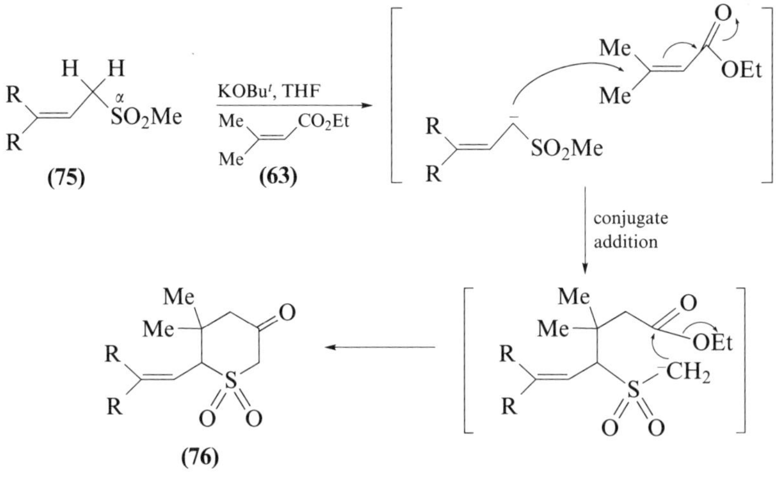

$$(\mathbf{75}) \xrightarrow{KOBu^t,\ THF,\ (\mathbf{63})} [\ ] \xrightarrow{\text{conjugate addition}} [\ ] \longrightarrow (\mathbf{76})$$

Scheme 34

Sulfonyl carbanions may also undergo the Smiles rearrangement, which is an intramolecular process and follows the general pattern shown in Scheme 35. A specific example is the rearrangement of 2-hydroxy-2′-nitrodiphenyl sulfone (**77**) to the diphenyl ether sulfinate (**78**) (Scheme 36). In this rearrangement, the ArO^- group acts as the nucleophile and the SO_2Ar is the leaving group, the *ortho*-position being activated by the electron-withdrawing nitro group (see the related Truce–Smiles rearrangement, Chapter 10, p. 202).

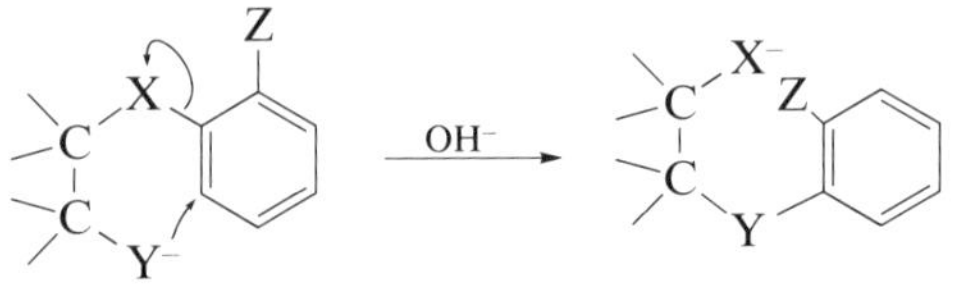

X is usually SO_2 but can be SO or S
Y is the conjugate base of OH, NH_2, NHR or SR
Z is generally NO_2

Scheme 35

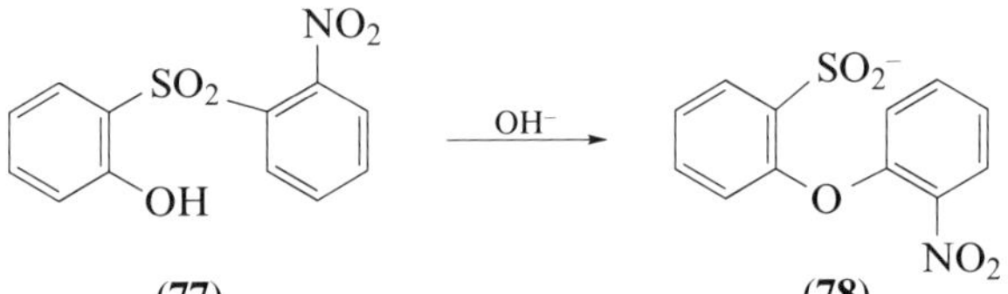

(**77**) (**78**)

Scheme 36

Sulfones and sulfoxides containing at least one β-hydrogen atom undergo a bimolecular elimination reaction (E2) an treatment with alkoxides, leading preferentially to the formation of the least substituted alkene (Hofmann's rule) (Scheme 37). The reaction resembles the analogous eliminations of quaternary ammonium and sulfonium salts (see Chapter 6, p. 83).

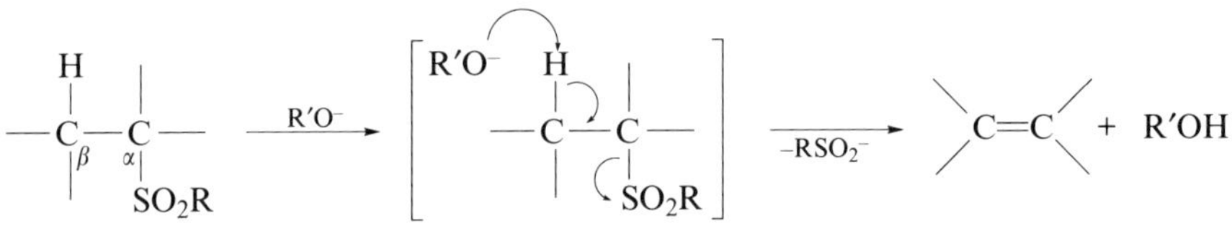

Scheme 37

Sulfoxides undergo thermal *syn*-elimination (see p. 69), but sulfones do not suffer an analogous reaction.

Sulfones are valuable in carbon–carbon bond formation (see Chapter 10, p. 197); alkylation occurs easily with allyl and benzyl halides, and this can be utilised in the formation of 2-alkenes. The reaction is illustrated by the conversion of allyl phenyl sulfone (**79**) to the alkene (**80**) (Scheme 38). The last step involves reductive elimination of the phenylsulfonyl group by treatment with potassium graphite.

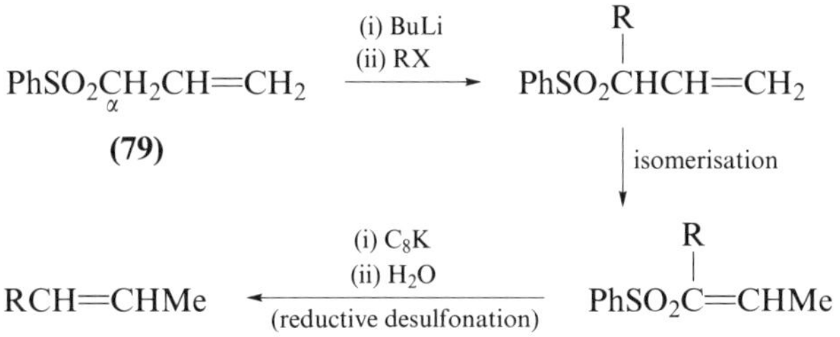

$$\underset{\textbf{(79)}}{PhSO_2\underset{\alpha}{C}H_2CH{=}CH_2} \xrightarrow[\text{(ii) RX}]{\text{(i) BuLi}} PhSO_2CH(R)CH{=}CH_2 \xrightarrow{\text{isomerisation}} PhSO_2C(R){=}CHMe \xrightarrow[\text{(reductive desulfonation)}]{\text{(i) } C_8K \text{, (ii) } H_2O} \underset{\textbf{(80)}}{RCH{=}CHMe}$$

Scheme 38

t-Butylphenyl sulfones (**81**) are useful for selective generation of *ortho*-carbanions e.g. (**82**) and (**83**). The unstable dicarbanion (**83**) readily eliminates the sulfinate anion to give the *o*-lithobenzyne (**84**) which finally affords the *para*-disubstituted benzene (**85**) (Scheme 39).

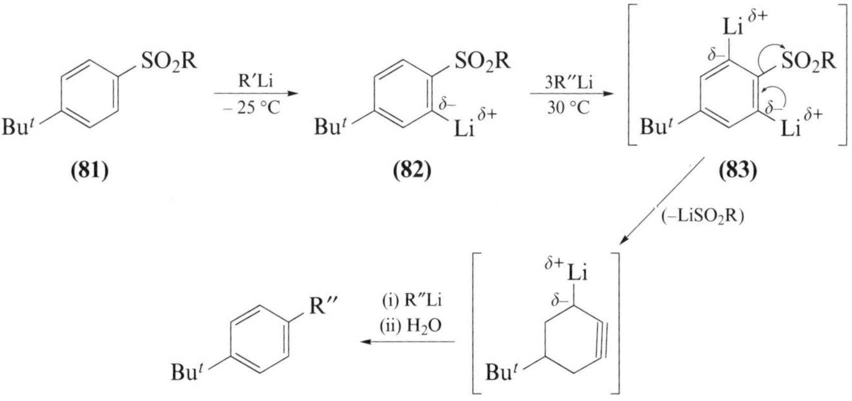

(81) $\xrightarrow[-25\ °C]{R'Li}$ **(82)** $\xrightarrow[30\ °C]{3R''Li}$ **(83)** $\xrightarrow{(-LiSO_2R)}$ **(84)** $\xrightarrow{\text{(i) } R''Li \text{, (ii) } H_2O}$ **(85)**

Scheme 39

Cyclic three-, four- and five-membered sulfones show a number of interesting special reactions which are described in Chapter 10 (see p. 210).

Biological activity of sulfoxides and sulfones[8]

Sulfides (**2**), sulfoxides (**1**) and sulfones (**55**) are interconvertible (Scheme 40), and sulfoxides and sulfones exist as metabolites of thioether-containing chemicals. The sulfoxidations can be achieved in plants, animals and microorganisms by essentially two microsomal systems: one involving cytochrome P-450, and the other flavin-containing monooxygenase.[9a] The reverse reaction, namely the reduction of sulfones (**55**), can be achieved by mammalian tissues and microsomal enzymes.[9a]

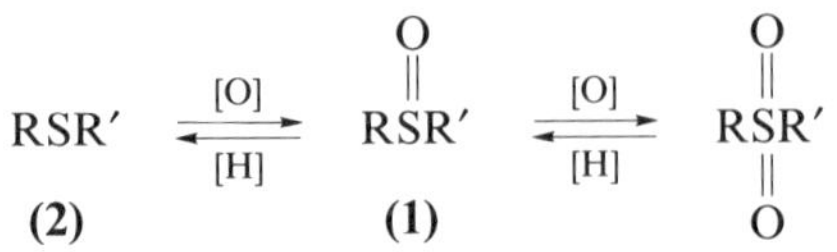

$$\underset{\textbf{(2)}}{RSR'} \underset{[H]}{\overset{[O]}{\rightleftharpoons}} \underset{\textbf{(1)}}{RS(=O)R'} \underset{[H]}{\overset{[O]}{\rightleftharpoons}} \underset{\textbf{(55)}}{RS(=O)_2R'}$$

Scheme 40

Alliins (**86**) (*S*-alkenyl- and *S*-alkylcysteine sulfoxides) occur in the plant genus *Allium* (onion, garlic, leek) (see chapter 4, p.7). The original alliin isolated from garlic (1951) was (+)-*S*-(prop-2-enyl)-L-cysteine sulfoxide (**86a**) (Figure 2). This was one of the first naturally occurring sulfoxides to be identified. Later work showed the presence of analogous compounds (**86b**)–(**86d**). The alkenylcysteine sulfoxides (**86**) are the immediate precursors of the thiol sulfinate allicins (**87**) (Figure 2), which are probably the major biologically active sulfur compounds in alliums (see Chapter 4, p. 56). The allicins probably derive from the biological oxidation of the corresponding dialkenyl disulfides (see Chapter 4, p. 60).[9b] Both symmetrical and unsymmetrical thiolsulfinates (**87**) have been found, and several of their metabolites, e.g. the sulfoxide (**88**) (Figure 2), have antithrombotic properties. Garlic may therefore be valuable in preventing heart disease, hence the increasing use of garlic capsules.

$RS(=O)CH_2CH(NH_2)CO_2H$ (**86**)

$RS(=O)SR'$ (**87**)

(**86a**) R = $CH_2{=}CHCH_2$
(**86b**) R = Me
(**86c**) R = $MeCH_2CH_2$
(**86d**) R = $MeCH{=}CH$

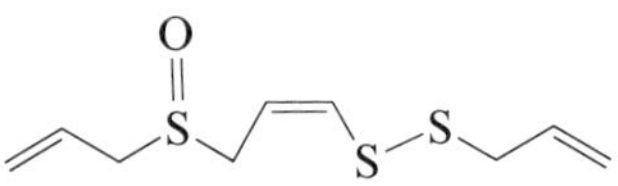

(**88**) (*Z*)–isomer

Figure 2

Several sulfoxides and sulfones are biologically active; examples include diaryl sulfones, e.g. dapson (**89**) (Figure 3). This is an antibacterial agent because it acts as a PABA antagonist like the sulfonamide drugs (see Chapter 11, p. 223). Other diaryl sulfones, e.g. tetradifon (**90**), are used as acaricides to control spider mites on plants.

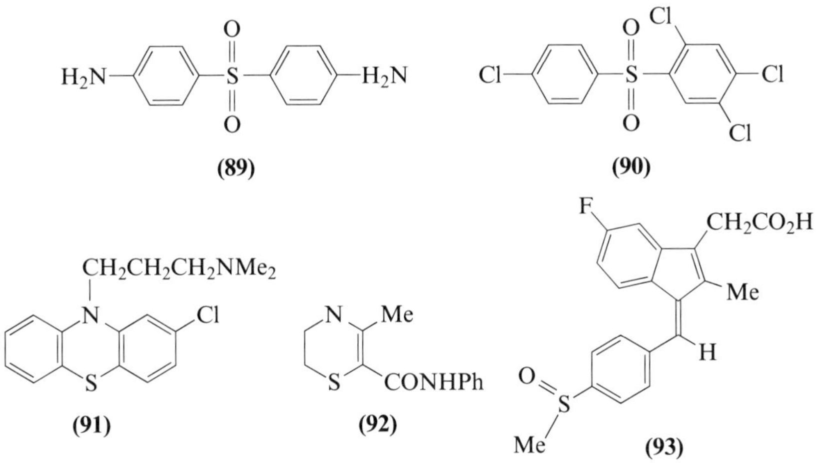

(89) **(90)**

(91) **(92)** **(93)**

Figure 3

As indicated previously (Scheme 40), sulfur centres are susceptible to oxidation; consequently, biologically active sulfur compounds may undergo metabolic oxidation to the corresponding sulfoxides and sometimes to the sulfones. In this oxidation, the sulfur may be part of an aliphatic chain or may be in a ring; thus, both the tranquiliser drug chlorpromazine (**91**) and the systemic agricultural fungicide carboxin (**92**) (Figure 3) are converted *in vivo* to the corresponding sulfoxides. Sulfoxides may also be partly reduced *in vivo* to the sulfides (Scheme 40), so that in many cases a redox equilibrium is established; this occurs with the antirheumatic sulfoxide drug sulindac (**93**) (Figure 3).

References

1. T. Durst, in *Comprehensive Organic Chemistry* (Eds D. H. R. Barton and W. D. Ollis), Vol. 3, Pergamon Press, Oxford, 1979, p. 121.
2. J. Drabowicz, P. Kielbasinski and M. Mikolajczyk, in *The Chemistry of Sulfones and Sulfoxides* (Eds S. Patai, Z. Rappoport and C. Sterling), Wiley, Chichester, 1988, p. 233.
3. M. C. Carreno, *Chem. Rev.*, **95**, 1717 (1995).
4. T. Tidwell, *Org. React.*, **39**, 297 (1990).
5. O. DeLucci, U. Miotti and G. Modena, *Org. React.*, **40**, 157 (1991).
6. P. G. Sammes, *Chem. Rev.*, **76**, 113 (1976).
7. N. S. Simpkins, *Sulfones in Organic Synthesis*, Pergamon Press, Oxford, 1993.
8. A. Kalir and H. H. Kalir, in *Chemistry of Sulfur-Containing Functional Groups*, (Eds S. Patai and Z. Rappoport), Suppl. S, Wiley, Chichester, 1993 p. 957.
9. *Sulfur-Containing Drugs and Related Compounds—Chemistry, Biology and Toxicology* (Ed. L. A. Damani), Ellis Horwood, Chichester, 1987: (a) A. G. Renwick, Vol. 2, Part B, Chap. 5, p 133; (b) G. R. Fenwick and A. B. Hanley, Vol. 2, Part B, Chap. 10, p. 269.

6 SULFONIUM AND OXOSULFONIUM SALTS, SULFUR YLIDES AND SULFENYL CARBANIONS

Sulfonium salts are compounds containing a tricoordinate, positively charged sulfur atom. Sulfonium salts (**1**) are prepared by the reaction of alkyl halides on dialkyl sulfides, the latter functioning as nucleophiles (Scheme 1).[1] The reaction is facilitated by the use of polar solvents like methanol. Methyl iodide is the most reactive alkyl halide, and in the alkyl iodides the reactivity decreases with increasing chain length.

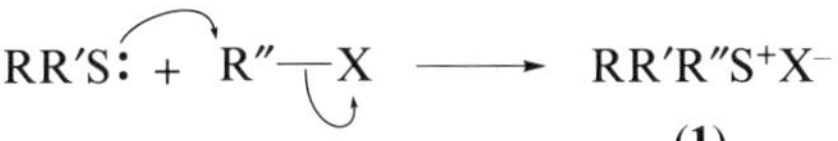

$$RR'S\!: + R''\!-\!X \longrightarrow RR'R''S^{+}X^{-}$$

(**1**)

Scheme 1

The sulfonium ion (**2**), like a saturated carbon atom, possesses a tetrahedral configuration. The unshared electrons on the sulfur atom, however, unlike those of nitrogen in quaternary ammonium salts, can hold their configuration at ordinary temperatures.[1,2] Sulfonium salts containing three different groups (R, R′ and R″ all different) consequently can exist as optical isomers, e.g. (**2**) and (**3**), owing to the presence of the chiral sulfur atom (Figure 1).[2]

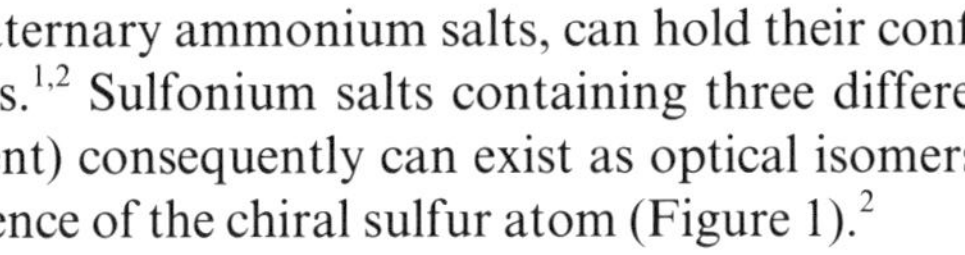

(**2**) (**3**) (**4**) $EtS^{+}(Me)CH_2CO_2H\ Br^{-}$

Figure 1

Ethyl(methyl)carboxymethylsulfonium bromide (**4**) was resolved into optical enantiomers by Pope and Peachey (1900), and since that time a large number of optically pure sulfonium salts have been obtained by resolution of racemic mixtures or by stereospecific syntheses. Chiral sulfonium salts can suffer stereomutation at sulfur by three major mechanisms (Scheme 2): (i) pyramidal inversion,

(ii) reversible dissociation into the sulfide and a carbocation via the S_N1 mechanism, and (iii) S_N2 attack at the α-carbon atom followed by re-formation of the sulfonium salt.

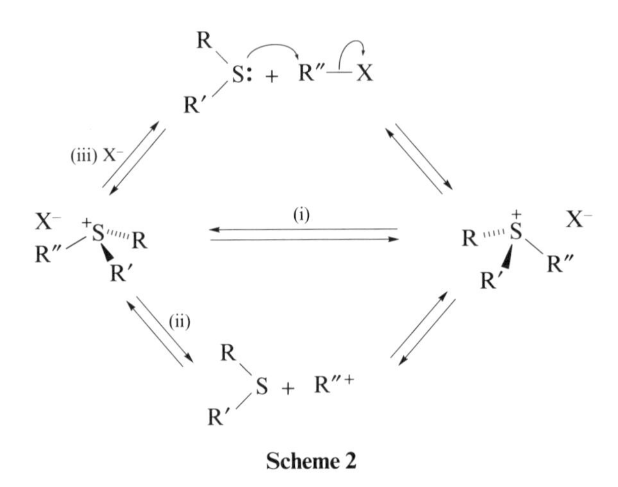

Scheme 2

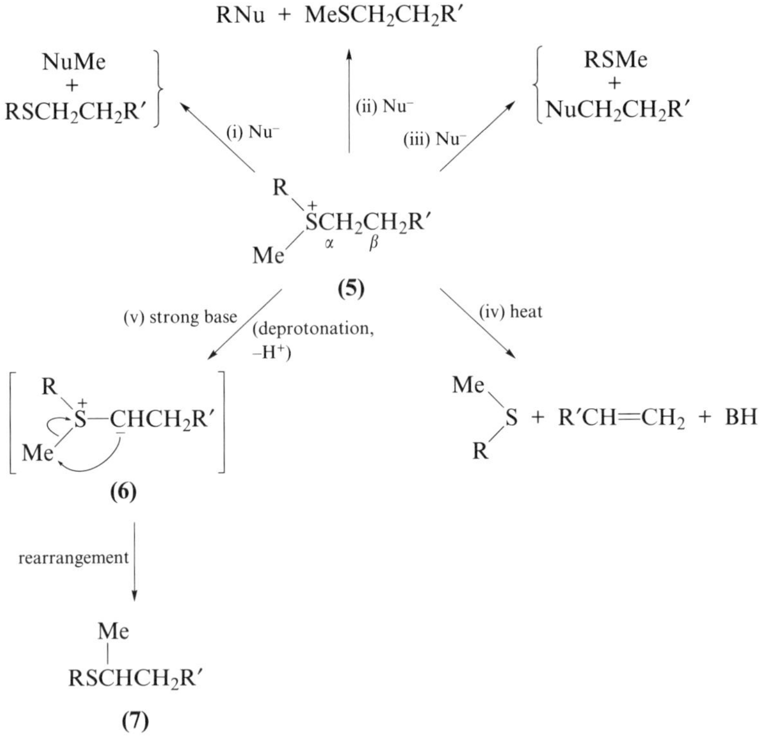

RNu + $MeSCH_2CH_2R'$

NuMe + $RSCH_2CH_2R'$

(i) Nu⁻

(ii) Nu⁻

(iii) Nu⁻

RSMe + $NuCH_2CH_2R'$

R(Me)$\overset{+}{S}CH_2CH_2R'$ (α, β)

(5)

(v) strong base (deprotonation, $-H^+$)

(iv) heat

[R(Me)$\overset{+}{S}$—$\overset{-}{C}HCH_2R'$]

(6)

rearrangement

$RS\underset{}{C}(Me)HCH_2R'$

(7)

Me(R)S + $R'CH{=}CH_2$ + BH

Scheme 3

Sulfonium salts undergo several synthetically valuable reactions. They are susceptible to carbon–sulfur bond fission via nucleophilic diplacements. Thus, the sulfonium salt (**5**) can react with a nucleophile (Nu^-) by attack of Nu^- on either the Me, R or CH_2CH_2R' group, namely route (i), (ii) or (iii) in Scheme 3.

Warming the sulfonium salt (**5**) may cause elimination of a β-hydrogen to yield a mixture of the sulfide and the alkene, i.e. route (iv) in Scheme 3. On the other hand, proton abstraction from (**5**) by treatment with a strong base gives the sulfur ylide (**6**) which can undergo molecular rearrangement to form the sulfide (**7**). Sulfonium salts are unstable in basic media as a result of their tendency to undergo nucleophilic substitution reactions.

Only four naturally occurring sulfonium salts are known. Two examples are dimethyl-β-propiothetin (**8**) and *S*-methyl-L-methionine (**9**) (Figure 2). Propiothetin (**8**) occurs in algae, plankton, fish and molluscs, and *S*-methyl-L-methionine (**9**) in cabbage, celery and other vegetables.

$$Me_2\overset{+}{S}CH_2CH_2CO_2H\ X^- \qquad Me_2\overset{+}{S}CH_2CH_2\underset{}{\overset{NH_2}{\overset{|}{C}}}HCO_2H$$

(**8**) (**9**)

Figure 2

Sulfonium salts by treatment with moist silver oxide, which reacts as silver hydroxide (AgOH), are converted into the corresponding sulfonium hydroxides (**10**), and the latter on heating undergo a bimolecular (E2) elimination reaction in which the Hofmann rule is obeyed, so that the major product is the least substituted alkene (**11**), or the Hofmann product (HP). The elimination resembles the analogous reaction of quaternary ammonium hydroxides, except that as the dialkyl sulfide leaving group is less bulky than the tertiary amine group, the relative yield of the HP is generally smaller (Scheme 4).[2,3] The substrate (**10**) contains two types of β-hydrogen atoms, so the elimination yields a mixture of the two alkenes (**11**) and (**12**) in which the former predominates (Scheme 4).

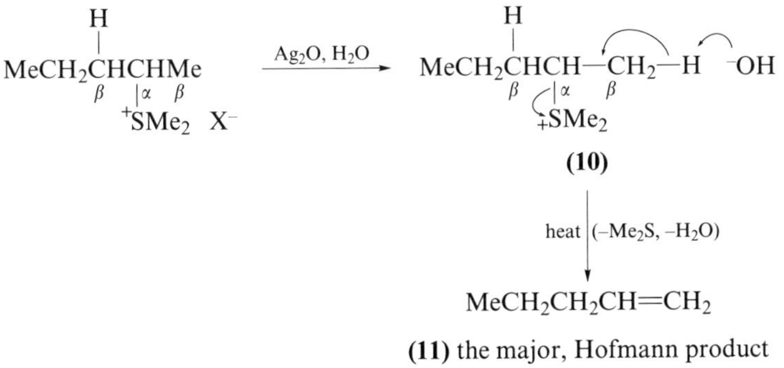

$$MeCH_2\overset{H}{\overset{|}{C}}HCHMe\ (\beta,\ \alpha,\ \beta)\ \text{with}\ {}^+SMe_2\ X^-\ \text{on}\ C_\alpha \xrightarrow{Ag_2O,\ H_2O} MeCH_2\overset{H}{\overset{|}{C}}HCH{-}CH_2{-}H\ \ {}^-OH\ \ (\beta,\ \alpha,\ \beta)\ \text{with}\ {}^+SMe_2\ \text{on}\ C_\alpha$$

(**10**)

$$\xrightarrow{\text{heat}\ (-Me_2S,\ -H_2O)} MeCH_2CH_2CH{=}CH_2$$

(**11**) the major, Hofmann product

Scheme 4 *(continued)*

Scheme 4 *(continued)*

or

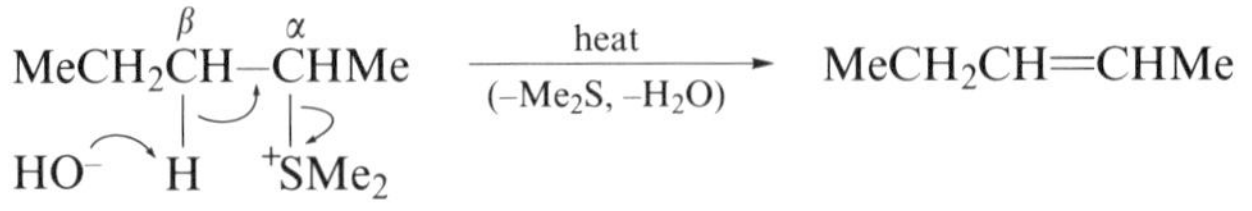

$$\underset{\substack{|\\ \mathrm{HO^- \quad H}}}{\mathrm{MeCH_2\overset{\beta}{C}H}}-\underset{\substack{|\\ \mathrm{^+SMe_2}}}{\overset{\alpha}{\mathrm{C}}\mathrm{HMe}} \xrightarrow[(-\mathrm{Me_2S},\ -\mathrm{H_2O})]{\text{heat}} \mathrm{MeCH_2CH{=}CHMe}$$

(10) **(12)** the minor, Saytzeff product

Scheme 4

The elimination reaction is not confined to the sulfonium hydroxides but occurs with a wide range of sulfonium salts. Studies of these eliminations have demonstrated that (i) increasing the size of the leaving group (Z) in the substrate **(13)** (Scheme 5) increases the yield of the HP **(11)** relative to the Saytzeff product (SP) **(12)**; (ii) increasing the degree of branching at the γ-carbon atom reduces the relative amount of the Saytzeff product (SP), because it makes attack by the anion on the β-carbon atom of the sulfonium hydroxide **(14)** more difficult (Scheme 6); and (iii) increasing the steric size of the anion (X^-) favours the formation of the HP as compared with the SP; for instance, with the sulfonium salt **(15)**, the yield of the HP markedly increases with the bulk of the anion (Scheme 7) because the HP results from attack of the anion on the least hindered β-hydrogen atom.

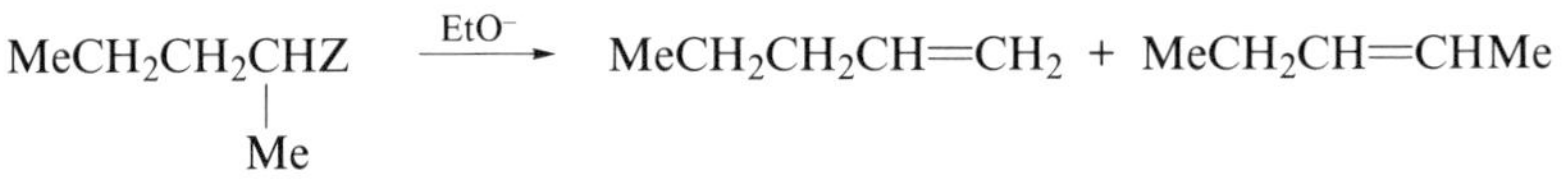

$$\mathrm{MeCH_2CH_2\underset{\substack{|\\ \mathrm{Me}}}{C}HZ} \xrightarrow{\mathrm{EtO^-}} \mathrm{MeCH_2CH_2CH{=}CH_2} + \mathrm{MeCH_2CH{=}CHMe}$$

(13) **(11)** HP **(12)** SP

Z (increasing size of Z ↓)	(11) HP yield (%)	(12) SP yield (%)
Br	31	69
OTs	48	52
$\overset{+}{\mathrm{S}}\mathrm{Me_2}$	87	13
$\overset{+}{\mathrm{N}}\mathrm{Me_3}$	98	2

Scheme 5

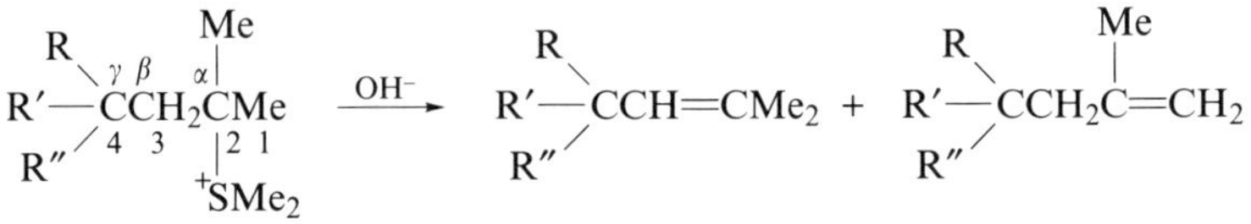

$$\mathrm{RR'R''\overset{\gamma}{C}\overset{\beta}{C}H_2\overset{\alpha}{C}(Me)(\overset{+}{S}Me_2)Me} \xrightarrow{\mathrm{OH^-}} \mathrm{RR'R''CCH{=}CMe_2} + \mathrm{RR'R''CCH_2C(Me){=}CH_2}$$

(carbons numbered 4, 3, 2, 1 from γ to Me)

(14)	SP (%)	HP (%)
R = R′ = R″ = H no branching	70	30
R = Me, R′ = R″ = H	50	50
R = R′ = Me, R″ = H	45	55
R = R′ = R″ = Me	15	85

Scheme 6

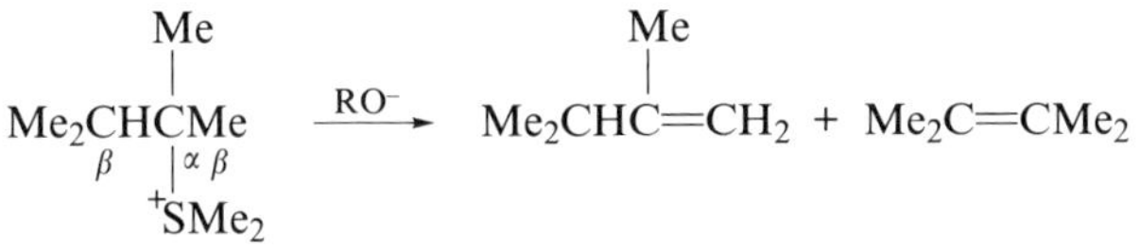

$$\underset{\substack{\beta\ \ |\alpha\,\beta \\ \overset{+}{S}Me_2}}{Me_2CH\overset{\overset{Me}{|}}{C}Me} \xrightarrow{RO^-} Me_2CH\overset{\overset{Me}{|}}{C}=CH_2 + Me_2C=CMe_2$$

(15)

RO^-	HP (%)	SP (%)
EtO^-	20	80
Me_3CO^-	73	27
Et_3CO^-	99	1

Scheme 7

Halosulfonium salts are unstable in the absence of Lewis acids. The instability arises from their liability to nucleophilic attack by the counterion, which results in bond cleavage (Scheme 8).

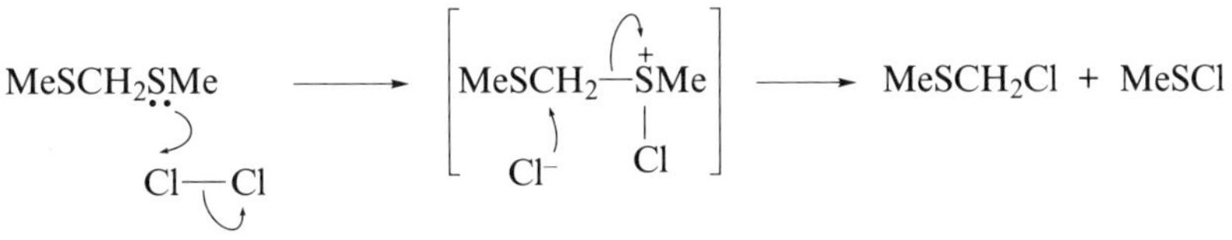

$$MeSCH_2\ddot{S}Me + Cl-Cl \longrightarrow \left[MeSCH_2-\overset{+}{S}(Cl)Me \quad Cl^-\right] \longrightarrow MeSCH_2Cl + MeSCl$$

Scheme 8

When sulfonium salts possessing at least one α-hydrogen atom are treated with strong bases e.g. butyllithium, the corresponding sulfonium or sulfur ylides (**16**) are formed (Scheme 9).

$$RR'\overset{+}{S}Me\ X^- \xrightarrow[(-HX)]{BuLi,\ THF} \left[RR'\overset{\delta-\ \delta+}{S=CH_2} \longleftrightarrow RR'\overset{+}{S}-\overset{-}{C}H_2\right]$$

(16)

Scheme 9

Sulfur ylides are of great importance in organic synthesis, especially in the preparation of epoxides and cyclopropanes (see Chapter 10, p. 188). Sulfonium ylides[1] are prone to suffer rearrangement reactions; a simple example is the Stevens rearrangement in which a sulfonium ylide (**17**) undergoes an intramolecular nucleophilic displacement (Scheme 10). The ylide (**17**) is formed by abstraction of the most acidic proton from the sulfonium salt (**18**) by treatment with a base (sodium methoxide). The closely related Sommelet rearrangement occurs with sulfonium salts containing aromatic nuclei; for example benzyldimethylsulfonium bromide (**19**) on treatment with sodium amide in liquid ammonia gives *o*-methylbenzyl methyl sulfide (**20**) (Scheme 11). The final stage involves a proton shift, the driving force for which is the regeneration of the stable aromatic

$PhCO\underset{\alpha}{C}H_2\overset{+}{S}(Me)CH_2Ph\ X^-$ **(18)** $\xrightarrow{MeO^-}$ $[PhCO\overset{-}{C}H\overset{+}{S}(Me)CH_2Ph]$ **(17)** $\xrightarrow{S_Ni}$ $PhCOCH(CH_2Ph)SMe$

Scheme 10

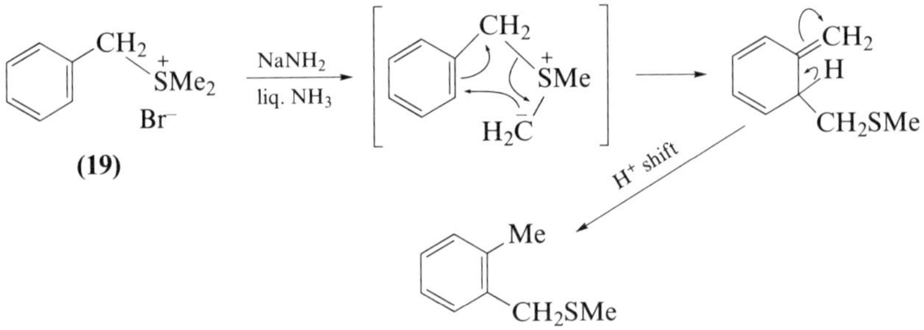

$PhCH_2\overset{+}{S}Me_2\ Br^-$ **(19)** $\xrightarrow[\text{liq. } NH_3]{NaNH_2}$ $[PhCH_2\overset{+}{S}(Me)\overset{-}{C}H_2]$ (CH_2, $\overset{+}{S}Me$, $H_2\overset{-}{C}$) → (CH_2, H, CH_2SMe) $\xrightarrow{H^+ \text{ shift}}$ o-$MeC_6H_4CH_2SMe$ (Me, CH_2SMe) **(20)**

Scheme 11

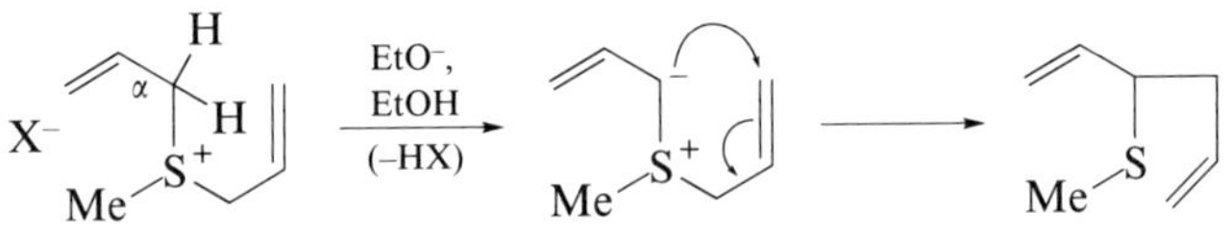

X^-, H, α, H, $Me-S^+$ **(21)** $\xrightarrow[(-HX)]{EtO^-,\ EtOH}$ $Me-S^+$ **(23)** → $Me-S$ **(22)**

Scheme 12

$Ph\underset{\alpha}{C}H_2\overset{+}{S}(Me)CH_2Ph\ X^-$ **(24)** $\xrightarrow[(-H^+)]{OH^-}$ $Ph\overset{-}{C}H\overset{+}{S}(Me)CH_2Ph$ $\xrightarrow{(i)}$ $PhCH(CH_2Ph)SMe$

or

$PhCH_2\overset{+}{S}(Me)\underset{\alpha}{C}H_2Ph$ (CH_2, $\overset{+}{S}Me$, α CH_2, Ph) **(24)** $\xrightarrow{OH^-}$ (CH_2, SMe, $\overset{-}{C}H$, Ph) $\xrightarrow{(ii)}$ (CH_2, H, CHSMe, Ph)

$\xrightarrow{H^+ \text{ shift}}$ (Me, SMe, CH, Ph)

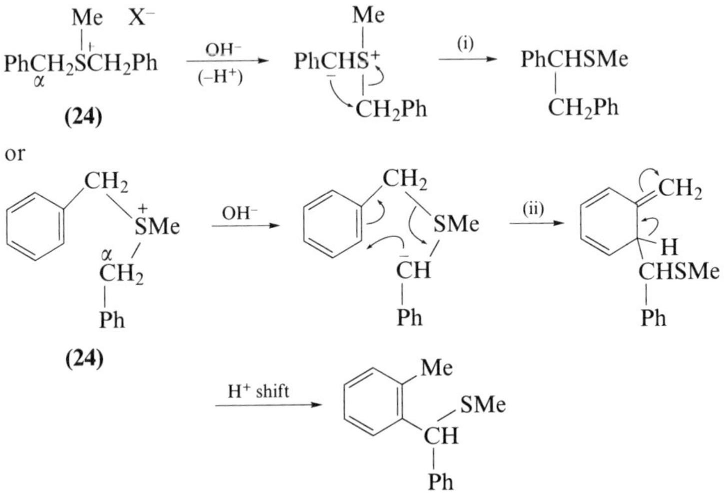

Scheme 13

nucleus. With alkylsulfonium salts like (**21**), the Sommelet rearrangement affords the sulfide (**22**) via the sulfur ylide (**23**) (Scheme 12).

These rearrangements involve 1,2-shifts, and benzylsulfonium salts (**24**) behave like the corresponding benzyl sulfides in undergoing 1,2-shifts (Scheme 13) (Stevens (i) or Sommelet (ii) rearrangement). The direction of the rearrangement is sensitive to the experimental conditions, the Stevens pathway (i) being favoured at higher temperatures (Scheme 13).

Sulfoxonium salts

Methylation of DMSO with methyl iodide yields trimethylsulfoxonium iodide (Scheme 14). This is an unusual reaction since other sulfoxides and halides form the products of *O*-alkylation. By treatment with strong bases, the trimethylsulfoxonium salt is converted into the sulfoxonium ylide (**25**) (Scheme 14).

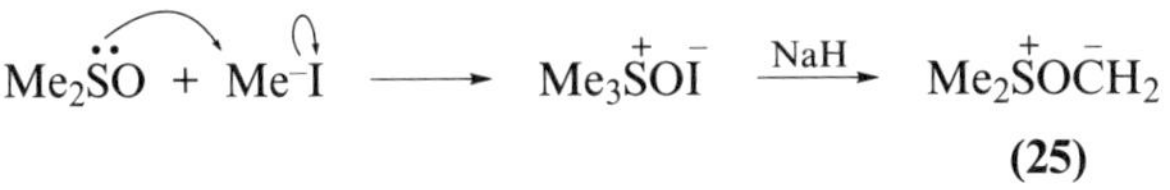

$$Me_2\ddot{S}O + Me{-}I \longrightarrow Me_3\overset{+}{S}O\overset{-}{I} \xrightarrow{NaH} Me_2\overset{+}{S}O\overset{-}{C}H_2$$

(25)

Scheme 14

Sulfoxonium ylides are useful in the preparation of cyclopropane derivatives by reaction with α,β-unsaturated carbonyl compounds (see Chapter 10, p. 189). Sulfoxonium salts (**26**) are intermediates in the Moffatt oxidation (Scheme 15a) (see Chapter 5, p. 66), and they also undergo several other useful reactions (Schemes 15b–15d). The Pummerer rearrangement involves an oxosulfonium ylide intermediate (see Chapter 5, p. 68).

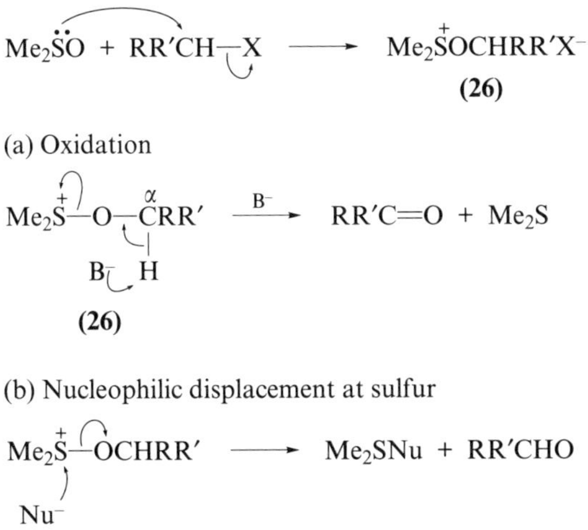

$$Me_2\ddot{S}O + RR'CH{-}X \longrightarrow Me_2\overset{+}{S}OCHRR'X^-$$

(26)

(a) Oxidation

$$Me_2\overset{+}{S}{-}O{-}\overset{\alpha}{C}RR'(H)\ \ B^- \xrightarrow{B^-} RR'C{=}O + Me_2S$$

(26)

(b) Nucleophilic displacement at sulfur

$$Me_2\overset{+}{S}{-}OCHRR' \ (Nu^-) \longrightarrow Me_2SNu + RR'CHO$$

Scheme 15 *(continued)*

Scheme 15 *(continued)*

(c) Nucleophilic displacement at carbon

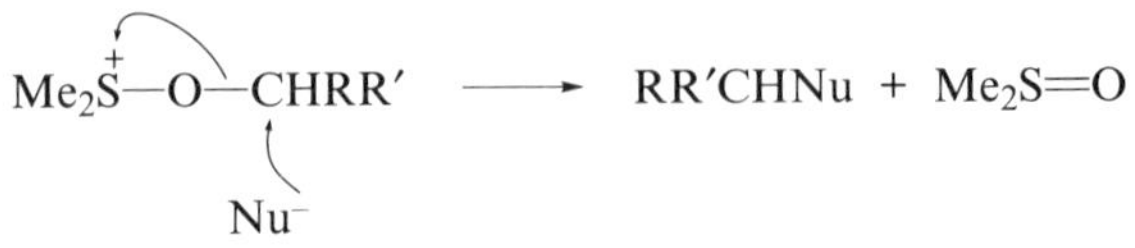

$$Me_2\overset{+}{S}{-}O{-}CHRR' + Nu^- \longrightarrow RR'CHNu + Me_2S{=}O$$

(d) Pummerer rearrangement

$$Me_2\overset{+}{S}OCHRR' \longrightarrow MeSCH_2OCHRR'$$

Scheme 15

Ylide formation in the presence of a base provides a rationalisation for the unusual reactions of unsaturated sulfonium salts, e.g. compound (**27**) (Scheme 16). The rearrangement occurs via the ylide (**28**) which finally collapses to give the sulfide (**29**). β-Alkynic sulfonium salts (**30**) tend to rearrange in solution to yield the isomeric allenic sulfonium salts (**31**) (Scheme 17).

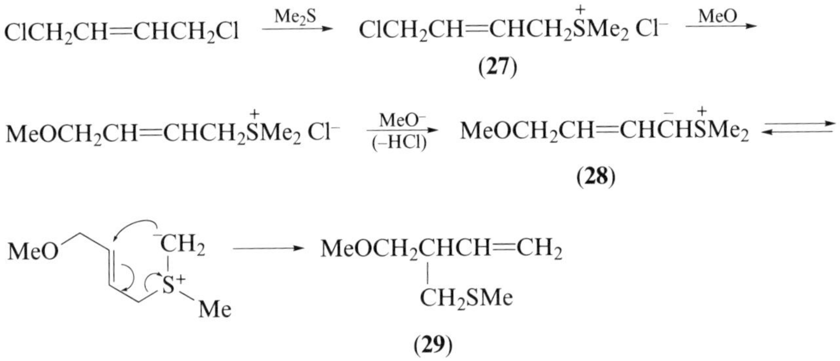

$$ClCH_2CH{=}CHCH_2Cl \xrightarrow{Me_2S} \underset{\textbf{(27)}}{ClCH_2CH{=}CHCH_2\overset{+}{S}Me_2\ Cl^-} \xrightarrow{MeO}$$

$$MeOCH_2CH{=}CHCH_2\overset{+}{S}Me_2\ Cl^- \xrightarrow[(-HCl)]{MeO^-} \underset{\textbf{(28)}}{MeOCH_2CH{=}CH\overset{-}{C}H\overset{+}{S}Me_2} \rightleftharpoons$$

$$MeO{\sim}CH{=}CH{-}CH_2{-}\overset{+}{S}(Me){-}\overset{-}{C}H_2 \longrightarrow \underset{\textbf{(29)}}{MeOCH_2CH(CH_2SMe)CH{=}CH_2}$$

Scheme 16

$$\underset{\textbf{(30)}}{R_2\overset{+}{S}CH_2C{\equiv}CH} \xrightarrow[\text{in solution}]{\text{standing}} \underset{\textbf{(31)}}{R_2\overset{+}{S}CH{=}C{=}CH_2}$$

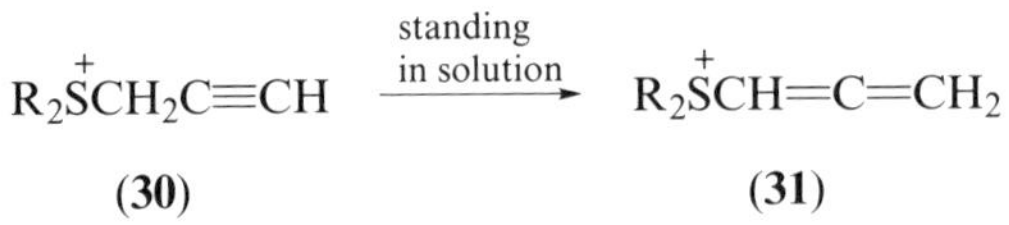

Scheme 17

This reaction illustrates the stabilising effect exerted by the sulfonium centre on the adjacent double bonds. Isomeric allenes (**32**) may be similarly obtained from propargylic sulfonium salts (**33**) (Scheme 18). The allene (**32**), by treatment with diethyl sodiomalonate (**34**), followed by addition and rearrangement of the derived sulfonium ylide (**35**), affords the methyl sulfide (**36**) (Scheme 18).

The reactions of sulfonium salts with organometallic reagents probably involve five-coordinate sulfur species. For example, dephenylation of triphenylsulfonium tetrafluoroborate (**37**) with vinyllithium and the stereospecific

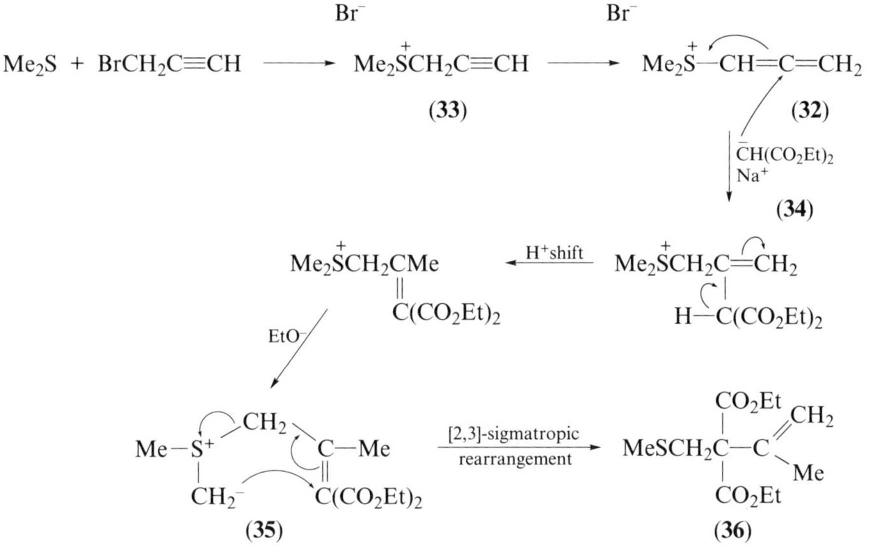

$Me_2S + BrCH_2C{\equiv}CH \longrightarrow Me_2\overset{+}{S}CH_2C{\equiv}CH\ Br^- \ (\mathbf{33}) \longrightarrow Me_2\overset{+}{S}{-}CH{=}C{=}CH_2\ Br^- \ (\mathbf{32})$

$\overline{C}H(CO_2Et)_2\ Na^+$ (**34**)

$Me_2\overset{+}{S}CH_2C(=CH_2)-CH(CO_2Et)_2$ (H—C(CO2Et)2) $\xrightarrow{H^+ shift}$ $Me_2\overset{+}{S}CH_2CMe{=}C(CO_2Et)_2$

EtO^-

$Me{-}S^+(CH_2^-){-}CH_2{-}C(Me){=}C(CO_2Et)_2$ (**35**) $\xrightarrow[\text{rearrangement}]{\text{[2,3]-sigmatropic}}$ $MeSCH_2C(CO_2Et)_2{-}C(=CH_2)Me$ (**36**)

Scheme 18

fragmentation of the *trans*-2,4-dimethylthietanium salt (**38**) to *cis*-dimethylcyclopropane (**39**) (Scheme 19) are in agreement with the primary *S*-alkylation step controlling the final stereochemistry of the product.[3]

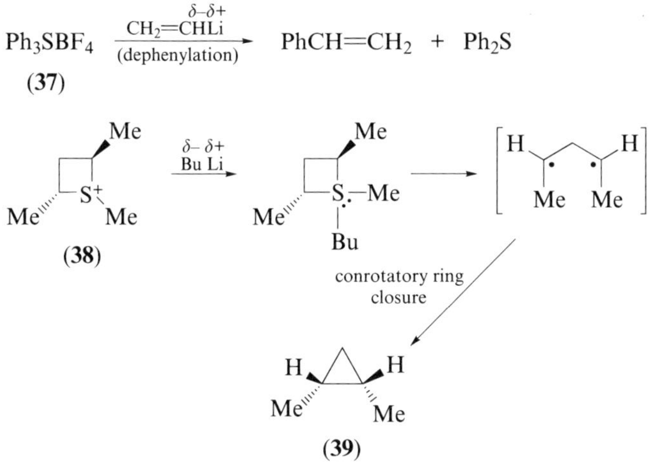

Ph_3SBF_4 (**37**) $\xrightarrow[\text{(dephenylation)}]{\overset{\delta-\ \delta+}{CH_2{=}CHLi}}$ $PhCH{=}CH_2 + Ph_2S$

(**38**) (Me, Me, S+, Me) $\xrightarrow{\overset{\delta-\ \delta+}{Bu\ Li}}$ (Me, Me, S–Me, Bu) $\longrightarrow$ [H, H, Me, Me diradical]

conrotatory ring closure

(**39**) (H, H, Me, Me)

Scheme 19

Sulfenyl carbanions

Sulfenyl carbanions (**40**) may be generated by metallation or deprotonation of the corresponding sulfides (**41**) (Scheme 20) (see Chapter 3, p. 30).

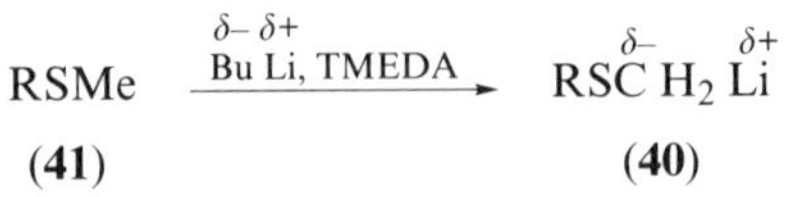

$$\underset{(\mathbf{41})}{RSMe} \xrightarrow[\text{Bu Li, TMEDA}]{\delta-\ \delta+} \underset{(\mathbf{40})}{RS\overset{\delta-}{C}H_2\overset{\delta+}{Li}}$$

Scheme 20

α-Sulfenyl carbanions (**40**) (R = Ph) will react with various electrophiles, e.g. with alkyl iodides (**42**); the reaction can be used to ascend the homologous series, so that compound (**42**) is converted into the homologue (**43**) (Scheme 21).

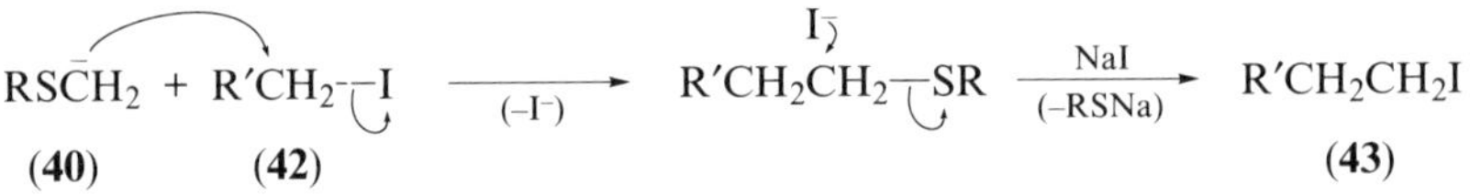

$$\underset{(\mathbf{40})}{RS\bar{C}H_2} + \underset{(\mathbf{42})}{R'CH_2{-}I} \xrightarrow[(-I^-)]{} R'CH_2CH_2{-}SR \xrightarrow[(-RSNa)]{NaI} \underset{(\mathbf{43})}{R'CH_2CH_2I}$$

Scheme 21

Sulfenyl carbanions can also be used in the formation of new carbon–carbon bonds under mild conditions. This is achieved by reaction of the carbanion (**40**) with a ketone and reductive desulfuration of the intermediate acetoxy sulfide (**44**) to produce the alkene (**45**) (Scheme 22).

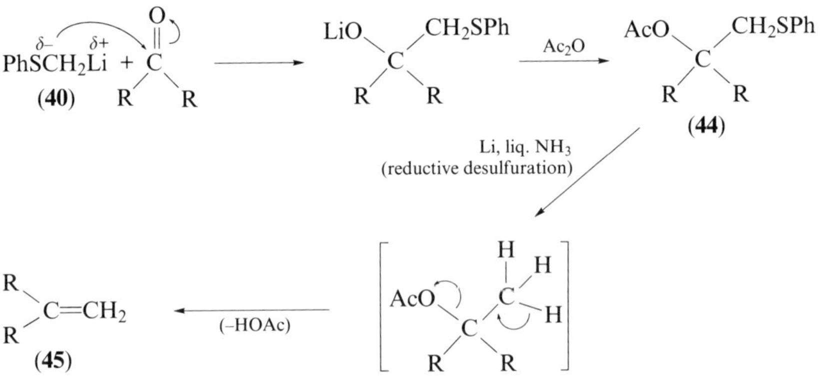

$$\underset{(\mathbf{40})}{Ph\overset{\delta-}{S}CH_2\overset{\delta+}{Li}} + R_2C{=}O \longrightarrow LiO(CH_2SPh)CR_2 \xrightarrow{Ac_2O} \underset{(\mathbf{44})}{AcO(CH_2SPh)CR_2}$$

$$\xrightarrow[\text{(reductive desulfuration)}]{\text{Li, liq. } NH_3} \left[AcO{-}CR_2{-}\bar{C}H_2\right] \xrightarrow[(-HOAc)]{} \underset{(\mathbf{45})}{R_2C{=}CH_2}$$

Scheme 22

The formation of sulfenyl carbanions is facilitated when the substrate contains a suitably positioned second sulfur atom or double bond, as in the case of dithians (see Chapter 3, p. 31) or allyl sulfides (**46**). The latter are thus easily converted into carbanions which can then be alkylated, and this provides a convenient procedure for coupling allyl groups; for example, the allyl sulfide (**46**) can be transformed into diallyl (**47**) (Scheme 23).

1,3-Dithians (thioketals) (**48**) are easily prepared by reaction of an aldehyde with propane-1,3-dithiol; the dithians readily form 2-carbanions owing to the stabilising effect of the two adjacent sulfur atoms on the anion. The carbanion can subsequently be alkylated by treatment with primary or secondary alkyl halides,

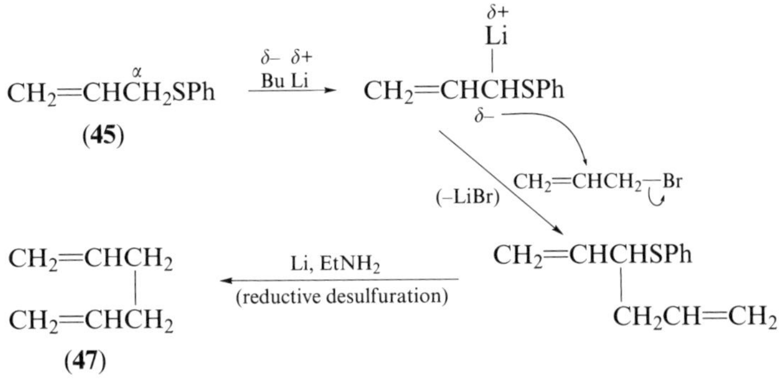

$$CH_2{=}CH\overset{\alpha}{C}H_2SPh \xrightarrow{\overset{\delta-\ \delta+}{Bu\ Li}} CH_2{=}CHCH(Li^{\delta+})SPh^{\delta-}$$

(45)

$$\xrightarrow[(-LiBr)]{CH_2{=}CHCH_2{-}Br} CH_2{=}CHCH(SPh)CH_2CH{=}CH_2 \xrightarrow[\text{(reductive desulfuration)}]{Li,\ EtNH_2} CH_2{=}CHCH_2{-}CH_2CH{=}CH_2$$

(47)

Scheme 23

the iodides being the most reactive halides. After reaction, the 1,3-dithian system can be reconverted to the carbonyl compound by acid hydrolysis (oxidative desulfuration) in the presence of the mercury (II) ion; this procedure provides a method of converting an aldehyde into a ketone (**49**) or a homologous aldehyde (**50**) (Scheme 24).

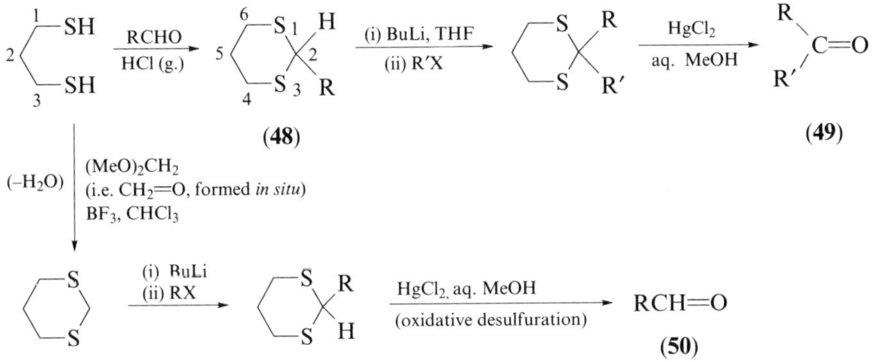

$$HS(CH_2)_3SH \xrightarrow[HCl\ (g.)]{RCHO} \text{2-R-1,3-dithian (48)} \xrightarrow[(ii)\ R'X]{(i)\ BuLi,\ THF} \text{2-R-2-R'-1,3-dithian} \xrightarrow[aq.\ MeOH]{HgCl_2} RR'C{=}O\ (49)$$

$$HS(CH_2)_3SH \xrightarrow[(-H_2O)]{(MeO)_2CH_2\ (i.e.\ CH_2{=}O,\ formed\ \textit{in situ}),\ BF_3,\ CHCl_3} \text{1,3-dithian} \xrightarrow[(ii)\ RX]{(i)\ BuLi} \text{2-R-1,3-dithian} \xrightarrow[\text{(oxidative desulfuration)}]{HgCl_2,\ aq.\ MeOH} RCH{=}O\ (50)$$

Scheme 24

In Chapter 3 (p. 32), such a sequence was shown for the preparation of cyclobutanone, and the procedure has been applied by Corey and Seebach (1968) for the synthesis of three- to seven-membered cyclic ketones. However, unlike the more reactive sulfinyl carbanions (see Chapter 5, p. 70) sulfenyl carbanions do not undergo Michael 1,4-addition to α,β-unsaturated carbonyl compounds and only add across the carbonyl group (1,2-addition). In these 1,3-dithian syntheses, the usual mode of addition of the aldehyde is reversed; normally, the carbon atom of the carbonyl group is partially positively charged and consequently reacts with nucleophiles. When the aldehyde is converted to a 1,3-dithian and reacted with butyllithium, the identical carbon atom becomes negatively charged and therefore reacts with electrophiles. The reversal of the polarity of the carbonyl carbon atom is termed 'umpolung' (German, meaning polarity reversal) (Figure 3).

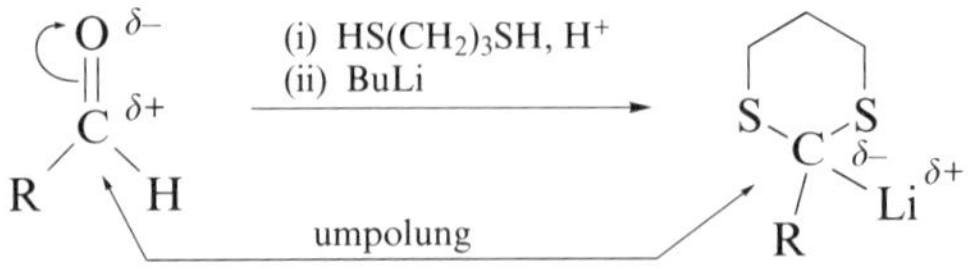

Figure 3

By treatment of the carbanion of (**48**) with an aldehyde, the procedure may be modified to yield hydroxy carbonyl compounds (**51**) (Scheme 25). Such dithian carbanions also react with epoxides e.g. (**52**), to give hydroxy ketones, e.g. (**53**). In this reaction, the less hindered end of the epoxide is preferentially attacked (Scheme 26).

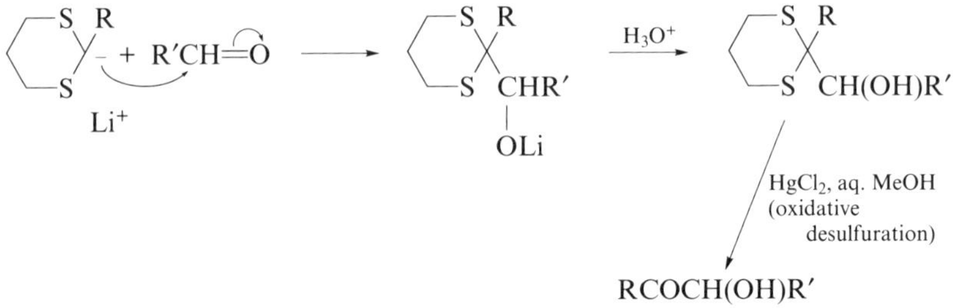

Scheme 25

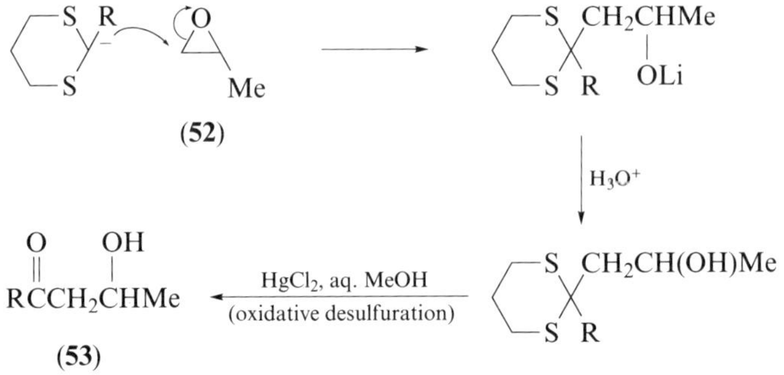

Scheme 26

References

1. *The Chemistry of the Sulfonium Group* (Eds C. J. M. Stirling and S. Patai), Wiley, Chichester, 1981.
2. C. J. M. Stirling, in *The Organic Chemistry of Sulfur* (Ed. S. Oae), Plenum Press, New York, 1977, p. 473.
3. G. C. Barrett, in *Comprehensive Organic Chemistry* (Eds D. H. R. Barton and W. D. Ollis), Vol. 3, Pergamon Press, Oxford, 1979, p. 105.

7 SULFINIC ACIDS, SULFONIC ACIDS AND DERIVATIVES; SULFENES

Sulfinic acids probably have the structure (**1**) rather than (**2**), and certainly any equilibrium between (**1**) and (**2**) lies on the side of the hydroxy isomer (**1**) (Figure 1).[1a] The free sulfinic acids are rather unstable and tend to disproportionate into the thiolsulfonate and the sulfinic acid; hence, they are generally used as the stable sodium salts. Owing to their instability, few sulfinic acids occur naturally, but they may exist as intermediates in the oxidation of thiols; thus, cysteinesulfinic acid is an intermediate in the oxidation of cysteine and 2-aminoethanesulfinic acid has been isolated from molluscs.

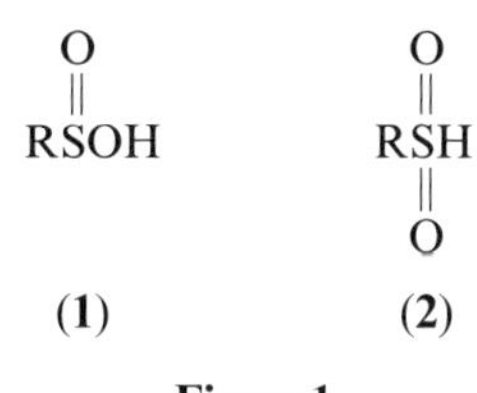

Figure 1

Preparation

Sulfinic acids are generally prepared by reduction of the readily available sulfonyl chlorides with zinc in neutral or basic aqueous solution.[2] The reduction may also sometimes be achieved by treatment with tin(II) chloride or sodium sulfite (Scheme 1). Another route involves reaction of a Grignard reagent with sulfur dioxide (Scheme 1).

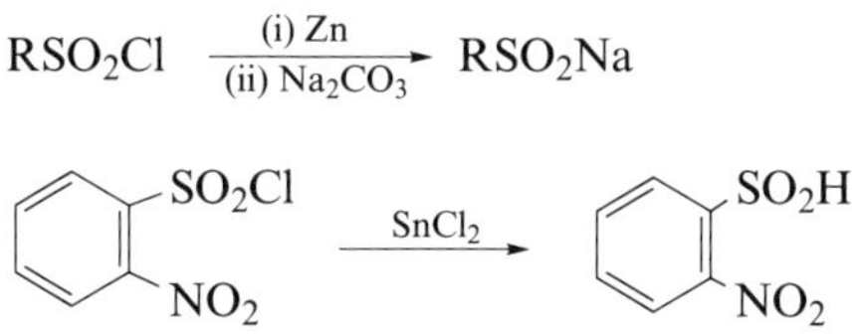

$$RSO_2Cl \xrightarrow[\text{(ii) } Na_2CO_3]{\text{(i) Zn}} RSO_2Na$$

SO_2Cl, NO_2 $\xrightarrow{SnCl_2}$ SO_2H, NO_2

Scheme 1 *(continued)*

Scheme 1 *(continued)*

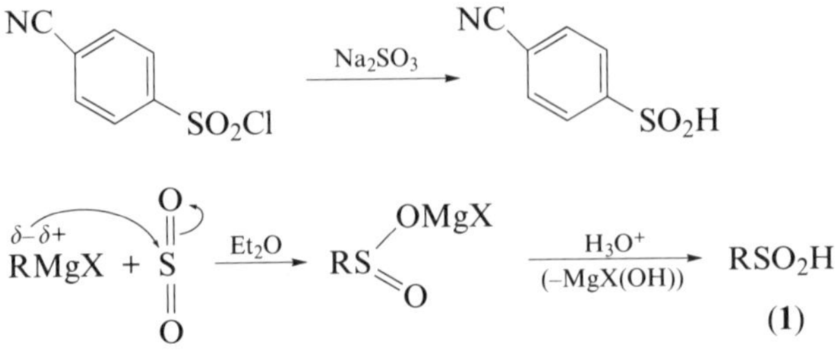

$$4\text{-}NC\text{-}C_6H_4\text{-}SO_2Cl \xrightarrow{Na_2SO_3} 4\text{-}NC\text{-}C_6H_4\text{-}SO_2H$$

$$\overset{\delta-\;\delta+}{RMgX} + O{=}S{=}O \xrightarrow{Et_2O} RS(=O)OMgX \xrightarrow[(-MgX(OH))]{H_3O^+} \underset{(\mathbf{1})}{RSO_2H}$$

Scheme 1

The preparation of sulfinic acids by the oxidation of thiols is difficult because of the danger of overoxidation to the sulfonic acid; however, in certain cases various oxidants, e.g. dilute hydrogen peroxide, halogens or *m*-chloroperbenzoic acid (MCPBA), have been successfully used (Scheme 2).

$$PhSH \xrightarrow{MCPBA} PhSO_2H$$

Scheme 2

Reactions

Aliphatic sulfinic acids (**1**) are much less stable than the aromatic derivatives; sulfinic acids decompose on heating (30–100 °C) into the corresponding thiol sulfonates (**3**) and sulfonic acids (**4**) (Scheme 3). Alphatic sulfinic acids can be

$$\underset{(\mathbf{1})}{3RSO_2H} \longrightarrow \underset{(\mathbf{3})}{RSO_2SR} + \underset{(\mathbf{4})}{RSO_3H} + H_2O$$

Scheme 3

oxidised or reduced. Oxidation, with, for instance, hydrogen peroxide, nitric acid, iodine, hypochlorite or alkaline potassium permanganate, yields the sulfonic acids. Reduction gives thiols, disulfides or thiol sulfonates depending on the reaction conditions; for example, zinc dust–acid yields thiols, lithium aluminium hydride gives disulfides, and iron(II) chloride in acetone or acetic acid gives thiolsulfonates. Sulfinic acids undergo nucleophilic addition to compounds containing activated alkenic double bonds like methyl acrylate, maleic anhydride and vinyl ketones. Some examples of the Michael additions of sulfinate anions to α,β-unsaturated carbonyl compounds are given in Scheme 4.

Sulfinic acids will also add to the carbonyl groups of aldehydes to give α-hydroxy sulfones (**5**) (Scheme 5).

Sulfinic acids will also open lactone rings by nucleophilic addition to form sulfonyl carboxylic acids; thus, the γ-lactone (**6**) yields the γ-sulfonyl carboxylic

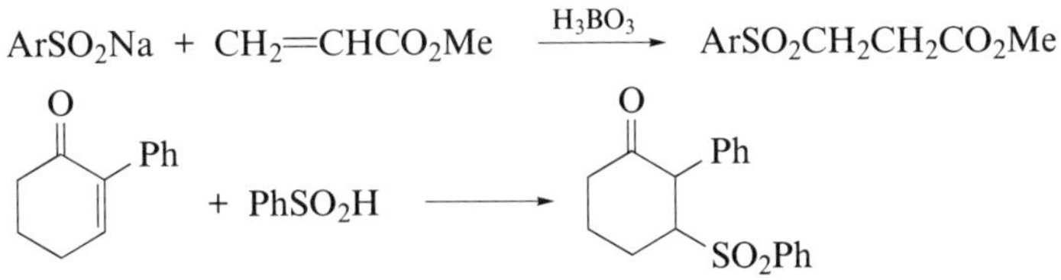

$$ArSO_2Na + CH_2{=}CHCO_2Me \xrightarrow{H_3BO_3} ArSO_2CH_2CH_2CO_2Me$$

2-Phenylcyclohex-2-en-1-one + $PhSO_2H$ ⟶ 2-phenyl-3-(phenylsulfonyl)cyclohexan-1-one (SO_2Ph)

Scheme 4

$$RSO_2H + R'CH{=}O \longrightarrow R'CH(OH)SO_2R$$

(5)

Scheme 5

(6) (γ-butyrolactone, ring positions α, β, γ) + RSO_2Na ⟶ $RSO_2CH_2CH_2CH_2CO_2Na$ (γ β α)

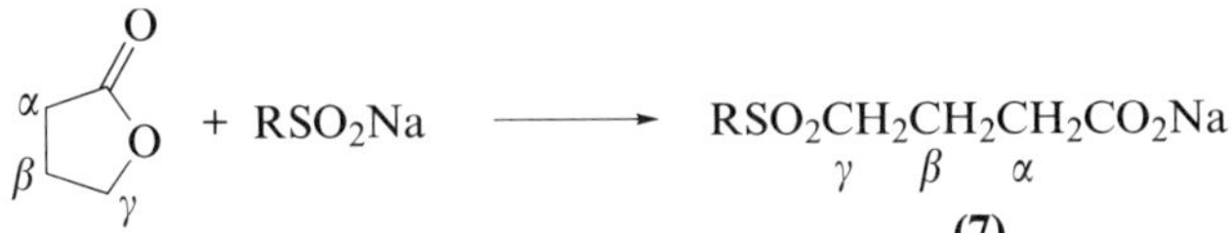

(6) **(7)**

Scheme 6

acid salt (**7**) (Scheme 6). The ring-opening reaction involves nucleophilic attack by the sulfinate anion on the γ-carbon atom of the lactone ring (**6**).

Sulfinic acids react easily with alkyl or benzyl halides to give the corresponding sulfones, e.g. (**8**), and (**9**) (Scheme 7). Sulfinic acids react similarly with sulfenyl, sulfinyl and sulfonyl chlorides to give thiolsulfonates (**10**), sulfinyl sulfones (**11**) and disulfones (**12**), respectively (Scheme 8).

$$RSO_2Na + EtI \longrightarrow RSO_2Et + NaI$$

(8)

$$RSO_2Na + PhCH_2Cl \longrightarrow RSO_2CH_2Ph + NaCl$$

(9)

Scheme 7

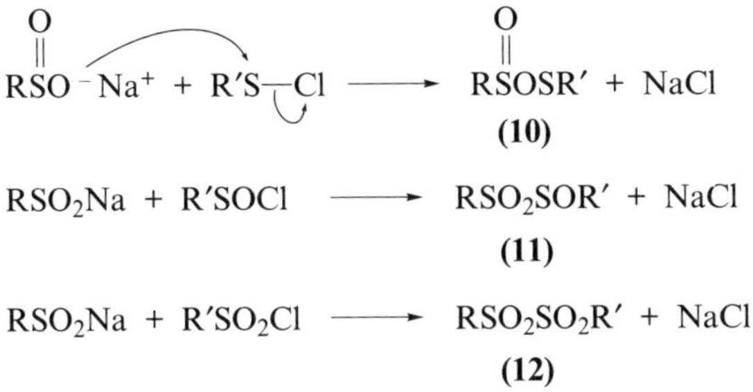

$$RS(=O)O^- Na^+ + R'S{-}Cl \longrightarrow RS(=O)OSR' + NaCl$$

(10)

$$RSO_2Na + R'SOCl \longrightarrow RSO_2SOR' + NaCl$$

(11)

$$RSO_2Na + R'SO_2Cl \longrightarrow RSO_2SO_2R' + NaCl$$

(12)

Scheme 8

Derivatives

Sulfinyl chlorides (**13**) may be obtained by treatment of the sulfinic acid or the sodium salt with thionyl chloride (Scheme 9). In most cases, the best synthetic route involves chlorination of a solution of the corresponding thiol ester (**14**) in acetic anhydride (Scheme 10).

$$RSO_2Na + SOCl_2 \longrightarrow \underset{\textbf{(13)}}{RSOCl} + NaCl + SO_2$$

Scheme 9

$$\underset{\textbf{(14)}}{PhCH_2SCOMe} \xrightarrow{Cl_2,\ Ac_2O} PhCH_2SOCl$$

Scheme 10

In sulfinyl chlorides (13), the chlorine atom can be easily replaced by various nucleophiles (Scheme 11). This substitution reaction can be used to obtain a range of sulfinyl derivatives. Condensation of the sulfinyl chloride (**13**) with amines, azide ion, hydrazine, alcohols (or phenols) and thiols thus affords the corresponding sulfinamides (**15**), azides (**16**), hydrazides (**17**), sulfinate esters (**18**) and thiolsulfinates (**19**) (Scheme 12).[1a, 2]

$$\underset{\textbf{(13)}}{RSOCl} + NuH \xrightarrow{base} RSONu + [HCl]$$

Scheme 11

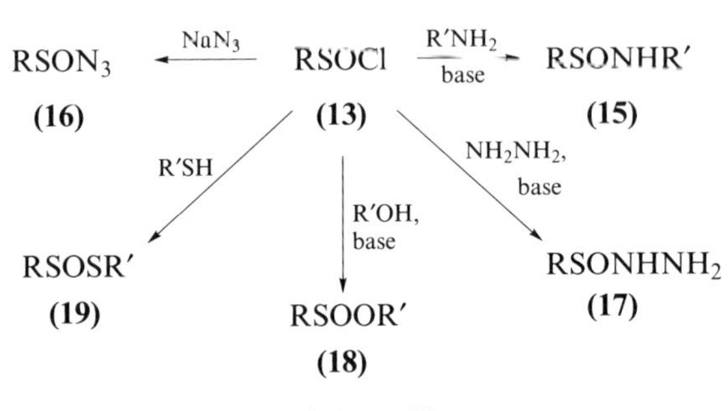

Scheme 12

Sulfinate esters (**18**) are isomeric with sulfones, but have very different properties. Cyclic sulfinate esters (sultines) (**21**) can be conveniently prepared by the action of sulfuryl chloride on *t*-2 butyl hydroxyalkyl sulfoxides (**20**) (Scheme 13). The most common reaction of sulfinate esters is nucleophilic substitution at sulfur with consequent sulfur–oxygen bond cleavage (Scheme 13).

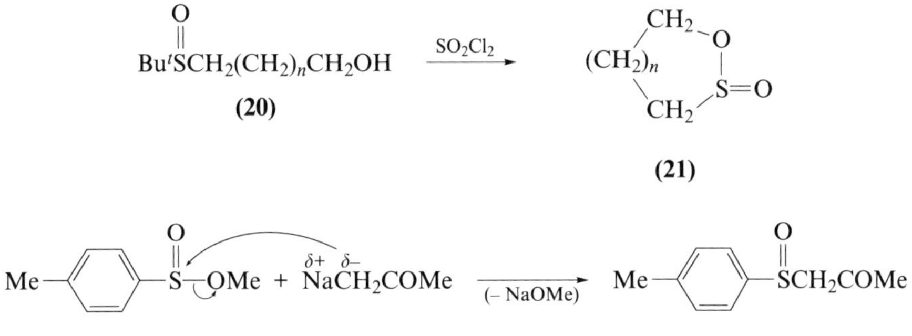

$$Bu^tS(O)CH_2(CH_2)_nCH_2OH \ \textbf{(20)} \xrightarrow{SO_2Cl_2} \textbf{(21)}$$

$$Me{-}C_6H_4{-}S(O){-}OMe + \overset{\delta+\ \ \delta-}{NaCH_2COMe} \xrightarrow[(-\,NaOMe)]{} Me{-}C_6H_4{-}S(O)CH_2COMe$$

Scheme 13

Sulfonic acids

Sulfonic acids (**22**) are strong acids comparable in strength with sulfuric acid; they are stable and do not decompose on heating. Sulfonic acids may be obtained by direct sulfonation of the appropriate aliphatic or aromatic substrate by treatment with sulfuric acid (Scheme 14). Direct sulfonation is generally used in the preparation of aromatic sulfonic acids.[1b, 3a]

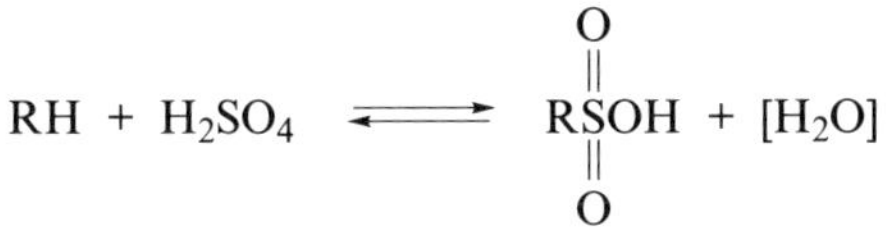

$$RH + H_2SO_4 \rightleftharpoons RSO_2OH \ \textbf{(22)} + [H_2O]$$

Scheme 14

Sulfonation is a bimolecular electrophilic substitution reaction (S_E2) and the electrophile is sulfur trioxide.[3a] Sulfur trioxide is a powerful electrophile because of the electron-withdrawing effect of the three double-bonded oxygen atoms. Consequently, oleum (fuming sulfuric acid), which contains approximately 10% of excess sulfur trioxide, is a much more powerful sulfonating agent than concentrated sulfuric acid. Sulfur trioxide is a sufficiently powerful electrophile to attack benzene (**23**) directly. The mechanism of the sulfonation of benzene by hot concentrated sulfuric acid to give benzenesulfonic acid (**24**) is shown in Scheme 15.[4a]

The reaction occurs via the σ-complex, the arenium carbocation, which is relatively unstable and reacts with the hydrogensulfate anion (HSO_4-) to regenerate the stable aromatic system (six π-electrons). In sulfonation, all four steps are reversible; consequently, sulfonation, unlike nitration, is a reversible process, so that by heating a sulfonic acid with dilute sulfuric acid the parent compound can be regenerated (desulfonation).

The reversibility of sulfonation enables the sulfonic acid group to be used in organic synthesis as a directing group. An illustrative example is provided by the synthesis of *o*-nitroaniline (**25**) from acetanilide (Scheme 16). In this sequence, the

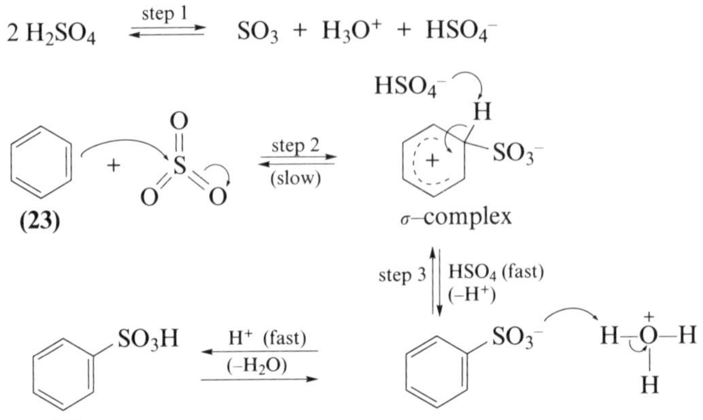

$$2\,H_2SO_4 \xrightleftharpoons{\text{step 1}} SO_3 + H_3O^+ + HSO_4^-$$

(23) step 2 (slow) σ-complex

step 3 HSO_4 (fast) $(-H^+)$

(24) SO_3H H^+ (fast) $(-H_2O)$

Scheme 15

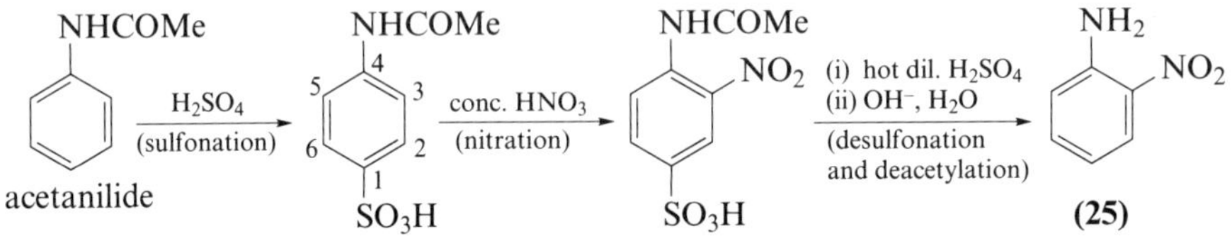

acetanilide NHCOMe H_2SO_4 (sulfonation) NHCOMe SO_3H conc. HNO_3 (nitration) NHCOMe NO_2 SO_3H (i) hot dil. H_2SO_4 (ii) OH^-, H_2O (desulfonation and deacetylation) NH_2 NO_2 (25)

Scheme 16

sulfonic acid group directs the subsequent nitration into the 2-position and blocks the 4-position; in the final stage, the sulfonic acid group is removed by hydrolysis. In the absence of sulfonation, the product formed would be mainly *p*-nitroaniline. Another analogous process is the use of sulfanilamide to obtain 2,6-disubstituted anilines (see p. 109).

In the sulfonation of monosubstituted benzenes, electron-donating substituents, such as alkoxy, acetamido, hydroxy or alkyl groups, facilitate the electrophilic reaction. In these cases, sulfonation occurs under comparatively mild conditions (0–35 °C) to give a mixture of the *o*- and *p*-sulfonic acids (**26**) and (**27**) (Figure 2). The relative amounts of these products depend on the steric size of the substituent. For instance, with acetanilide, the large size of the acetamido group results in mainly *para*-sulfonation, as shown in Scheme 16. On the other hand, the sulfonation of benzenes containing electron-withdrawing substituents, such as nitro, carbonyl or carboxy groups, is more difficult and requires more forcing conditions (temperatures above 100 °C and a large excess of the sulfonating reagent), and yields the *m*-sulfonic acid (**28**) (Figure 2).

The sulfonation of halobenzenes is anomalous because although the halogen atom exerts a strong (−I) inductive (electronic withdrawal) effect with deactivation, sulfonation occurs in the *ortho*-position and *para*-position owing to the electron donation involving the electromeric effect (+E) from the lone electron pairs on the halogen atom in the presence of the reagent.

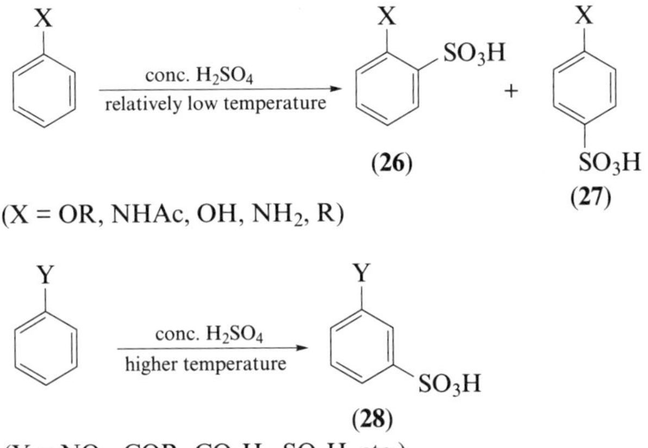

conc. H_2SO_4, relatively low temperature → **(26)** + **(27)**

(X = OR, NHAc, OH, NH_2, R)

conc. H_2SO_4, higher temperature → **(28)**

(Y = NO_2, COR, CO_2H , SO_3H, etc.)

Figure 2

Sulfonation has a very broad scope; thus, aromatic hydrocarbons, halides, ethers, ketones, carboxylic acids, amines, phenols and nitro compounds can be used without damage to the functional group during sulfonation.

With highly reactive substrates, di-, tri- or polysulfonation may occur. In such cases, monosulfonation may often require the use of an inert diluent such as chloroform or carbon tetrachloride to moderate the vigour of the reaction.

In the sulfonation of polyalkylbenzenes such as 1,2,4,5-tetramethylbenzene (**29**) intramolecular migration of an alkyl group (Jacobsen rearrangement, 1886) may occur prior to sulfonation, so that the product is 2,3,4,5-tetramethylbenzenesulfonic acid (**30**) in which the alkyl groups have moved closer together (Scheme 17). The Jacobsen rearrangement is restricted to tetra- and pentaalkylbenzenes.

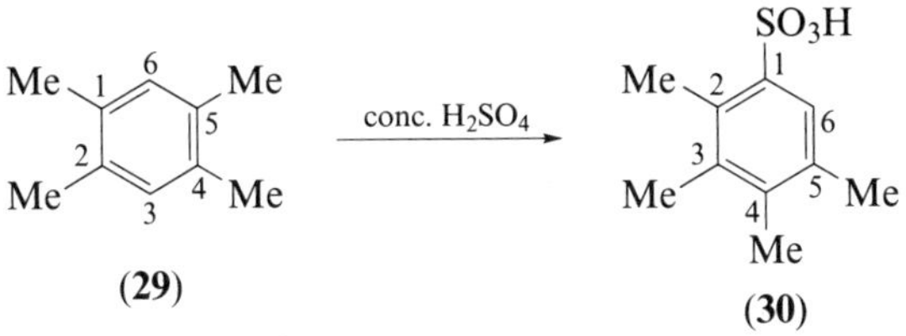

(29) conc. H_2SO_4 → **(30)**

Scheme 17

In the sulfonation of aromatic substrates containing a free amino group with hot concentrated sulfuric acid, the initially formed amine sulfate may rearrange to yield the required sulfonic acid. Thus, when aniline (**31**) is heated with concentrated sulfuric acid at 200 °C (the 'baking process'), the amine sulfate (**32**) is probably converted into phenylsulfamic acid (**33**) which subsequently rearranges by migration of the sulfonic acid group into orthanilic acid (**34**) and finally to the most thermodynamically stable sulfanilic acid (**35**) (Scheme 18).

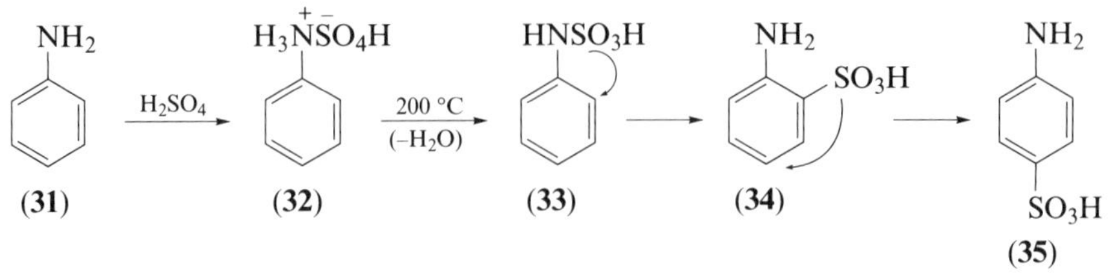

NH_2 (31) $\xrightarrow{H_2SO_4}$ $H_3\overset{+}{N}\overset{-}{S}O_4H$ (32) $\xrightarrow[(-H_2O)]{200\,°C}$ $HNSO_3H$ (33) $\longrightarrow$ NH_2, SO_3H (34) $\longrightarrow$ NH_2, SO_3H (35)

Scheme 18

The majority of aliphatic compounds tend to be resistant to direct sulfonation and under forcing conditions often form a complex mixture of products. Consequently, aliphatic sulfonic acids are generally prepared by other methods, such as treatment of an alkyl halide with sodium sulfite (see p. 101). However, in certain cases direct sulfonation works well, namely with terminal alkynes (**36**) and tertiary alkanes (**37**) (Scheme 19).

$$\underset{(\mathbf{36})}{RC{\equiv}CH} + H_2SO_4 \longrightarrow RC{\equiv}CSO_3H + [H_2O]$$

$$\underset{(\mathbf{37})}{R_3CH} + H_2SO_4 \longrightarrow R_3CSO_3H + [H_2O]$$

Scheme 19

Sulfur trioxide is a more powerful sulfonating agent than concentrated sulfuric acid, and consequently it will sulfonate benzene at low temperatures (0–10 °C). Sulfur trioxide also reacts with terminal alkenes (**38**) to yield the sulfonic acids (**39**) (Scheme 20).

$$\underset{(\mathbf{38})}{RCH{=}CH_2} + SO_3 \longrightarrow \underset{(\mathbf{39})}{RCH{=}CHSO_3H}$$

Scheme 20

The pyridine–sulfur trioxide adduct is a valuable reagent for the sulfonation of acid sensitive substrates like the heterocycles furan and pyrrole, which are polymerised by strong acids like sulfuric acid. Pyridine–sulfur trioxide thus converts furan (**40**) and pyrrole (**41**) into the corresponding 2-sulfonic acids (Scheme 21).

The highly reactive sulfur trioxide–dioxan complex is effective for the terminal sulfonation of alkyl aryl ketones, for instance the reagent converts acetophenone (**42**) into the ω-sulfonic acid (**43**) (Scheme 22) (see Chapter 2, p. 24).

The reaction of an aromatic compound with chlorosulfonic acid (**44**) (one molar equivalent) in an inert solvent, e.g. chloroform, can be used to prepare aromatic sulfonic acids (Scheme 23). Chlorosulfonic acid is a powerful sulfonating agent approximately equivalent to fuming sulfuric acid (oleum), and care must be taken with this reagent to avoid the formation of sulfones or sulfonyl chlorides (see p. 103).[5] Aliphatic or benzylic sulfonic acids can be obtained in good yields

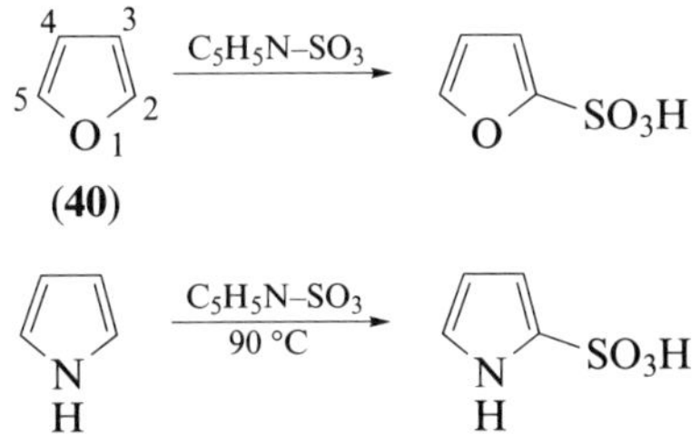

$$\text{(40) furan} \xrightarrow{C_5H_5N\text{–}SO_3} \text{furan-2-}SO_3H$$

$$\text{(41) pyrrole} \xrightarrow[90\ °C]{C_5H_5N\text{–}SO_3} \text{pyrrole-2-}SO_3H$$

Scheme 21

$$\underset{(\mathbf{42})}{PhCOMe} \xrightarrow[\text{room temperature}]{SO_3\text{–dioxan}} \underset{(\mathbf{43})}{PhCOCH_2SO_3H}$$

Scheme 22

$$ArH + \underset{(\mathbf{44})}{ClSO_3H} \xrightarrow{CHCl_3,\ 0\text{–}25\ °C} ArSO_3H + HCl$$

Scheme 23

by reaction of an alkyl or benzylic halide (**45**) or an epoxide with sodium sulfite or sodium hydrogen sulfite (the Strecker reaction, 1868) (Scheme 24) (see Chapter 2, p. 22). The sulfite anion is nucleophilic at sulfur, which accounts for these reactions which go best with primary alkyl halides.

$$RCH_2X + Na_2SO_3 \xrightarrow[\text{EtOH}]{\text{reflux in aq.}} RCH_2SO_3Na + NaX$$

$$\underset{(\mathbf{45})}{PhCH_2X} + NaHSO_3 \longrightarrow PhCH_2SO_3H + NaX$$

$$RCH(O)CH_2 \text{ (epoxide)} \xrightarrow[HSO_3^-]{NaHSO_3} RCH(OH)CH_2SO_3Na$$

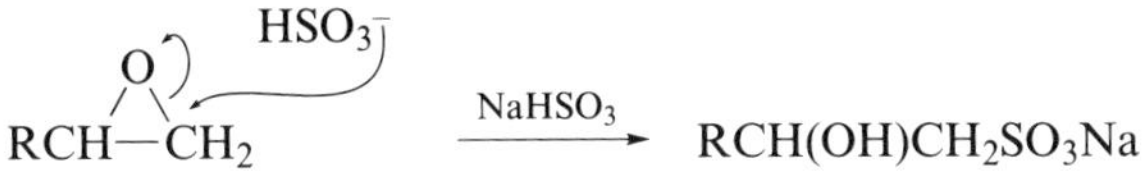

Scheme 24

The Strecker reaction is also successful with aromatic compounds containing reactive halogen atoms, e.g. 2,4-dinitrochlorobenzene (**46**) (Scheme 25). In this compound, the chlorine atom is activated with respect to nucleophilic attack by the strongly electron-withdrawing (−I, −M) nitro groups. Aliphatic sulfonic acids may also be prepared by addition of sodium hydrogen sulfite to alkenes in the presence of oxygen or peroxides (Scheme 26). This sulfonation involves anti-Markownikoff free radical addition of the reagent to the alkene double bond, so that the sulfite anion preferentially adds to the carbon atom with the most hydrogen atoms.

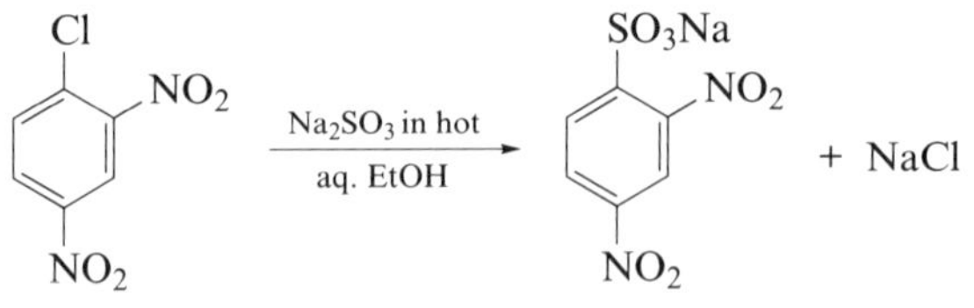

$$\text{(46) 1-chloro-2,4-dinitrobenzene} \xrightarrow[\text{aq. EtOH}]{Na_2SO_3 \text{ in hot}} \text{2,4-}(NO_2)_2C_6H_3SO_3Na + NaCl$$

Scheme 25

$$RCH{=}CH_2 + NaHSO_3 \xrightarrow{O_2 \text{ or } H_2O_2} RCH_2CH_2SO_3Na$$

Scheme 26

Sulfonic acids may also be obtained by oxidation of the appropriate thiols (see Chapter 4, p. 43). The oxidants may be halogens, hydrogen peroxide, nitric acid, potassium permanganate or chromic anhydride (Scheme 27). An illustrative example is provided by the oxidation of 4-chloro-2-methylbutane-2-thiol (**47**) to the corresponding sulfonic acid (**48**) (Scheme 27).

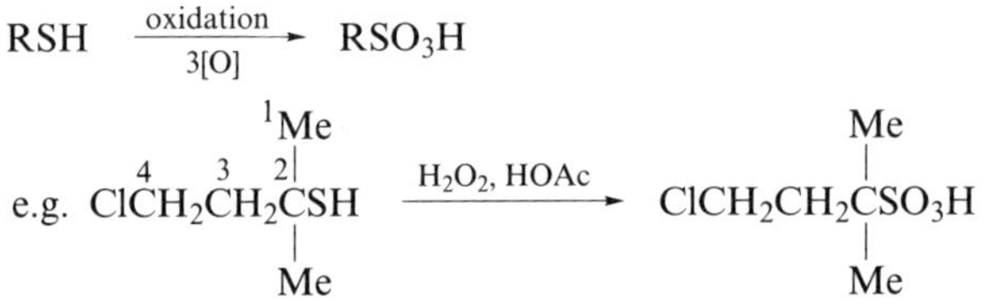

$$RSH \xrightarrow[3[O]]{\text{oxidation}} RSO_3H$$

$$\text{e.g. } \overset{4}{Cl}CH_2\overset{3}{C}H_2\overset{2}{C}(\overset{1}{Me})(Me)SH \; \textbf{(47)} \xrightarrow{H_2O_2,\ HOAc} ClCH_2CH_2C(Me)_2SO_3H \; \textbf{(48)}$$

Scheme 27

The method can also be applied to the conversion of 4-chloropyridine (**49**) to pyridine-4-sulfonic acid (**50**) (Scheme 28). In compound (**49**), the chlorine atom is activated to nucleophilic attack by the electron-withdrawing nitrogen atom. Scheme 28 is a useful synthetic sequence since direct sulfonation of pyridine only yields the 3-sulfonic acid as a consequence of the electrophilic nature of the nitrogen atom. Sulfonic acids can also be obtained by oxidation of sulfides, disulfides and sulfinic acids. The introduction of a sulfonic acid group into an organic molecule provides a useful method of increasing the aqueous solubility of the

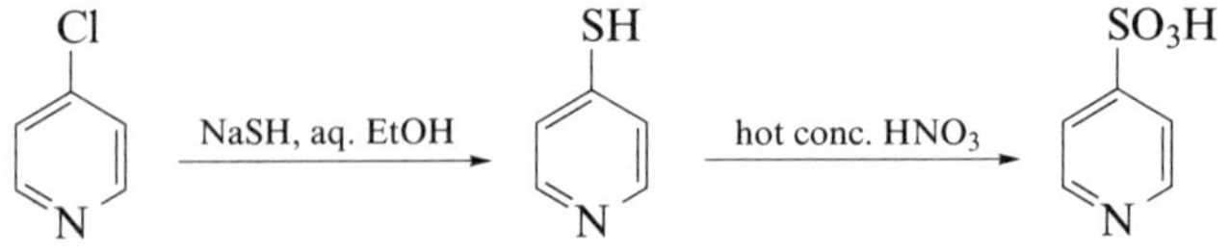

$$\text{4-chloropyridine } \textbf{(49)} \xrightarrow{\text{NaSH, aq. EtOH}} \text{4-mercaptopyridine (SH)} \xrightarrow{\text{hot conc. } HNO_3} \text{pyridine-4-}SO_3H \; \textbf{(50)}$$

Scheme 28

compound. Many sulfonic acids as their sodium salts are important detergents and dyes (see Introduction, p. 5 and Chapter 11, p. 220).

Sulfonyl chlorides[3a]

Aromatic sulfonyl chlorides (**51**) are conveniently prepared by treatment of the aromatic compound with an excess of chlorosulfonic acid (Scheme 29). The reaction occurs via the sulfonic acid (step 1) which is then chlorinated by the excess reagent (step 2) (Scheme 29). The mechanism probably involves the chlorosulfonic acid acting as the electrophile, as shown for the reaction with benzene (**23**) in Scheme 30.[5]

$$ArH + 2\,ClSO_3H \rightleftharpoons \underset{\textbf{(51)}}{ArSO_2Cl} + HCl + H_2O + SO_3$$

$$ArH + ClSO_3H \rightleftharpoons ArSO_3H + HCl \text{ (step 1)}$$

$$ArSO_3H + ClSO_3H \rightleftharpoons ArSO_2Cl + H_2O + SO_3 \text{ (step 2)}$$

Scheme 29

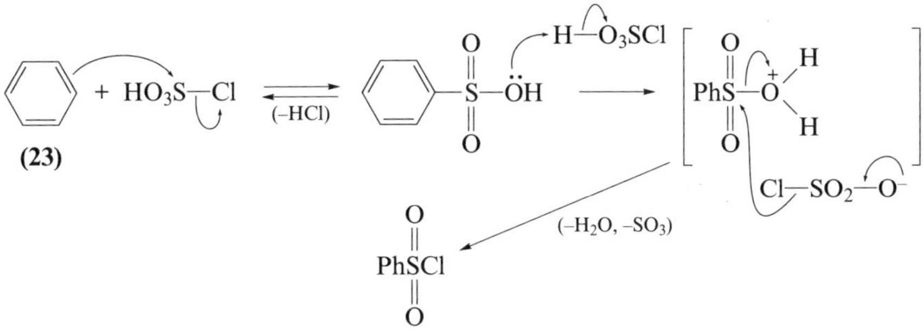

$$\textbf{(23)}\ C_6H_6 + HO_3S{-}Cl \underset{(-HCl)}{\rightleftharpoons} C_6H_5{-}S(=O)_2{-}\ddot{O}H \xrightarrow{H{-}O_3SCl} \left[PhS(=O)_2{-}\overset{+}{O}H_2 \quad Cl{-}SO_2{-}O^-\right] \xrightarrow{(-H_2O,\ -SO_3)} PhS(=O)_2Cl$$

Scheme 30

Since chlorosulfonation is a reversible reaction, a large excess of the reagent (approximately six molar equivalents) is often required to obtain the maximum yield of the sulfonyl chloride and to avoid the formation of sulfones. However, by using a mixture of chlorosulfonic acid and thionyl chloride, the amount of the acid can often be substantially reduced without affecting the yield of the sulfonyl chloride.

Sulfonyl chlorides (**51**) may also be obtained by heating the sulfonic acid or its sodium salt with thionyl chloride or phosphorus pentachloride (Scheme 31). The reaction with thionyl chloride is catalysed by the addition of a few drops of DMF. The chlorosulfonation of organic compounds may also be achieved by reaction with sulfur dioxide and chlorine; this is an important industrial process (the Reed reaction) (Scheme 32).

In the chlorosulfonation of 4-chloro-2-methylbutane (**52**), the electrophilic chlorine radical abstracts a hydrogen atom from the most electron-rich site to

$$RSO_3H + SOCl_2 \longrightarrow \underset{\textbf{(51)}}{RSO_2Cl} + HCl + SO_2$$

$$RSO_3H + PCl_5 \longrightarrow \underset{\textbf{(51)}}{RSO_2Cl} + POCl_3 + HCl$$

Scheme 31

$$RH + SO_2 + Cl_2 \xrightarrow{h\nu} \underset{\textbf{(51)}}{RSO_2Cl}$$

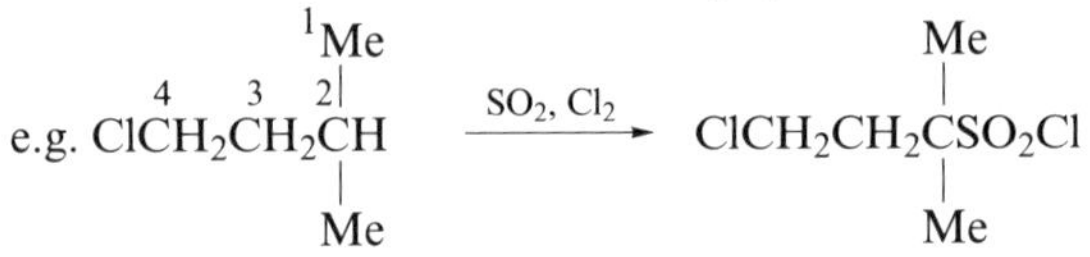

$$\text{e.g. } \underset{\textbf{(52)}}{\overset{4}{Cl}CH_2\overset{3}{C}H_2\overset{2}{C}H(\overset{1}{Me})Me} \xrightarrow{SO_2,\ Cl_2} \underset{\textbf{(53)}}{ClCH_2CH_2C(Me)_2SO_2Cl}$$

Scheme 32

yield the sulfonyl chloride (**53**) (Scheme 32). Organic compounds may sometimes be converted to the corresponding sulfonyl chlorides by treatment with sulfuryl chloride (54), but the method often gives complex mixtures of products owing to competing chlorination. However alkyl and benzyl halides give good yields of the sulfonyl chlorides (**55**) via the Grignard reagents (Scheme 33).

$$RCH_2Cl \xrightarrow{Mg} RCH_2MgCl \xrightarrow[\textbf{(54)}]{SO_2Cl_2} \underset{\textbf{(55)}}{RCH_2SO_2Cl} + MgCl_2$$

Scheme 33

Aromatic diazonium salts (**56**) may be converted to the sulfonyl chlorides by treatment with sulfur dioxide in the presence of copper(II) salts as catalysts (Scheme 34).

$$ArNH_2 \xrightarrow[\substack{<5\ ^\circ C \\ \text{(diazotisation)}}]{NaNO_2,\ HCl} \underset{\textbf{(56)}}{Ar\overset{+}{N_2}\overset{-}{Cl}} \xrightarrow[\substack{\text{warm} \\ (-N_2)}]{SO_2,\ CuCl_2} ArSO_2Cl$$

Scheme 34

Sulfonyl chlorides (**51**) are important intermediates in the synthesis of a range of sulfonyl derivatives since the chlorine atom is readily replaced by nucleophilic reagents such as amines, alcohols and hydrazine to give the corresponding sulfonamides (**57**), sulfonates (**58**) and hydrazides (**59**) (Scheme 35).

Sulfonyl chlorides (**51**) are used in the manufacture of many biologically active compounds like the sulfonamide antibacterial drugs (see Chapter 2, p. 25 and Chapter 11, pp. 222–6) and pesticides such as the acaricide tetradifon (**60**) (see Chapter 11, p. 235), which is synthesised from 1,2,4-trichlorobenzene (**61**) (Scheme 36).

$$\underset{\textbf{(51)}}{RSO_2Cl} + 2\,R'NH_2 \longrightarrow \underset{\textbf{(57)}}{RSO_2NHR'} + R'NH_3Cl$$

$$RSO_2Cl + R'OH \xrightarrow[C_5H_5N]{\text{base, e.g.}} \underset{\textbf{(58)}}{RSO_3R'} + [HCl]$$

$$RSO_2Cl + 2\,N_2H_4 \longrightarrow \underset{\textbf{(59)}}{RSO_2NHNH_2} + NH_2\overset{+}{N}H_3\overset{-}{Cl}$$

Scheme 35

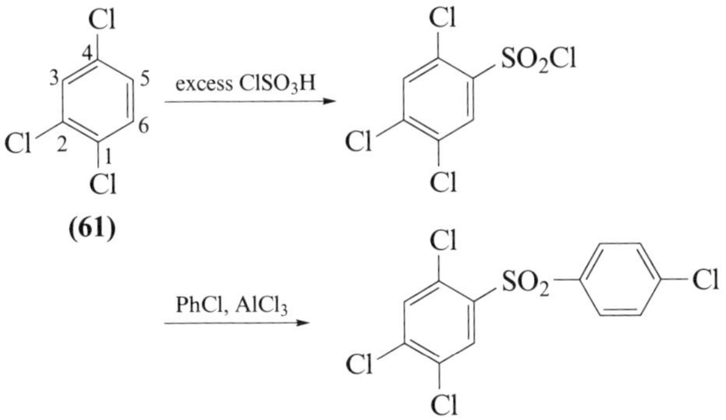

(61) **(60)**

Scheme 36

Tetradifon is a valuable acaricide for the control of phytophagous mites on a wide range of crops; the last step in the synthesis is an example of the Friedel–Crafts reaction, and can be likened to the reaction between *t*-butylbenzene (**62**) and a sulfonyl chloride (**51**) in the presence of aluminium trichloride catalyst to give the sulfone (**63**) (see Chapter 10, p. 196) (Scheme 37). This reaction is a

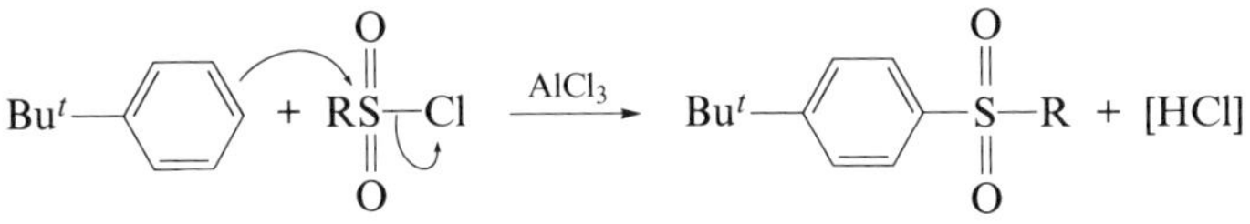

(62) **(51)** **(63)**

Scheme 37

bimolecular electrophilic substitution reaction (S_E2) and may also be regarded as a nucleophilic displacement of the chlorine atom by the carbon moiety. Another related acaricide is the sulfonate, chlorfenson (**64**), obtained from the chlorobenzene (**65**) as indicated in Scheme 38.

Benzenesulfonyl chloride (**66**) is used in organic qualitative analysis for the separation of primary (**67**), secondary (**68**) and tertiary (**69**) amines by the Hinsberg method (Scheme 39). The alkaline reaction mixture is first filtered, which removes

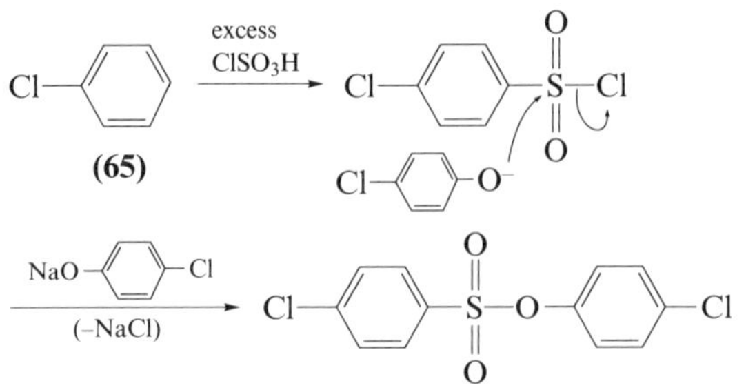

Scheme 38

$$RNH_2 + PhSO_2Cl \xrightarrow[(-HCl)]{aq.\ NaOH} [RNHSO_2Ph] \longrightarrow R\bar{N}SO_2Ph\ Na^+ \xrightarrow{H_3O^+} RNHSO_2Ph \downarrow$$

(67) (66) (71)

$$R_2NH + PhSO_2Cl \xrightarrow{aq.\ NaOH} R_2NSO_2Ph \downarrow$$

(68) (70)

R_3N does not react with $PhSO_2Cl$

(69)

Scheme 39

the precipitate of the dialkyl sulfonamide (**70**), and subsequent distillation of the filtrate yields the volatile tertiary amine (**69**). The residual liquid is acidified to precipitate the monoalkyl sulfonamide (**71**) (Scheme 39). The free primary and secondary amines are finally regenerated from their sulfonamides by boiling with dilute hydrochloric acid. The separation depends on the fact that the hydrogen atom of the sulfonamido (RSO_2NH) group is strongly acidic, and hence only the sulfonamide (**71**) derived from the primary amine (**67**) forms a water soluble salt.

Kinetic and substituent effects indicate that the majority of nucleophilic substitution reactions of sulfonyl chlorides follow the bimolecular S_N2-type mechanism[4b,6] involving the linear transition state (**72**) (Scheme 40) (see Chapter 3, p. 29).

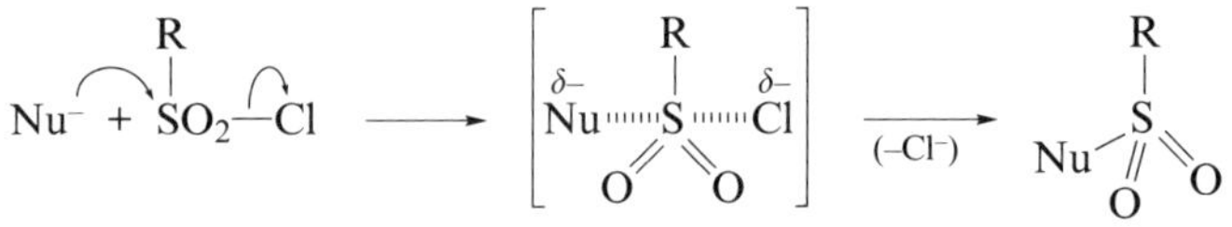

(72)

Scheme 40

On the other hand, in certain cases when the sulfonyl chloride contains an α-hydrogen atom, there is strong evidence indicating that the substitution reaction occurs by an elimination–addition (E–A) mechanism involving a highly reactive sulfene intermediate.[3b] For instance, the reaction of methanesulfonyl chloride (**73**) with aniline to form the *N*-phenylsulfonamide (**74**) probably occurs via the sulfene (**75**) (Scheme 41).

$$\underset{(\mathbf{73})}{MeSO_2Cl} \xrightarrow[(-HCl)\ (E)]{NEt_3} \underset{(\mathbf{75})}{\overset{\delta-}{H_2C}=\overset{\delta+}{SO_2}} \xrightarrow[(A)]{PhNH_2} \underset{(\mathbf{74})}{MeSO_2NHPh}$$

Scheme 41

Sulfonates (**58**) are obtained by reaction of a sulfonyl chloride with an alcohol or phenol in the presence of a base, e.g. pyridine, at low temperatures to avoid side reactions (Scheme 42). In the reaction, yields of the sulfonate (**58**) may be enhanced under anhydrous conditions or by the use of phase transfer catalysts, and it is found that primary alcohols react must faster than secondary alcohols and the primary sulfonates are more stable. Tertiary alcohols only form alkenes. The selective sulfonation of a primary hydroxy group in the presence of a secondary hydroxy group is therefore possible and is a useful procedure for the selective protection of a primary alcohol group (Scheme 43).

$$\underset{(\mathbf{51})}{RSO_2Cl} + R'OH \xrightarrow{C_5H_5N} \underset{(\mathbf{58})}{RSO_3R'} + C_5H_5\overset{+}{N}H\overset{-}{Cl}$$

Scheme 42

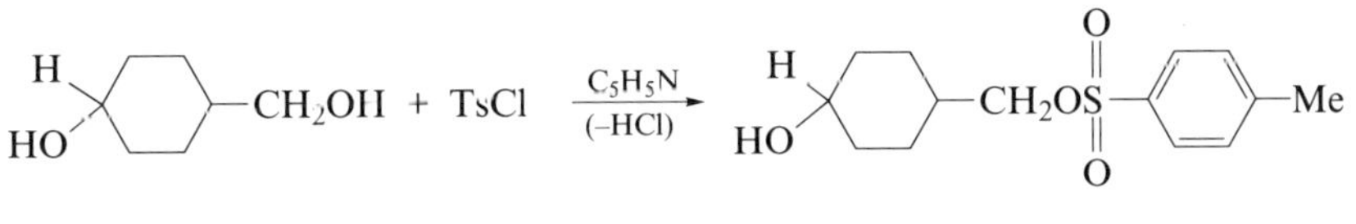

$$\text{H(HO)C}_6\text{H}_{10}\text{-CH}_2\text{OH} + \text{TsCl} \xrightarrow[(-HCl)]{C_5H_5N} \text{H(HO)C}_6\text{H}_{10}\text{-CH}_2\text{OS(=O)}_2\text{-C}_6\text{H}_4\text{-Me}$$

Scheme 43

Alkyl sulfonates (**58**; R′ = alkyl) tend to undergo nucleophilic substitution with preferential R′—O bond cleavage rather than sulfur–oxygen or sulfur–carbon bond cleavage owing to the excellent leaving group properties of the alkanesulfonate anion (RSO_2O^-) (Scheme 44).

On the other hand, with aryl sulfonates (**58**) (R′ = aryl), nucleophilic substitution occurs with preferential sulfur–oxygen bond cleavage since aryl substituents show little tendency to undergo nucleophilic attack (Scheme 44). The relative order of nucleophilicity towards the sulfur atom is similar to that obtaining at a carbonyl carbon atom[4b] and is reported to be as shown in Figure 3.

In the majority of substitution and elimination reactions, the sulfonate moiety functions as the leaving group, and certain sulfonates, e.g. methanesulfonate (mesylate), *p*-toluenesulfonate (tosylate) and trifluoromethanesulfonate (triflate), are exceptionally good leaving groups. The introduction of these sulfonate groups

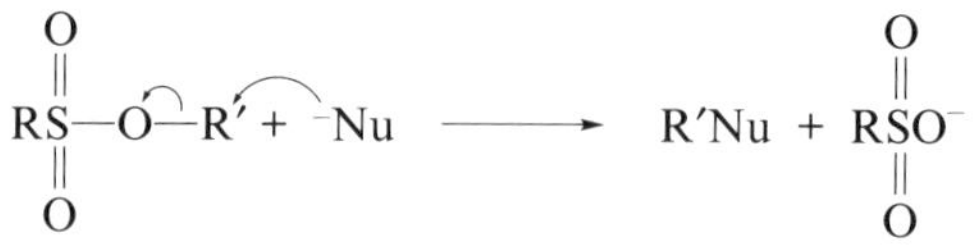

$$RSO_2{-}O{-}R' + {}^{-}Nu \longrightarrow R'Nu + RSO_3^{-}$$

(**58**) R′ = alkyl

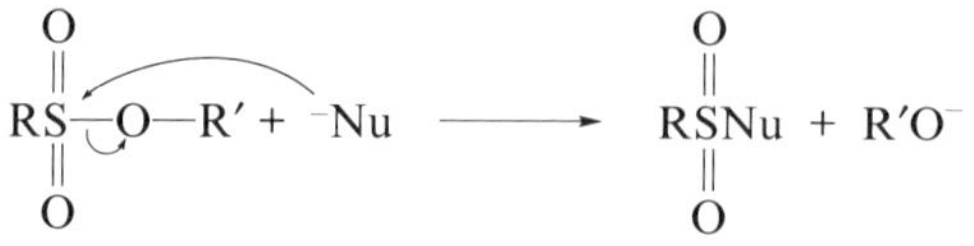

$$RSO_2{-}O{-}R' + {}^{-}Nu \longrightarrow RSO_2Nu + R'O^{-}$$

(**58**) R′ = aryl

Scheme 44

$$HO^{-} > RNH_2 > N_3^{-} > F^{-} > OAc^{-} > Cl^{-} > H_2O > I^{-}$$

decreasing nucleophilicity ⟶

Figure 3

therefore greatly facilitate the nucleophilic substitution reaction; the triflate is the best leaving group and will undergo acetolysis 10^4 times faster than the analogous tosylate. The hydroxy group of an alcohol is a poor leaving group and sulfonation affords a useful method of activating the group towards nucleophilic substitution reactions. Thus, an alcohol can be activated by conversion to the tosylate derivative (**76**) by treatment with *p*-toluenesulfonyl chloride (Scheme 45).

$$ROH \xrightarrow[C_5H_5N]{TsCl} R{-}OTs \xrightarrow{^{-}NuH} RNu + TsOH$$

(**76**)

Scheme 45

Sulfonates react with various organometallic reagents and with chiral substrates. The reaction occurs with stereochemical inversion of configuration (see Chapter 3, p. 30 and Chapter 5, p. 63) (Scheme 46a). On the other hand, if the reaction at the carbon atom is sterically hindered or elimination is difficult, then attack at sulfur occurs (Scheme 46b).

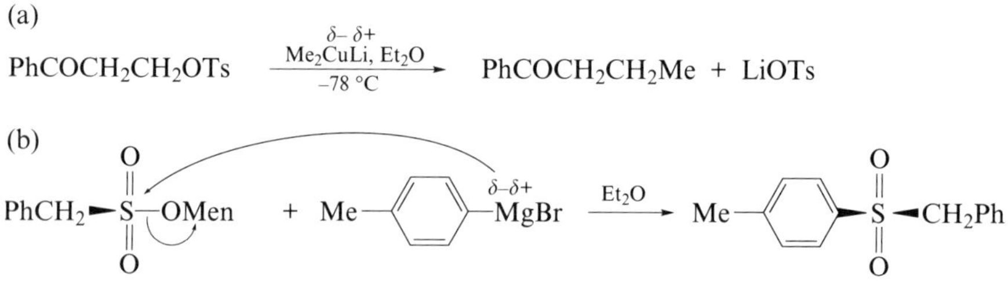

(a)

$$PhCOCH_2CH_2OTs \xrightarrow[-78\ ^\circ C]{\overset{\delta- \ \delta+}{Me_2CuLi},\ Et_2O} PhCOCH_2CH_2Me + LiOTs$$

(b)

$$PhCH_2{-}SO_2{-}OMen + Me{-}C_6H_4{-}\overset{\delta-\delta+}{MgBr} \xrightarrow{Et_2O} Me{-}C_6H_4{-}SO_2{-}CH_2Ph$$

Scheme 46

Sulfonamides are important as drugs and pesticides (see Chapter 11, p. 225); they are readily prepared by condensation of the appropriate sulfonyl chloride with an excess of ammonia or a primary or secondary amine or by using 1 equivalent of the amine in the presence of a suitable base such as a tertiary amine (Scheme 47).

$$RSO_2Cl + 2\,R'R''NH \longrightarrow RSO_2NR'R'' + R'R''\overset{+}{N}H_2\overset{-}{Cl}$$

Scheme 47

Sulfanilamide (**77**) is a useful intermediate in the synthesis of 2,6-disubstituted anilines (Scheme 48). The last step is the hydrolytic removal of the sulfamoyl group, which occurs as a result of the reversibility of the sulfonation reaction (see p. 97).

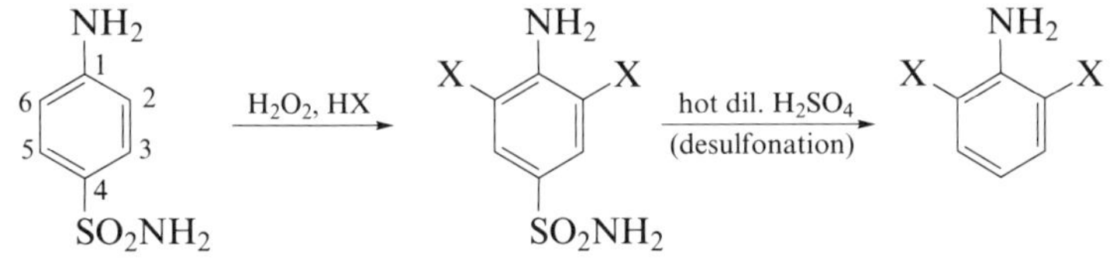

(77) 4-$NH_2C_6H_4SO_2NH_2$ (ring positions 1–6) $\xrightarrow{H_2O_2,\ HX}$ 4-amino-3,5-dihalo-benzenesulfonamide (NH_2, X, X, SO_2NH_2) $\xrightarrow[\text{(desulfonation)}]{\text{hot dil. } H_2SO_4}$ 2,6-$X_2C_6H_3NH_2$

Scheme 48

Sulfonamides of type RSO_2NH_2 or RSO_2NHR' containing the SO_2NH moiety are weakly acidic and form water soluble sodium salts (see the Hinsberg separation of amines, p. 105). In such cases, removal of the acidic proton allows alkylation or acylation at nitrogen (Scheme 49).

$$RSO_2NH_2 \xrightarrow[\text{heat}]{Ac_2O,\ AcCl} RSO_2NHAc$$

Scheme 49

Sulfonamides can be hydrolysed to the corresponding sulfonic acids by heating in strongly acidic solution (concentrated hydrochloric acid), but they resist alkaline hydrolysis.

N,*N*-Diaryl or *N*-alkyl-*N*-aryl benzenesulfonamides (**78**) rearrange to *o*-aminoaryl arylsulfones (**79**) by reaction with organometallic reagents; the rearrangement involves *ortho*-metallation of both aryl rings (Scheme 50).

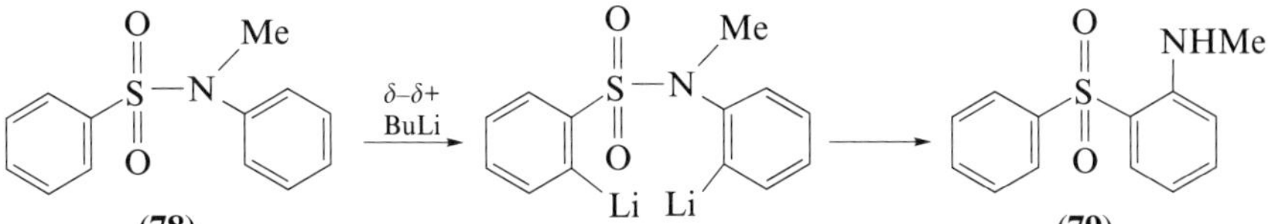

(78) $PhSO_2N(Me)Ph$ $\xrightarrow{\delta-\delta+ \text{ BuLi}}$ dilithiated intermediate (Li ortho on both rings) $\longrightarrow$ (79) $PhSO_2C_6H_4$-2-NHMe

Scheme 50

N-chloro- and *N*,*N*-dichlorosulfonamides are stable, water soluble compounds which are used as antiseptics. They are prepared by chlorination of an alkaline solution of the appropriate sulfonamide; thus, chloramine T (**81**) and dichloramine T (**82**) are obtained from *p*-toluenesulfonamide (**80**) (Scheme 51). Chloramine T (**81**) cleaves 1,3-dithians like (**83**) (see Chapter 6, p. 90) to the corresponding carbonyl compounds, as indicated in Scheme 52.

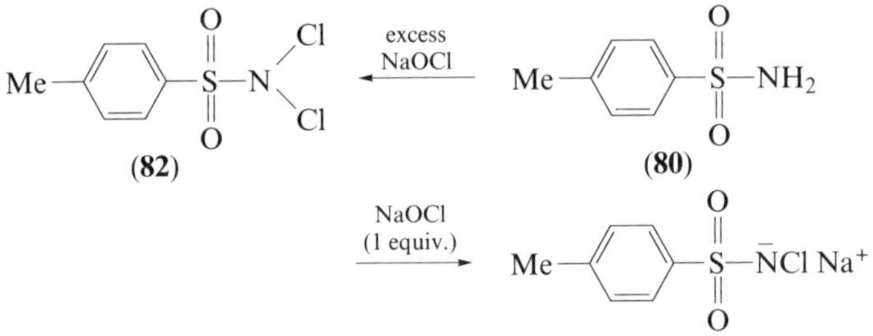

$$p\text{-MeC}_6H_4SO_2NCl_2\ (\mathbf{82}) \xleftarrow{\text{excess NaOCl}} p\text{-MeC}_6H_4SO_2NH_2\ (\mathbf{80}) \xrightarrow{\text{NaOCl (1 equiv.)}} p\text{-MeC}_6H_4SO_2\bar{N}Cl\ Na^+\ (\mathbf{81})$$

Scheme 51

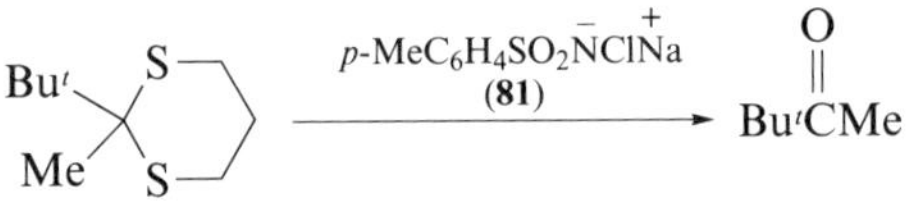

$$\text{2-Bu}^t\text{-2-Me-1,3-dithiane}\ (\mathbf{83}) \xrightarrow[(\mathbf{81})]{p\text{-MeC}_6H_4SO_2\bar{N}Cl\overset{+}{Na}} Bu^tC(=O)Me$$

Scheme 52

Sulfonyl hydrazides

These derivatives (**59**) are generally obtained by condensation of the appropriate sulfonyl chloride (**51**) with an excess of hydrazine (Scheme 53). Sulfonyl hydrazides (**59**) react with nitrous acid (sodium nitrite–dilute acid) to form the corresponding sulfonyl azides (**84**), and with carbonyl compounds they yield the sulfonyl hydrazones (**85**) (Scheme 54).

$$\underset{(\mathbf{51})}{RSO_2Cl} + 2NH_2NH_2 \longrightarrow \underset{(\mathbf{59})}{RSO_2NHNH_2} + NH_2\overset{+}{N}H_3\overset{-}{Cl}$$

Scheme 53

$$\underset{(\mathbf{59})}{RSO_2NHNH_2} + HNO_2 \longrightarrow \underset{(\mathbf{84})}{RSO_2N_3} + H_2O$$

$$\underset{(\mathbf{59})}{RSO_2NHNH_2} + R'C(=O)R'' \xrightarrow[\text{EtOH}]{\text{Warm}} \underset{(\mathbf{85})}{RSO_2NHN{=}CR'R''} + H_2O$$

Scheme 54

p-Toluenesulfonyl hydrazones are useful synthons (see Chapter 10, p. 215); for instance, ketone sulfonyl hydrazones may be used to obtain alkenes by the Bamford–Stevens reaction (Scheme 55). Aldehyde tosylhydrazones may be similarly used to prepare alkanes.[7]

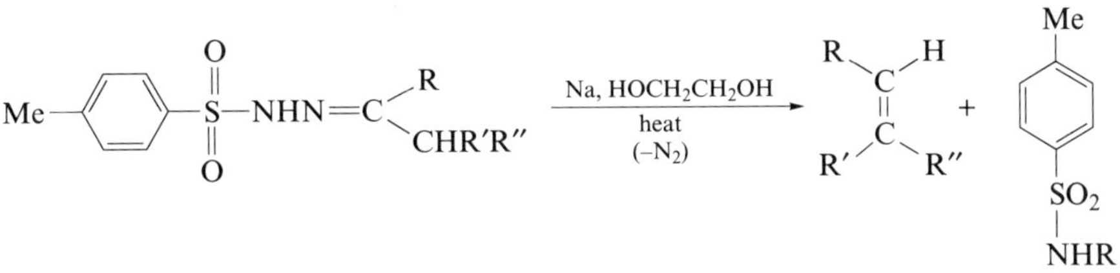

$$\text{Me}-C_6H_4-SO_2-NHN{=}C(R)(CHR'R'') \xrightarrow[\text{heat}\ (-N_2)]{\text{Na, }HOCH_2CH_2OH} RHC{=}CR'R'' + \text{Me}-C_6H_4-SO_2NHR$$

Scheme 55

The thermal instability of sulfonyl hydrazides allows them to be used for the reduction of carbon–carbon double bonds; the reaction depends on the *in situ* formation of the highly reactive diimide intermediate (**86**), which is a good reducing agent for alkenes. Tosylhydrazide can be used for this reduction, but triisopropylbenzenesulfonyl hydrazide (**87**) is preferred because it decomposes to the diimide at low temperature. In the final step, the diimide reduces the alkene to the alkane, as indicated in Scheme 56.

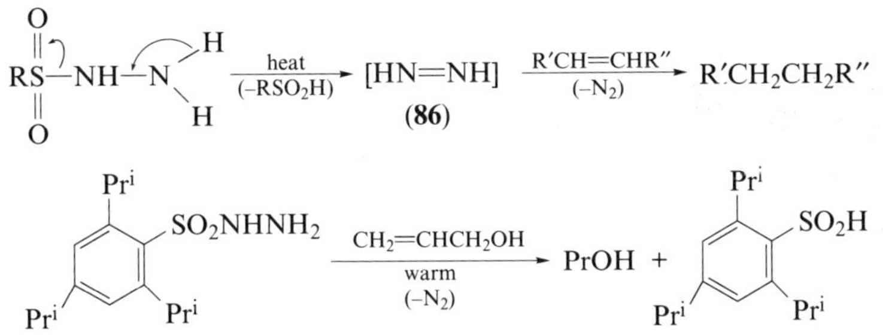

$$RSO_2-NH-NH_2 \xrightarrow[(-RSO_2H)]{\text{heat}} \underset{(\mathbf{86})}{[HN{=}NH]} \xrightarrow[(-N_2)]{R'CH{=}CHR''} R'CH_2CH_2R''$$

$$\underset{(\mathbf{87})}{2,4,6\text{-}Pr^i_3C_6H_2SO_2NHNH_2} \xrightarrow[\text{warm}\ (-N_2)]{CH_2{=}CHCH_2OH} PrOH + 2,4,6\text{-}Pr^i_3C_6H_2SO_2H$$

Scheme 56

Sulfonyl azides

Sulfonyl azides (**84**) are generally prepared by reaction of the sulfonyl chloride (**51**) with sodium azide in aqueous acetone (Scheme 57). Sulfonyl azides are generally reasonably stable at room temperature, especially the aromatic deriva-

$$\underset{(\mathbf{51})}{RSO_2Cl} + NaN_3 \xrightarrow{\text{aq. }Me_2CO} \underset{(\mathbf{84})}{RSO_2N_3} + NaCl$$

Scheme 57

tives which are crystalline solids. However, when sulfonyl azides (**84**) are heated (120–150 °C), they decompose with loss of nitrogen and the transient formation of the sulfonyl nitrene (**88**) (Scheme 58), which is the basis of many of their reactions.

$$RSO_2N_3 \xrightarrow[h\nu]{\text{heat or}} [RSO_2N:] + N_2$$

(**84**) (**88**)

Scheme 58

The electron-deficient sulfonyl nitrene (**88**) can insert into electron-rich carbon–hydrogen bonds, abstract hydrogen atoms, and add to double bonds and aromatic rings. These reactions may be initiated by acids, heat, light and transition metals. The reactions are illustrated by heating methanesulfonyl azide (**89**) with benzene (**23**) (Scheme 59). Here, the electrophilic sulfonyl nitrene (**90**) adds to the electron-rich aromatic double bond, but the kinetically favoured azepine (**91**) rearranges to give the thermodynamically favoured *N*-phenyl sulfonamide (**92**) (Scheme 59).

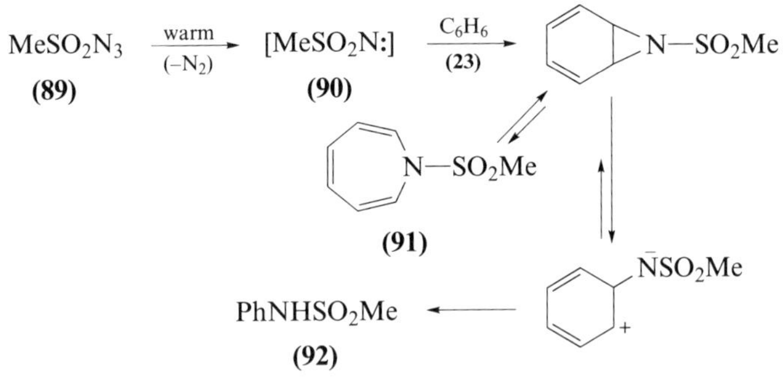

$$MeSO_2N_3 \xrightarrow[(-N_2)]{\text{warm}} [MeSO_2N:] \xrightarrow[(23)]{C_6H_6} \text{(benzene-fused aziridine)}N{-}SO_2Me$$

(**89**) (**90**)

N—SO_2Me (azepine)

(**91**)

$\bar{N}SO_2Me$ (+)

PhNHSO$_2$Me ←

(**92**)

Scheme 59

Some *ortho*-substituted aromatic sulfonyl azides, like *o*-toluenesulfonyl azide (**93**), on thermolysis are converted into the corresponding sultams (**94**) (Scheme 60). The reaction probably involves intramolecular C—H bond insertion by the reactive *o*-toluenesulfonyl nitrene intermediate. The azide (N_3^-) group also

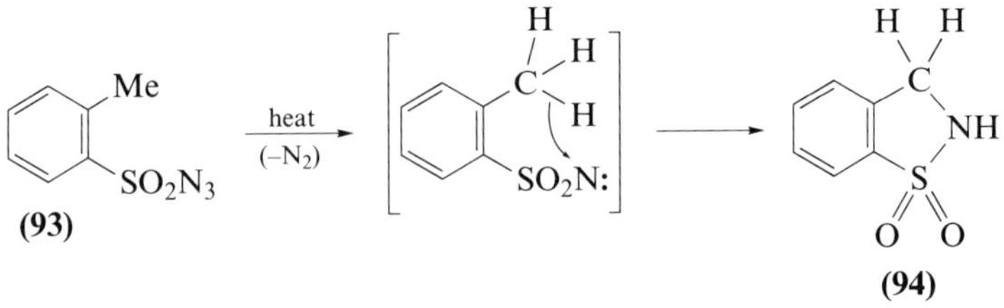

Me, SO_2N_3 (**93**) $\xrightarrow[(-N_2)]{\text{heat}}$ [C(H)(H)(H), $SO_2N:$] → H H C, NH, S, O O (**94**)

Scheme 60

functions as a pseudohalide; consequently, sulfonyl azides can participate in relatively low temperature bimolecular (S_N2-type) substitution reactions with nucleophilic substrates such as alcohols and amines (Scheme 61). Under these mild conditions, decomposition to the sulfonyl nitrene does not occur and it is the intact sulfonyl azide that reacts. *p*-Toluenesulfonyl azide is an efficient diazo transfer reagent; for example, diazo transfer to a reactive methylene group, such as that in dimedone (**95**), is a facile reaction (Scheme 62). The direct diazo transfer reaction fails with compounds containing less acidic methylene groups such as alicyclic ketones; however, a modification using 2,4,6-triisopropylbenzenesulfonyl azide under phase transfer conditions is often successful.

$$R'\ddot{N}H_2 + RSO_2{-}N_3 \longrightarrow RSO_2NHR' + HN_3$$

(**84**)

Scheme 61

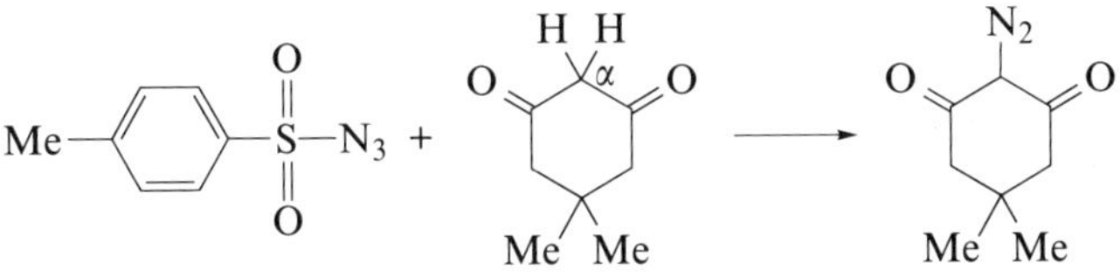

(**95**)

Scheme 62

Aromatic sulfonyl azides react with enol ethers of cyclic ketones to form arenesulfonyl imidate esters (**96**) with ring contraction, the addition–rearrangement process is very stereospecific (Scheme 63).

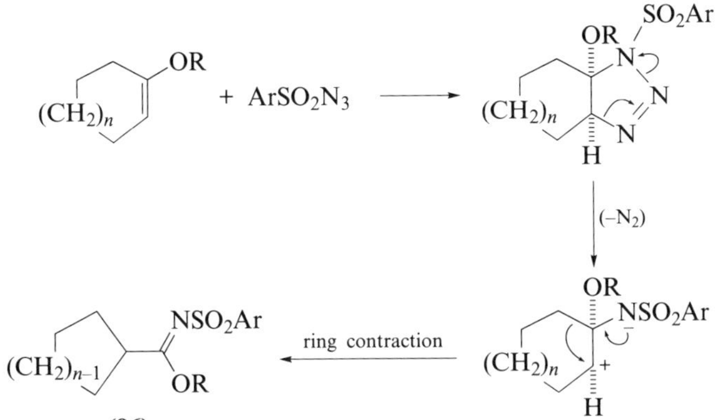

(**96**)

Scheme 63

Sulfonyl azides add to trivalent phosphorus compounds, strained alkenes, e.g. norbornene, enamines and vinyl ethers at comparatively low temperatures. In each case, the initial adduct decomposes on warming with loss of nitrogen, as indicated in Scheme 64.

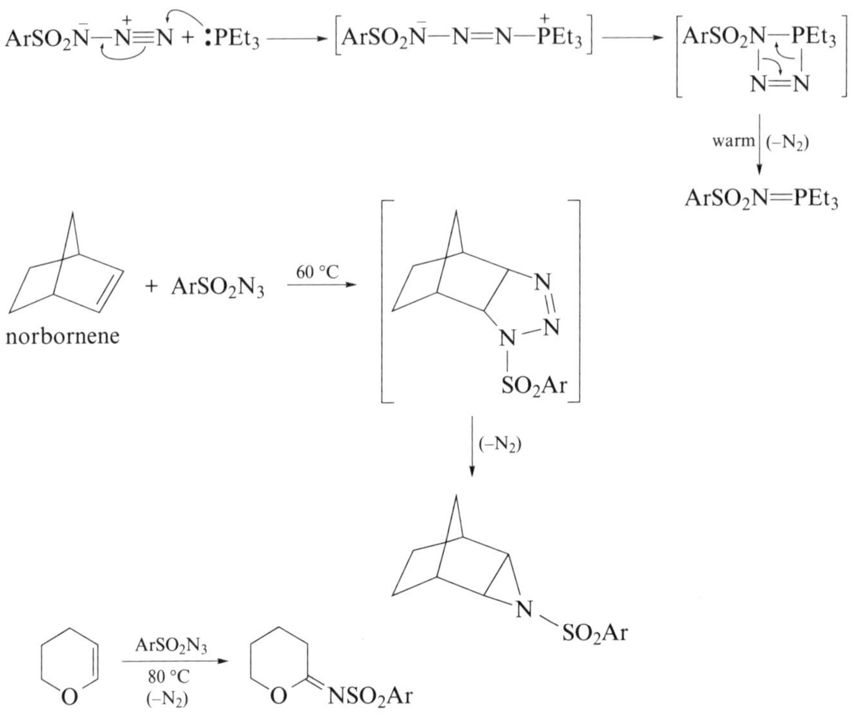

$$ArSO_2\bar{N}-\overset{+}{N}\equiv N + :PEt_3 \longrightarrow [ArSO_2\bar{N}-N=N-\overset{+}{P}Et_3] \longrightarrow [ArSO_2N-PEt_3 / N=N] \xrightarrow[(-N_2)]{warm} ArSO_2N=PEt_3$$

norbornene + $ArSO_2N_3$ $\xrightarrow{60\ °C}$ [triazoline, N–SO_2Ar] $\xrightarrow{(-N_2)}$ aziridine N–SO_2Ar

dihydropyran $\xrightarrow[80\ °C,\ (-N_2)]{ArSO_2N_3}$ tetrahydropyran-2-ylidene NSO_2Ar

Scheme 64

Sulfenes

Sulfenes are very reactive and occur as intermediates in several synthetically important reactions (see p. 115).[3b] They have not been isolated and are formed by base-induced eliminations; for instance, a sulfene is formed in the reaction of a sulfonyl chloride containing at least one α-hydrogen atom with a tertiary amine. As an illustration, when phenylmethanesulfonyl chloride (**97**) is treated with pyridine, phenylsulfene (**98**) is formed as a transient species (Scheme 65).

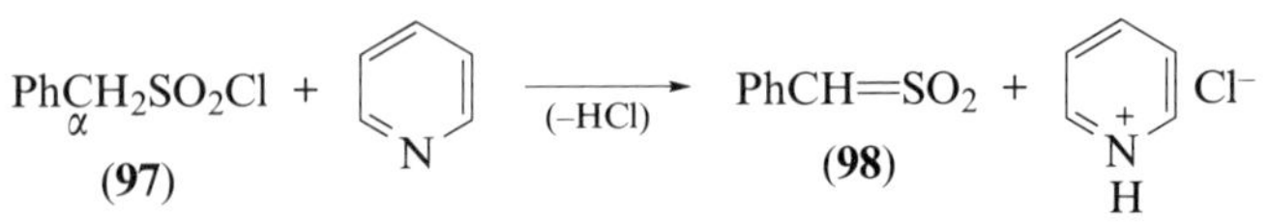

$$\underset{(\mathbf{97})}{PhC_{\alpha}H_2SO_2Cl} + \text{pyridine} \xrightarrow{(-HCl)} \underset{(\mathbf{98})}{PhCH=SO_2} + \text{pyridinium (N–H)}^+\ Cl^-$$

Scheme 65

Evidence for sulfene formation is obtained by carrying out the reaction in the presence of suitable trapping agents, namely substrates which undergo nucleophilic additions, like alcohols and amines (Scheme 66).

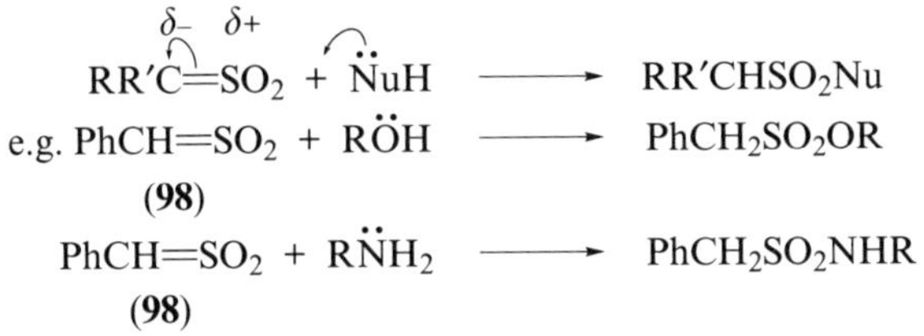

$$\overset{\delta-}{RR'C}{=}\overset{\delta+}{SO_2} + \ddot{N}uH \longrightarrow RR'CHSO_2Nu$$

$$\text{e.g. } \underset{\textbf{(98)}}{PhCH{=}SO_2} + R\ddot{O}H \longrightarrow PhCH_2SO_2OR$$

$$\underset{\textbf{(98)}}{PhCH{=}SO_2} + R\ddot{N}H_2 \longrightarrow PhCH_2SO_2NHR$$

Scheme 66

Sulfenes are important in synthetic organic chemistry because they participate in a range of useful cycloaddition reactions, leading to the formation of four-membered and other cyclic compounds. Sulfenes will form cycloadducts with the following types of substrates.

ENAMINES AND YNAMINES

Sulfene (**75**), generated from methanesulfonyl chloride and triethylamine, reacts with the cyclohexanone enamine (**99**) to give the four-membered cyclic amino sulfone (**100**) (Scheme 67) (see Chapter 10, p. 210).

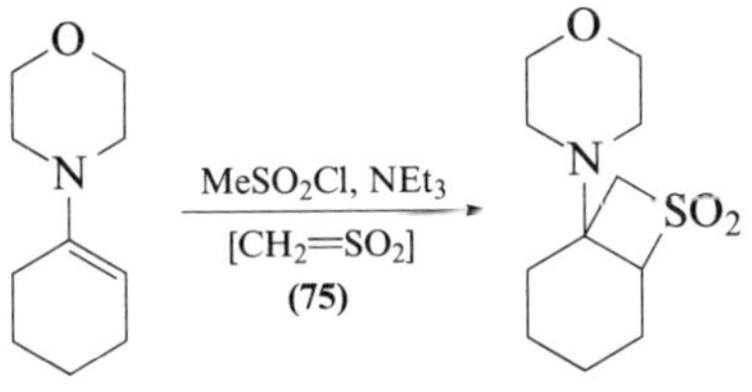

O N SO₂ MeSO₂Cl, NEt₃ [CH₂=SO₂] (75)

(**99**) (**100**)

Scheme 67

There is evidence indicating that the [2+2] cycloadditions of sulfene (**75**) to enamines (**101**) to give the cycloadducts (**102**) are stepwise rather than concerted processes, as indicated in Scheme 68.

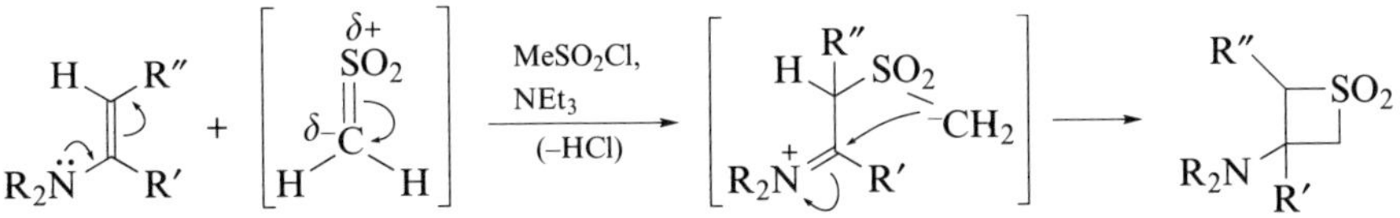

H R″ R₂N̈ R′ + [δ+ SO₂ δ− C H H] MeSO₂Cl, NEt₃ (−HCl) [H R″ SO₂ CH₂ R₂N⁺ R′] ⟶ R″ SO₂ R₂N R′

(**101**) (**75**) (**102**)

Scheme 68

With dienamines, [4+2] cycloaddition may compete with the generally favoured [2+2] cycloaddition, as occurs with the acyclic dienamine (**103**), which yields a mixture of the [4+2] adduct (104) and the [2+2] adduct (**105**) (Scheme 69). On the other hand, when the cyclic cisoid dienamine (**106**) is used, the [4+2] cycloadduct (**107**) is the sole product (Scheme 70).

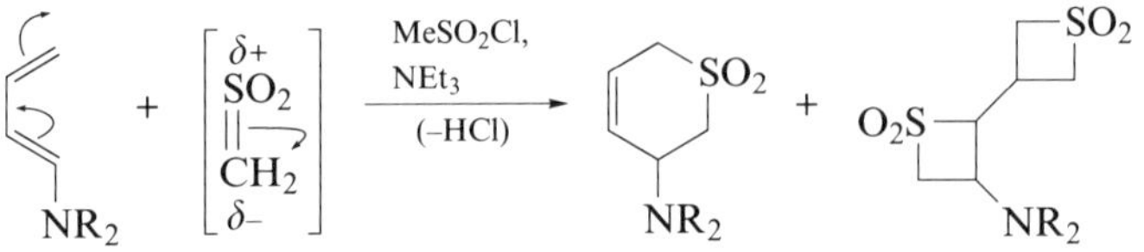

NR_2 (**103**) + [δ+ SO_2 = CH_2 δ−] (**75**) → $MeSO_2Cl$, NEt_3 (−HCl) → (**104**) + (**105**)

Scheme 69

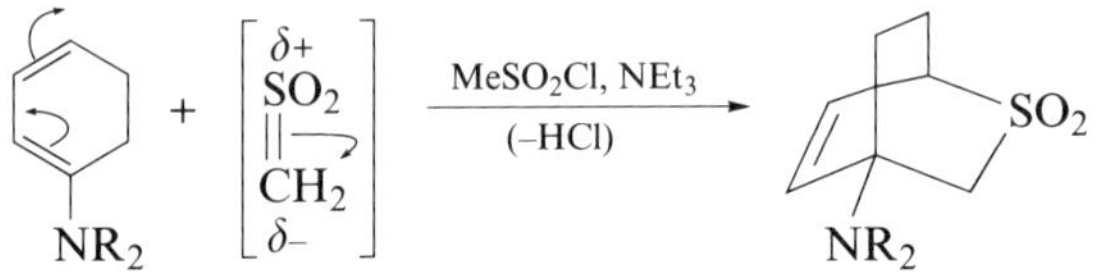

(**106**) + (**75**) → $MeSO_2Cl$, NEt_3 (−HCl) → (**107**)

Scheme 70

Sulfenes (**75**) act as dienophiles towards ketoenamines (**108**) to give the [4 + 2] cycloadducts (**109**); the yields are excellent when R′ = H and R is piperidino (Scheme 71).

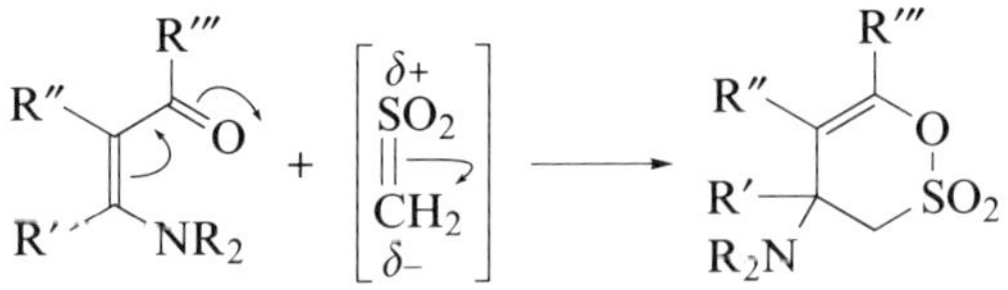

(**108**) + (**75**) → (**109**)

Scheme 71

Sulfenes, e.g. phenylsulfene (**98**), also react with the alkynic analogues of enamines termed 'ynamines', e.g. (**110**), to give the four-membered cyclic sulfones, e.g. (**111**), by [2 + 2] cycloaddition (Scheme 72).

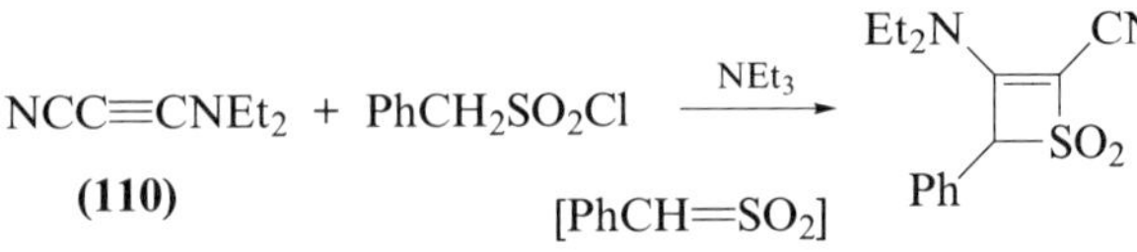

NCC≡CNEt$_2$ + PhCH$_2$SO$_2$Cl →(NEt_3) [PhCH=SO$_2$] (**98**) → (**111**)

(**110**)

Scheme 72

ALKENES AND DIENES

Sulfenes generated from sulfonyl chlorides and tertiary amines do not react with simple alkenes or dienes; however, the sulfene generated from (trimethylsilyl) methanesulfonyl chloride (**112**) and caesium fluoride affords the [4 + 2] cycloadduct (**113**) with cyclopentadiene (**114**) (Scheme 73).

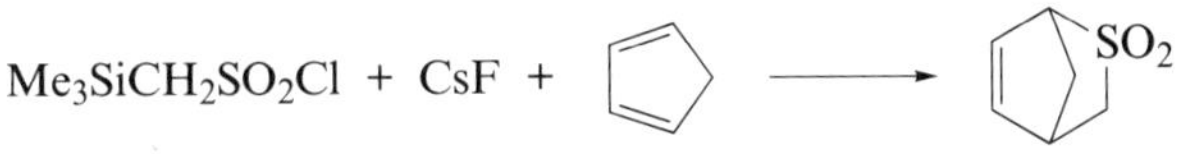

$Me_3SiCH_2SO_2Cl$ + CsF + (**114**) ⟶ (**113**) SO_2

(**112**) (**114**) (**113**)

Scheme 73

Vinylsulfenes (**115**), generated by thermolysis of thiete-1,1-dioxides (**116**), react with the strained cycloalkene norbornene (**117**) to give [4 + 2] cycloadducts (**118**) (Scheme 74).

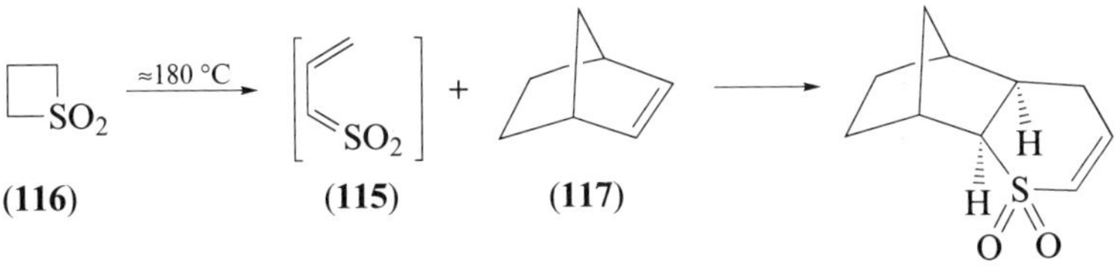

SO_2 (**116**) —≈180 °C→ [SO_2] (**115**) + (**117**) ⟶ (**118**) H, H, S, O, O

Scheme 74

CARBON–NITROGEN DOUBLE-BONDED COMPOUNDS

The first addition of a sulfene to a carbon–nitrogen double bond was reported by Staudinger and Pfenninger (1916), who demonstrated that diphenylsulfene (**119**) generated from diphenyldiazomethane (**120**) and sulfur dioxide, reacts with benzylideneaniline (**121**) to give the four-membered [2 + 2] cycloadduct (**122**) (Scheme 75).

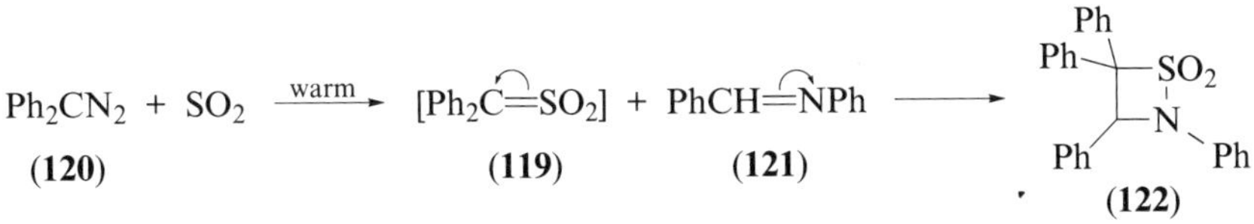

$Ph_2CN_2 + SO_2$ —warm→ $[Ph_2C{=}SO_2]$ + $PhCH{=}NPh$ ⟶ (**122**) Ph, Ph, SO_2, N, Ph, Ph

(**120**) (**119**) (**121**) (**122**)

Scheme 75

Phenylsulfene (**98**) was later shown to add to Schiff bases (**123**) to give a mixture of the *cis*-[2 + 2] cycloadduct (**124**) and the *trans*-[2 + 2] cycloadduct (**125**) (Scheme 76). The predominant formation of the *cis*-product (**124**) may indicate the operation of a concerted [2 + 2] cycloaddition mechanism. Sulfene (**75**) reacts with benzylideneanilines to give the [4 + 2] cycloadducts (**126**), which may arise from rearrangement of the initial [2 + 2] adducts (**127**) (Scheme 77).

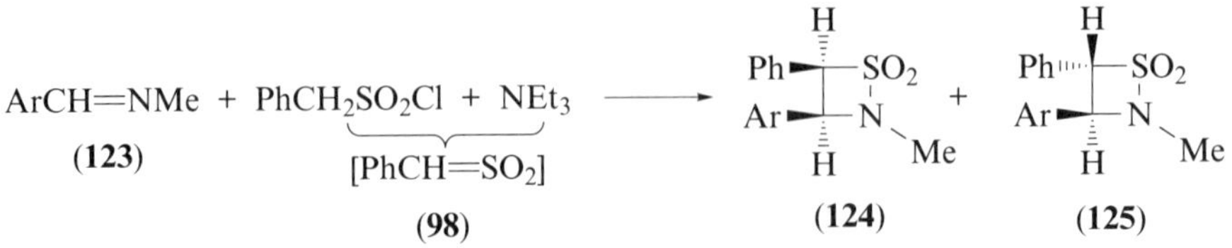

ArCH=NMe (**123**) + $PhCH_2SO_2Cl$ + NEt_3 [$PhCH{=}SO_2$] (**98**) ⟶ (**124**) + (**125**)

Scheme 76

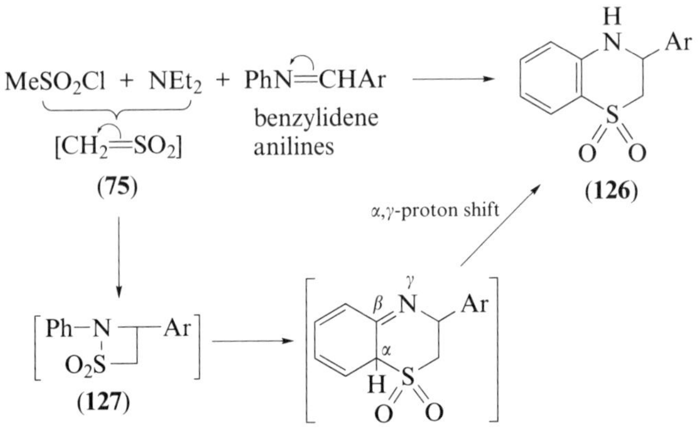

$MeSO_2Cl$ + NEt_3 [$CH_2{=}SO_2$] (**75**) + PhN=CHAr benzylidene anilines ⟶ (**126**)

(**127**) ⟶ [ ] α,γ-proton shift ⟶ (**126**)

Scheme 77

Sulfenes react with 1,3-diazabutadienes (**128**) to give excellent yields of the [4 + 2] cycloadducts (**129**) (Scheme 78). Sulfenes also react with diazoalkanes to form episulfones (see Chapter 10, p. 210).

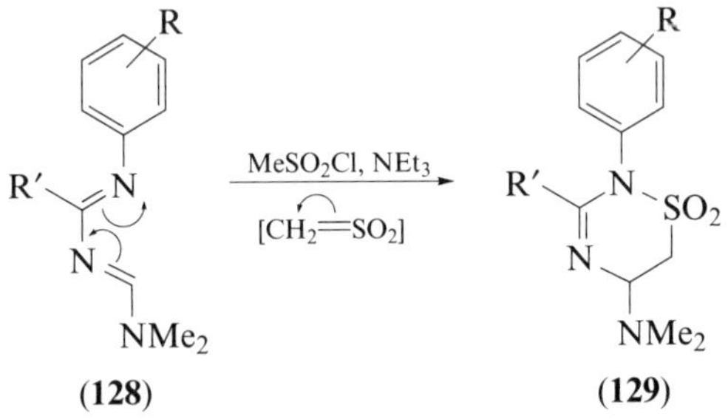

(**128**) $MeSO_2Cl$, NEt_3 [$CH_2{=}SO_2$] ⟶ (**129**)

Scheme 78

CARBON–OXYGEN DOUBLE BONDS

Sulfenes often undergo a [2 + 2] cycloaddition reaction with halocarbonyl compounds; thus, chloral (**130**) adds to sulfene (**75**) to give the β-sultone (**131**) (Scheme 79).

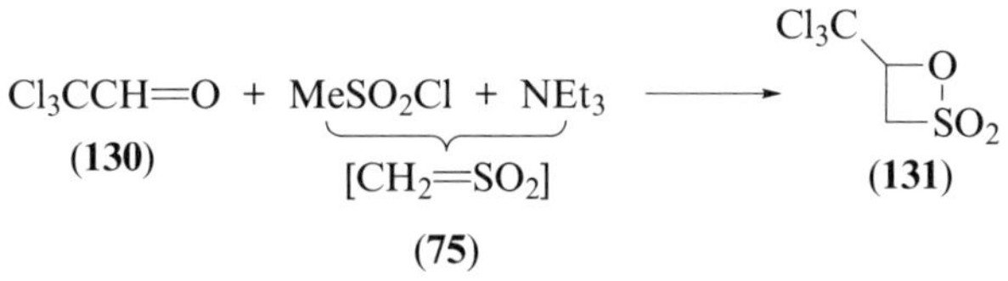

$Cl_3CCH{=}O + MeSO_2Cl + NEt_3 \longrightarrow$ (131)

(130) $[CH_2{=}SO_2]$ (75)

Scheme 79

1,3-DIPOLES

Sulfenes function as dipolarophiles towards *C,N*-diphenylnitrone (**132**) to give the 1:1 adduct (**133**), which probably arises from rearrangement of the initial [3 + 2] cycloadduct (**134**), as shown in (Scheme 80).

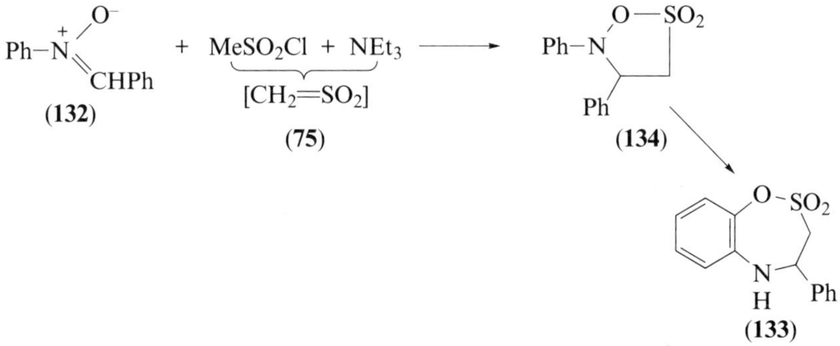

(132) + $MeSO_2Cl + NEt_3$ $[CH_2{=}SO_2]$ (75) $\longrightarrow$ (134) $\longrightarrow$ (133)

Scheme 80

Highly reactive dipoles like azomethine imines (**135**) form [3 + 2] cycloadducts (**136**) with sulfenes (Scheme 81).

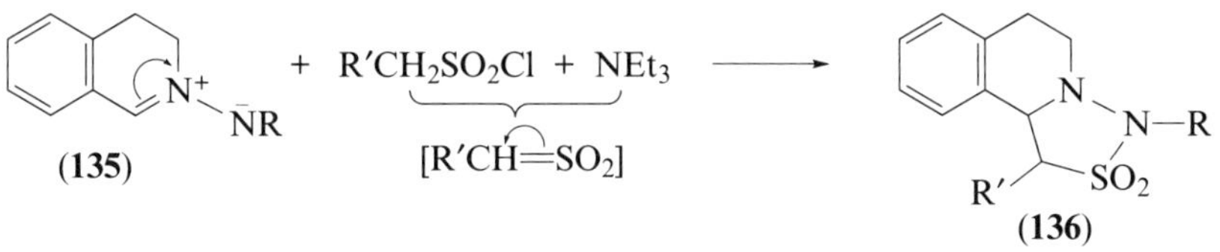

(135) + $R'CH_2SO_2Cl + NEt_3$ $[R'CH{=}SO_2]$ $\longrightarrow$ (136)

Scheme 81

References

1. K. Andersen, in *Comprehensive Organic Chemistry* (Eds D. H. R. Barton and W. D. Ollis), Vol. 3, Pergamon Press, Oxford, 1979: (a) p. 317; (b) p. 331.
2. S. Oae and N. Kunieda, in *Organic Chemistry of Sulfur* (Ed. S. Oae), Plenum Press, New York, 1977, Chap. 11, p 603.
3. *The Chemistry of Sulfonic Acids, Esters and their Derivatives* (Eds S. Patai and Z. Rappoport), Wiley, Chichester 1991: (a) J. Hoyle, Chap. 10, p. 351; (b) J. F. King and R. Rathore, Chap. 17, p. 697.

4. J. March, *Advanced Organic Chemistry*, 4th Edn, Wiley, New York, 1992: (a) p. 528; (b) p. 496.
5. R. J. Cremlyn and P. Bassin, *Phosphorus, Sulfur and Silicon*, **56**, 245 (1991).
6. I. M. Gordon, H. Maskell and M.-F. Ruasse, *Chem. Soc. Rev.*, **18**, 123 (1989).
7. A. H. Chamberlain and S. H. Bloom, *Org. React.*, **39**, 1 (1990).

8 THIOCARBONYL COMPOUNDS

The carbon–sulfur double bond (**1**), unlike the carbon–oxygen double bond, is comparatively weak; for instance, the bond energies of C=S and C=O in carbon disulfide and carbon dioxide are 128 kcal mol^{-1} and 192 kcal mol^{-1}, respectively. The difference is due to the failure of sulfur to participate in efficient $p\pi$–$p\pi$ bonding; that is, the overlap of the carbon $2p$-orbital and the sulfur $3p$-orbital is less effective than the $2p$–$2p$ overlap in the carbon–oxygen double bond owing to differences in spatial symmetry and electron density distribution between the involved orbitals. Thiocarbonyl compounds are consequently more reactive and less stable than their oxygen analogues; they show a marked tendency to give carbon–sulfur single bonds by enolisation.[1–4]

Thioaldehydes and thioketones

Monomeric thioaldehydes have not been isolated, although several, e.g. thioformaldehyde, have been identified by photoelectron spectroscopy. The thione group is relatively unstable in the monomeric form (**1**), and thioaldehydes can only be isolated as dimers (**2**), trimers (**3**), polymers (**4**) or enol derivatives (**5**) (Scheme 1).

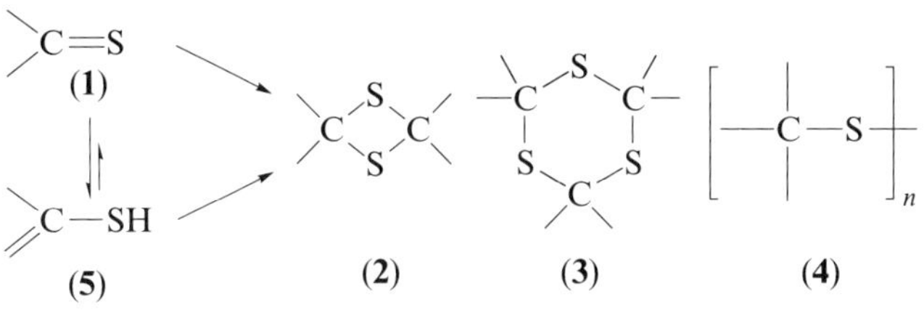

C=S
(**1**)
C—SH
(**5**)
(**2**) (**3**) (**4**)

Scheme 1

Simple thioketones (**6**) are unstable compounds, and until comparatively recently they were known almost exclusively as the cyclic trimers (1,3,5-trithians), (**7**). However, a number of simple aliphatic and alicyclic thioketones have now been isolated as red or violet liquids; they are easily oxidised and polymerised. In contrast, polycyclic thioketones, e.g. thiocamphor (**8**), are relatively stable, red crystalline solids (Figure 1); thiobenzophenone (**6**) (R = R′ = Ph) is also a stable

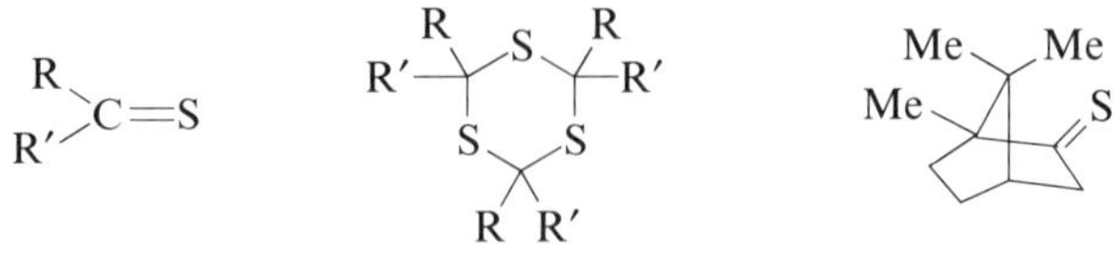

(6) (7) (8)

Figure 1

solid. In general, thioketones are more sensitive to the stabilising effect of neighbouring groups than ketones.

PREPARATION

Thioketones (**6**) can be obtained by the acid-catalysed reaction of ketones with hydrogen sulfide (Scheme 2). The course of the reaction is dependent on the reaction temperature, the nature of the solvent, the concentration of the ketone and the stability of the thioketone (**6**), especially in relation to enolisation. This appears to be the most generally useful preparative route to thioketones, and many simple aliphatic derivatives are obtained by performing the reaction in ethanol at low temperature (−80 °C to −55 °C). The *gem*-dithiol (**9**) may also be converted into the thioketone (**6**) by heating it at approximately 200 °C in the presence of a basic catalyst (Scheme 2). Reasonably stable thioketones, e.g. aromatic and heterocyclic derivatives like (**10**) and (**11**), can be prepared by heating the corresponding ketones with phosphorus pentasulfide in boiling toluene, pyridine or xylene (see Chapter 2, p. 21) (Scheme 3).

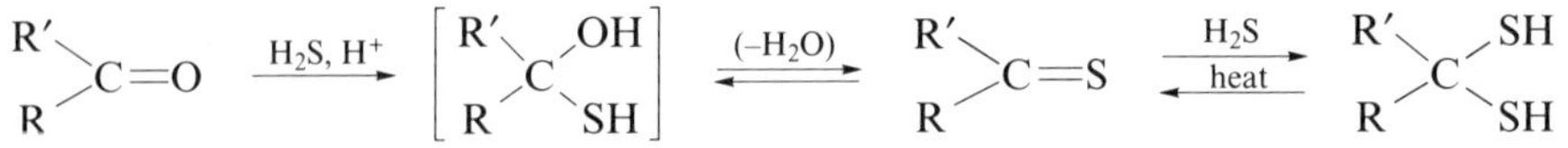

$$R'RC{=}O \xrightarrow{H_2S,\ H^+} [R'RC(OH)(SH)] \underset{}{\overset{(-H_2O)}{\rightleftharpoons}} R'RC{=}S\ (\mathbf{6}) \underset{heat}{\overset{H_2S}{\rightleftharpoons}} R'RC(SH)_2\ (\mathbf{9})$$

Scheme 2

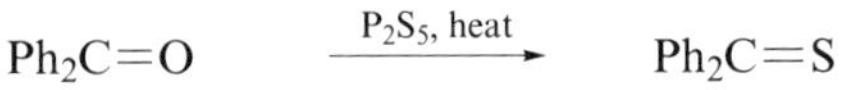

$$Ph_2C{=}O \xrightarrow{P_2S_5,\ heat} Ph_2C{=}S\ (\mathbf{10})$$

P₂S₅, heat → (11)

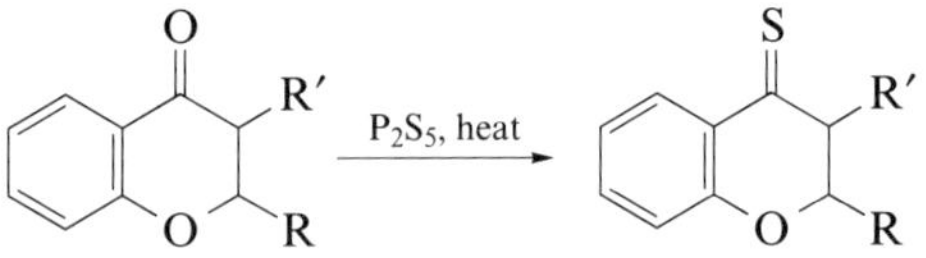

Scheme 3

The procedure has recently been extended to the preparation of less stable thioketones on the discovery that ketones will react with phosphorus pentasulfide in very polar solvents, e.g. acetonitrile, THF or diglyme, at much lower temperatures (approximately 30 °C); the reaction is performed in the presence of a basic catalyst and probably follows the mechanism depicted in Scheme 4.

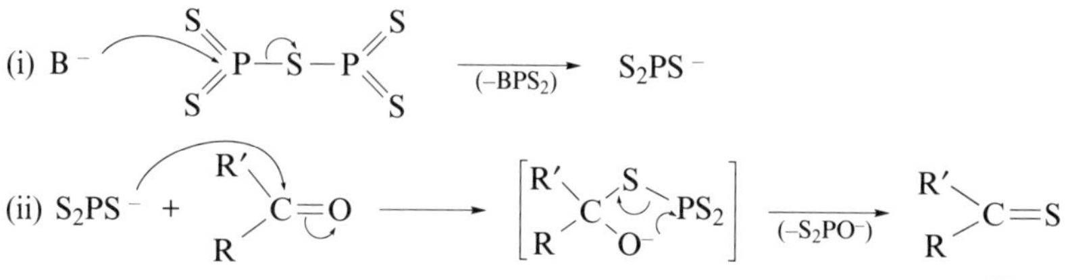

$$\text{(i)}\ B^- + S_2P{-}S{-}PS_2 \xrightarrow{(-BPS_2)} S_2PS^-$$

$$\text{(ii)}\ S_2PS^- + RR'C{=}O \longrightarrow \left[RR'C(S)(O^-)PS_2 \right] \xrightarrow{(-S_2PO^-)} RR'C{=}S\ \textbf{(6)}$$

Scheme 4

Thioketones (**6**) may also be prepared in high yield by treatment of acyclic ketones with Lawesson's reagent.[5] This reagent is easily obtained by heating anisole and phosphorus pentasulfide together in a 10:1 molar ratio (Scheme 5). The ketone is heated with Lawesson's reagent in either boiling benzene or toluene until the evolution of hydrogen sulfide gas has ceased (Scheme 5). The reaction is also successful with amides, esters, thioesters, lactones and lactams, and many substituents, e.g. halogen, nitro or dimethylamino, are not affected; R and R′ may be alkyl or aryl groups. Aromatic thioketones (**12**) may be conveniently prepared by the Friedel–Crafts reaction using thiophosgene; the analogous reaction using a thioacyl chloride affords the corresponding alkyl aryl thioketone (**13**) (Scheme 6).

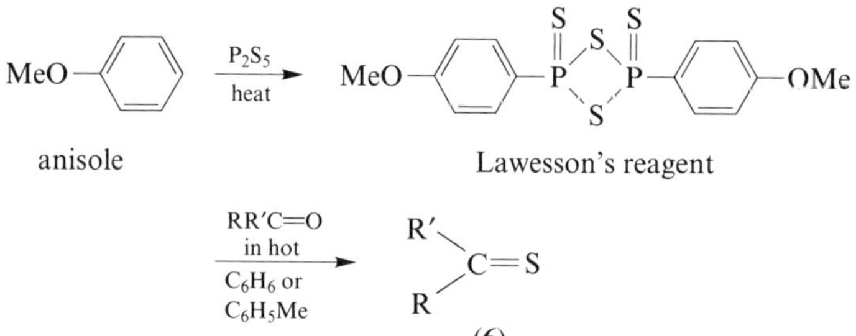

$$MeO{-}C_6H_5 \xrightarrow[\text{heat}]{P_2S_5} MeO{-}C_6H_4{-}P(S)(\mu{-}S)_2P(S){-}C_6H_4{-}OMe$$

anisole — Lawesson's reagent

$$\xrightarrow[\text{C}_6\text{H}_6\text{ or C}_6\text{H}_5\text{Me}]{\text{RR'C=O in hot}} RR'C{=}S\ \textbf{(6)}$$

Scheme 5

$$2\,ArH + CSCl_2 \xrightarrow{AlCl_3} Ar_2C{=}S + [2HCl]$$

thiophosgene — (**12**)

$$ArH + RCSCl \xrightarrow{AlCl_3} ArC(=S)R + [HCl]$$

(**13**)

Scheme 6

Aromatic and heteroaromatic thioketones may be obtained by the action of hydrogen sulfide on diarylimines (**14**); in this reaction, carbon disulfide may be substituted for hydrogen sulfide (Scheme 7). The mechanisms for these reactions are outlined in Scheme 7.

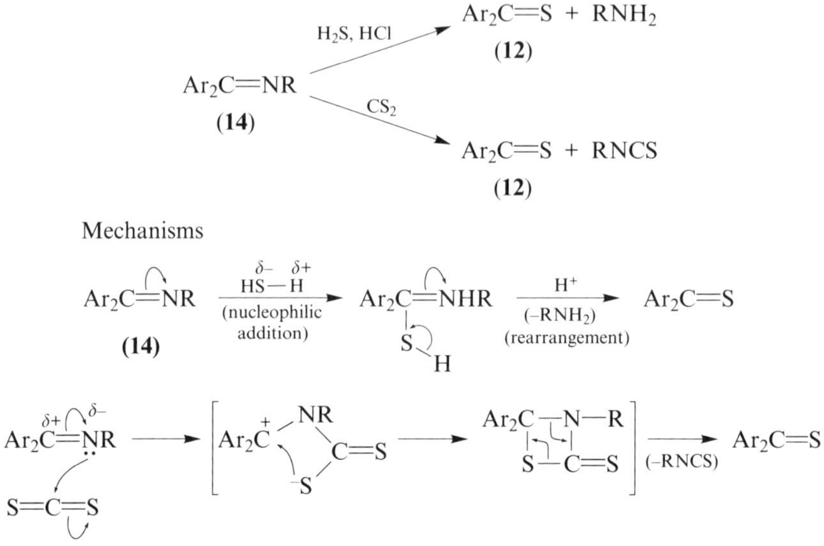

$$Ar_2C{=}NR \ (\mathbf{14}) \xrightarrow{H_2S,\ HCl} Ar_2C{=}S \ (\mathbf{12}) + RNH_2$$

$$Ar_2C{=}NR \ (\mathbf{14}) \xrightarrow{CS_2} Ar_2C{=}S \ (\mathbf{12}) + RNCS$$

Mechanisms

$$Ar_2C{=}NR \ (\mathbf{14}) \xrightarrow[\text{(nucleophilic addition)}]{\overset{\delta-}{HS}-\overset{\delta+}{H}} Ar_2C(SH){=}\overset{+}{N}HR \xrightarrow[\text{(rearrangement)}]{H^+,\ (-RNH_2)} Ar_2C{=}S$$

$$\overset{\delta+}{Ar_2C}{=}\overset{\delta-}{NR} + S{=}C{=}S \longrightarrow \left[Ar_2\overset{+}{C}\text{–}NR\text{–}C({=}S)\text{–}S^-\right] \longrightarrow \left[\begin{array}{c} Ar_2C-N-R \\ | \quad\ \ | \\ S-C{=}S \end{array}\right] \xrightarrow{(-RNCS)} Ar_2C{=}S$$

Scheme 7

REACTIONS

Thioketones containing α-hydrogen atoms exist as equilibrium mixtures with the enethiols. For instance, with thiocyclohexanone (**15**) at low temperatures ($-40\,°C$), the red thione tautomer (**15**) predominates, but on warming, the colourless enethiol (**16**) becomes the dominant species (Scheme 8); the latter may undergo dimerisation.

$$\text{cyclohexanethione } (\mathbf{15}) \rightleftharpoons \text{cyclohexene-1-thiol (–SH) } (\mathbf{16})$$

Scheme 8

At low temperatures, the monomeric thioketones are stable for some time, especially in the presence of a minute quantity of an antioxidant (hydroquinone). In a homologous series of aliphatic thioketones, the stability increases with the number of carbon atoms. Aliphatic thioketones often yield stable dimers, but do not polymerise. When thioketones are reacted with nucleophiles they generally behave similarly to ketones, often yielding identical products. However, thioketones are usually more reactive and catalysts are not required. The thioketone (**6**) thus reacts with water, alcohols, thiols and ammonia derivatives to give the ketone and the compounds (**17**), (**18**) and (**19**), respectively (Scheme 9). In these reactions, enethiolisation does not appear to have much influence. The reaction leading to (**19**) involves initial nucleophilic attack on the electrophilic thiono carbon atom followed by elimination of hydrogen sulfide and formation of the strong carbon–nitrogen double bond, as shown in Scheme 10.

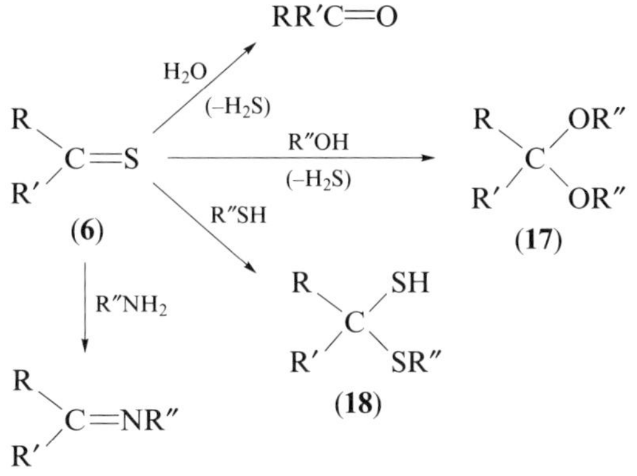

RR′C=O

H_2O ($-H_2S$)

R, R′ C=S (**6**) — R″OH ($-H_2S$) → R, R′ C(OR″)(OR″) (**17**)

R″SH → R, R′ C(SH)(SR″) (**18**)

$R''NH_2$ → R, R′ C=NR″

(**19**) (R″ = Ar, OH, NH_2, NHPh, $NHCONH_2$)

Scheme 9

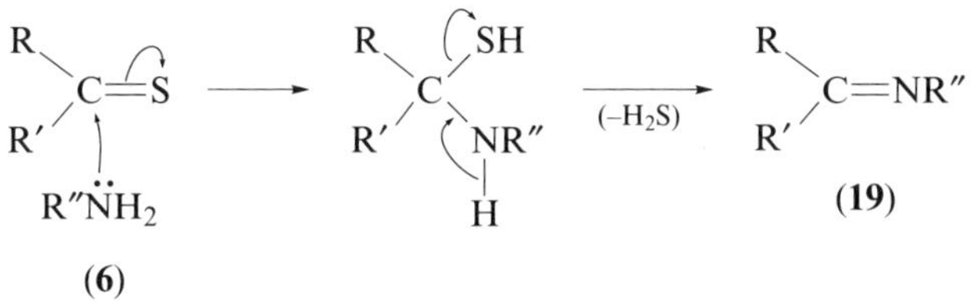

R, R′ C=S + $R''\ddot{N}H_2$ (**6**) → R, R′ C(SH)(NR″H) — ($-H_2S$) → R, R′ C=NR″ (**19**)

Scheme 10

Simple aliphatic thioketones (**6**) readily condense with active methylene compounds in a Knoevenagel-type reaction (Scheme 11). The reaction in Scheme 11 proceeds more easily than with ketones, often without catalysts, and involves nucleophilic attack by the $X(NC)HC^-$ moiety.

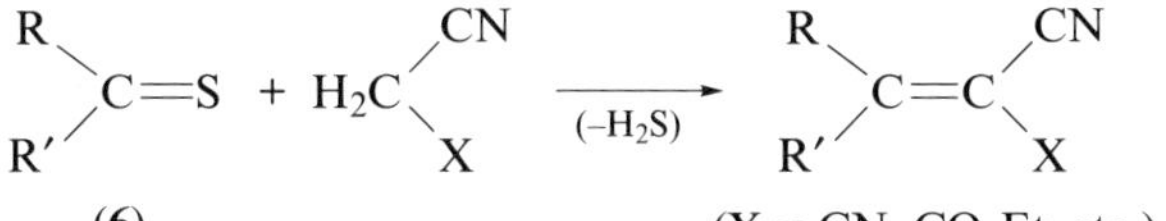

R, R′ C=S (**6**) + H_2C(CN)(X) — ($-H_2S$) → R, R′ C=C(CN)(X)

(X = CN, CO_2Et, etc.)

Scheme 11

Thioketones (**6**) in which R and R′ are electron-withdrawing groups react with organometallic reagents (RM), e.g. organolithium or Grignard reagents, by thiophilic addition. In this reaction, the attacking nucleophile becomes attached to the sulfur atom rather than to the thiocarbonyl carbon atom, leading to the sulfide (**20**) (Scheme 12). On the other hand, when the thioketone (**6**) contains electron-donating R and R′ groups, as occurs in aliphatic thioketones such as di-*t*-butyl thioketone, the nucleophile attacks both the carbon and sulfur atoms, yielding a mixture of the products (**21**) and (**22**) (Scheme 13).

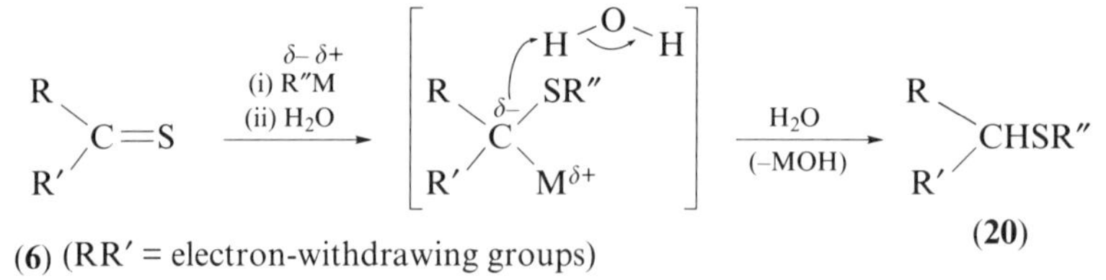

δ− δ+ (i) R″M (ii) H_2O; H_2O (−MOH); SR″; $M^{\delta+}$; CHSR″

(**6**) (RR′ = electron-withdrawing groups) (**20**)

Scheme 12

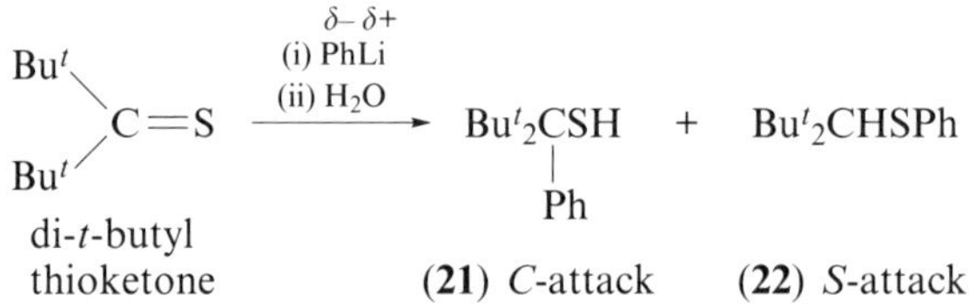

δ− δ+ (i) PhLi (ii) H_2O; $Bu^t_2C(Ph)SH$ + Bu^t_2CHSPh

di-*t*-butyl thioketone (**21**) *C*-attack (**22**) *S*-attack

Scheme 13

Enolisable thioketones (i.e. those possessing an α-hydrogen atom) like thiocyclohexanone (**15**) are easily methylated by treatment with diazomethane (Scheme 14a). However, non-enolisable thioketones (**6**) behave differently and yield the episulfides (**23**) (Scheme 14b). The first reaction (Scheme 14a) is the normal methylation of the acidic enethiol proton by diazomethane. The second reaction (Scheme 14b) involves initial nucleophilic attack by the diazomethane molecule

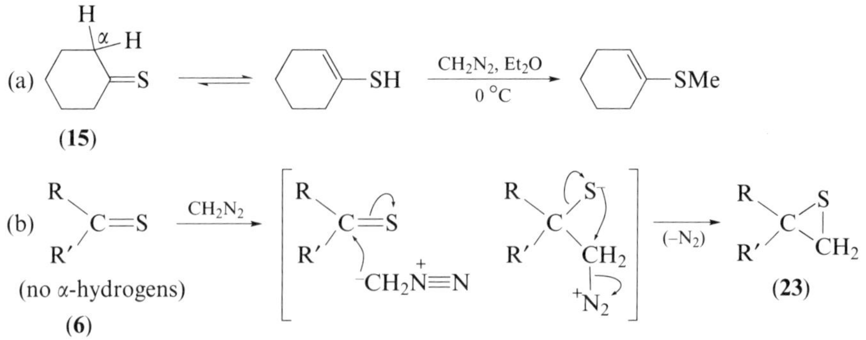

(a) (**15**) ⇌ SH; CH_2N_2, Et_2O, 0 °C; SMe

(b) CH_2N_2; $CH_2\overset{+}{N}\equiv N$; CH_2; $^+N_2$; (−N_2); CH_2

(no α-hydrogens) (**6**) (**23**)

Scheme 14

on the electrophilic thiocarbonyl carbon atom followed by elimination of nitrogen to give the episulfide (**23**).

Thioketones markedly differ from ketones in their reactions with electrophilic reagents.[1,4] Ketones react chiefly at the α-carbon atom, whereas with thioketones the electrophile preferentially attacks the sulfur atom (thiophilic addition), leading to derivatives of the enethiol. The reactions are often performed in the presence of a base (Scheme 15). Thus, alkylation and acylation yield the enethiol derivatives (**24**) and (**25**). Similarly, reactions with an arenesulfenyl chloride, with carbon disulfide followed by an alkyl iodide, and with a carbonyl compound afford the compounds (**26**), (**27**) and (**28**), respectively (Scheme 15). The mechanisms of formation of the last two products are depicted in Scheme 16.

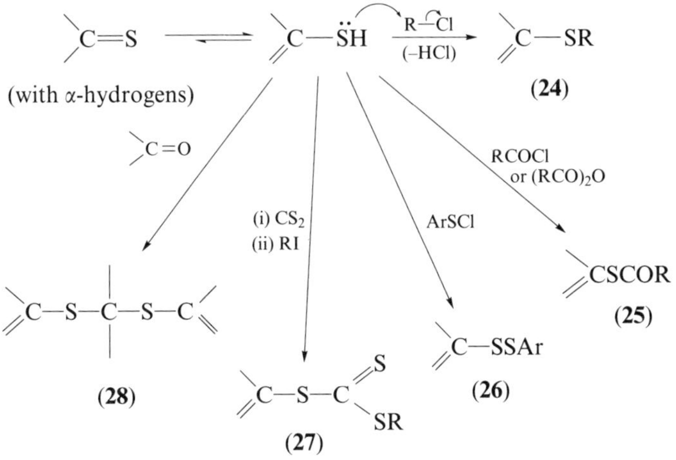

Scheme 15

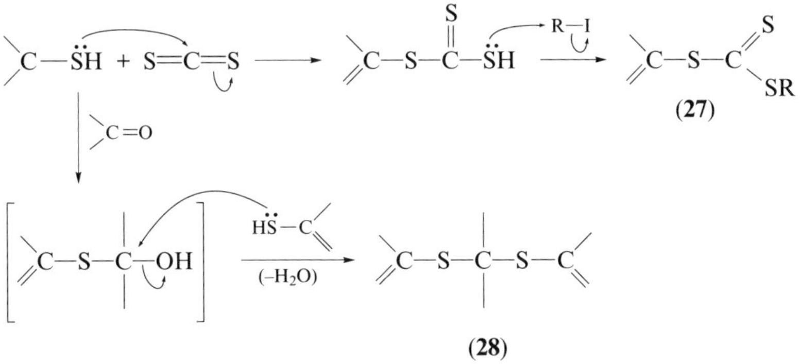

Scheme 16

These reactions may also be described as nucleophilic additions in which the nucleophilic enethiol sulfur atom attacks the electrophilic carbon atom. In sharp contrast to the analogous keto compounds, methylene groups adjacent to the thiocarbonyl moiety display only limited reactivity. This is probably a reflection of the lower acidity of the methylene α-hydrogen atoms in the CH_2CS system as compared with the CH_2CO system, and appears to constitute a fundamental difference between ketones and thioketones.

Thioketones (**6**) are slowly oxidised in air to the corresponding ketones, and they are reduced by the majority of common reducing agents to the thiols (**29**). Sodium borohydride often gives the optimum yield of the thiol. On the other hand, reduction with zinc–hydrochloric acid affords the hydrocarbon (**30**); the latter reaction is analogous to the Clemmensen reduction of ketones (Scheme 17).

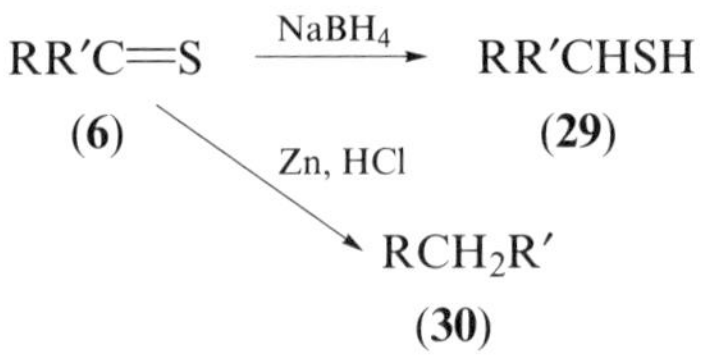

$$\underset{(\mathbf{6})}{RR'C{=}S} \xrightarrow{NaBH_4} \underset{(\mathbf{29})}{RR'CHSH}$$

$$\underset{(\mathbf{6})}{RR'C{=}S} \xrightarrow{Zn,\ HCl} \underset{(\mathbf{30})}{RCH_2R'}$$

Scheme 17

Thioketones, unlike ketones, undergo a coupling reaction on heating with certain powdered metals such as copper, iron, zinc or silver; thus, the diaryl thioketone (**6**) yields the tetraarylalkene (**31**) (Scheme 18).

$$\underset{(\mathbf{6})}{2\,Ar_2C{=}S} + 2Cu \xrightarrow{heat} \underset{(\mathbf{31})}{Ar_2C{=}CAr_2} + CuS$$

Scheme 18

Non-enolisable thioketones behave as dienophiles in Diels–Alder reactions and are considerably more reactive than the analogous ketones in these reactions.

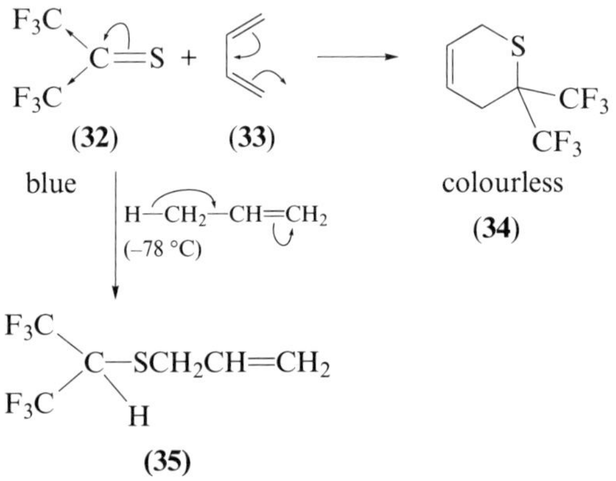

$(F_3C)_2C{=}S$ (**32**) blue + diene (**33**) ⟶ 2,2-bis(trifluoromethyl)thiacyclohex-4-ene (S, CF_3, CF_3) colourless (**34**)

(**32**) $\xrightarrow[(-78\ °C)]{H{-}CH_2{-}CH{=}CH_2}$ $(F_3C)_2C(H){-}SCH_2CH{=}CH_2$ (**35**)

Scheme 19

In particular, hexafluorothioacetone (**32**) is exceedingly reactive and reacts with butadiene (**33**) almost instantaneously to give the [4 + 2] cycloadduct (34) (Scheme 19). The thioketone (**32**) also readily adds to alkenes like propene (Scheme 19) to give the adduct (**35**) (Scheme 19).

Sulfines (thiocarbonyl *S*-oxides)[4]

Sulfines derived from simple aldehydes and ketones are rather unstable and generally appear as transient species. Sulfine (**36**) has been generated by flash vacuum thermolysis of, for instance, 1,3-dithietan-1-oxide (**37**) or methane-sulfinyl chloride (**38**) (Scheme 20). However, the dehydrochlorination of a sulfinyl chloride, like (**39**), by a tertiary amine is a common route to obtain a stable sulfine, like (**40**) (Scheme 21).

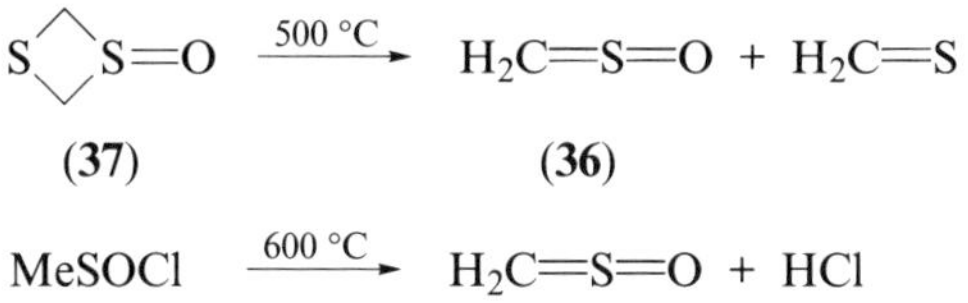

$$\text{(37)} \xrightarrow{500\,^{\circ}\text{C}} H_2C{=}S{=}O + H_2C{=}S \quad \text{(36)}$$

$$\underset{\textbf{(38)}}{MeSOCl} \xrightarrow{600\,^{\circ}\text{C}} \underset{\textbf{(36)}}{H_2C{=}S{=}O} + HCl$$

Scheme 20

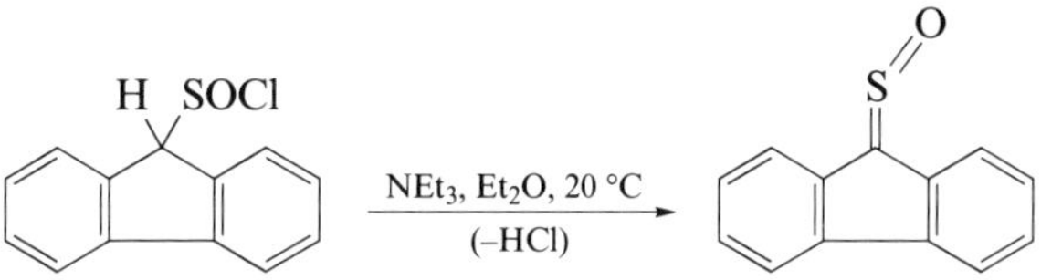

$$\textbf{(39)} \xrightarrow[(-HCl)]{NEt_3,\ Et_2O,\ 20\,^{\circ}C} \textbf{(40)}$$

H SOCl

Scheme 21

The most important synthesis of sulfines (**41**) is by oxidation of thiocarbonyl compounds (**42**) with peracids, e.g. MCPBA (Scheme 22). In this reaction, the thiocarbonyl sulfur atom is usually more susceptible to oxidation than the single-bonded sulfur, so that under the optimum conditions further oxidation to (**43**) and (**44**) can be largely avoided.

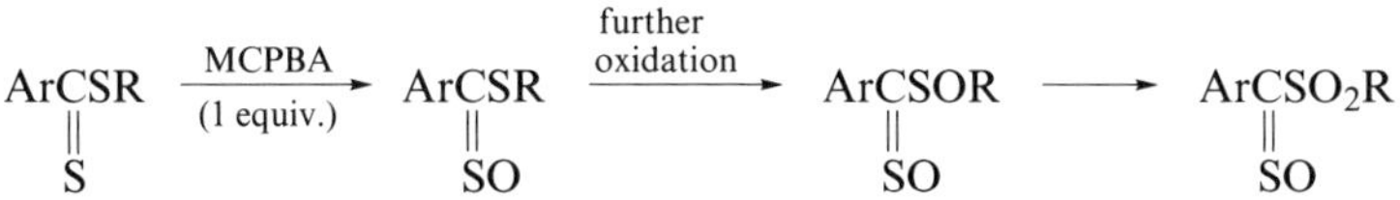

$$\underset{\textbf{(42)}}{ArC(=S)SR} \xrightarrow[(1\ equiv.)]{MCPBA} \underset{\textbf{(41)}}{ArC(=SO)SR} \xrightarrow{further\ oxidation} \underset{\textbf{(43)}}{ArC(=SO)SOR} \longrightarrow \underset{\textbf{(44)}}{ArC(=SO)SO_2R}$$

Scheme 22

The sulfine group has a bent structure, so that sulfines (**45**) in which R and R′ are different may exist as the geometric isomers (**45a**) and (**45b**). Some of these have been isolated, e.g. the *syn*- and *anti*-isomers of phenyl *o*-tolyl sulfine (Figure 2). The fact that some sulfines can be isolated is an important distinction from sulfenes (thiocarbonyl *S*-dioxides) (see Chapter 7, p. 114), which are only known as transient intermediates.

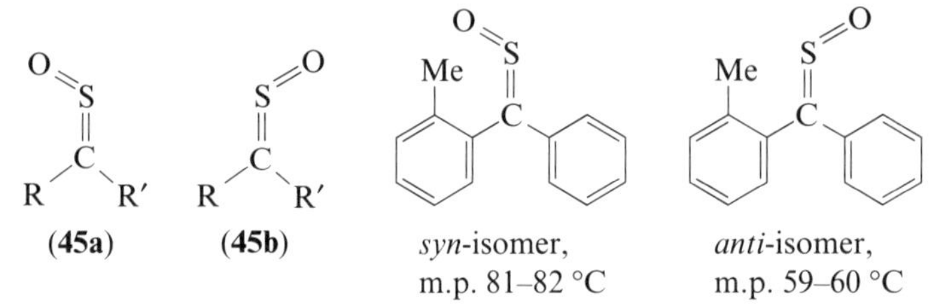

(**45a**) (**45b**) *syn*-isomer, m.p. 81–82 °C *anti*-isomer, m.p. 59–60 °C

Figure 2

Sulfines (**45**) may be hydrolysed to ketones (Scheme 23).

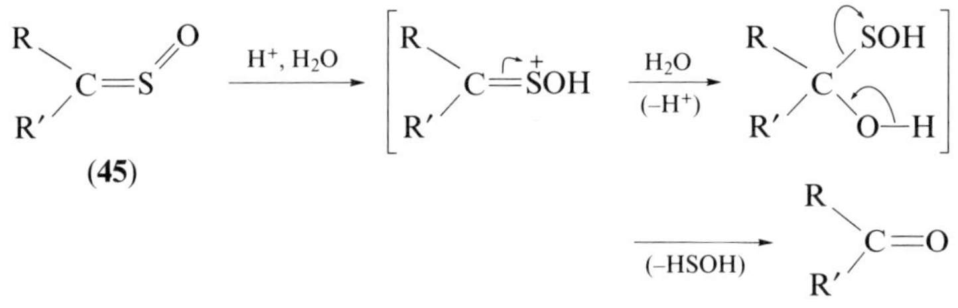

Scheme 23

Organolithium reagents react with diaryl sulfines (**46**) by a mechanism involving thiophilic attack by the nucleophile to give the diaryl sulfoxides (**47**) (Scheme 24).

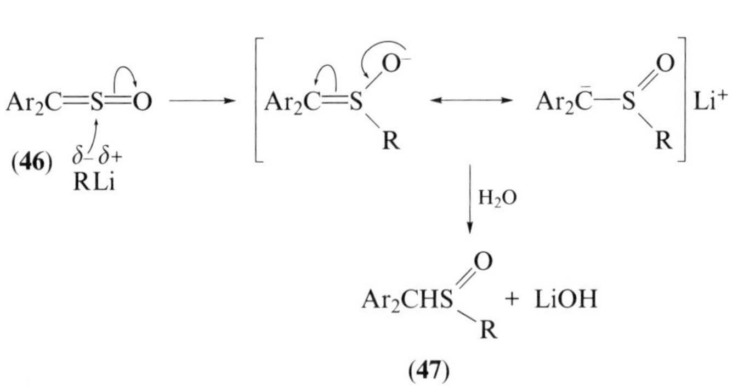

Scheme 24

With 1,3-dienes like 2,3-dimethylbutadiene (**48**), sulfines (**45**) undergo the Diels–Alder reaction to form the [4 + 2] cycloadducts (**49**) (Scheme 25). In this reaction, sulfines behave similarly to thioketones and certain sulfenes (see Chapter 7, p. 117), although the latter generally show a greater tendency to form [2 + 2] cycloadducts.

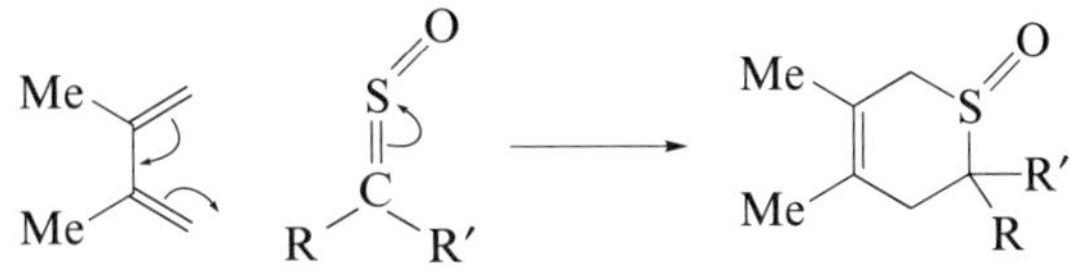

(**48**) (**45**) → (**49**)

Scheme 25

Thio- and dithiocarboxylic acids and derivatives

Monothioacids (**50**) and dithioacids (**51**) are stable and can be isolated. The monothioacid exists predominantly as the *S*-acid or thioloacid (**50**) rather than as the *O*-acid or thionoacid tautomer (**52**). The latter cannot be isolated, since as soon as it is generated (**52**) tautomerises to the thiolo form (**50**) (Figure 3). The derived thioloesters (**53a**) and the dithioesters (**53b**) are stable and can be readily isolated (Figure 3).

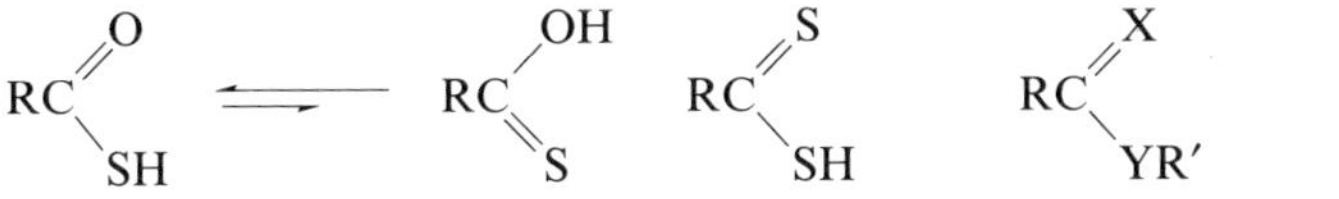

$RC(=O)SH$ (**50**) ⇌ $RC(OH)=S$ (**52**) $RC(=S)SH$ (**51**) $RC(=X)YR'$ (**53a**) X = O, Y = S; (**53b**) X = Y = S

Figure 3

Thioacids have disagreeable odours and slowly decompose on standing in air; they generally have lower boiling points and aqueous solubilities as compared with the corresponding oxygen compounds but are soluble in the majority of organic solvents. The thioloacids (**50**) and the thionoesters (**54a**) and dithioesters (**54b**) can be prepared by treatment of the oxygen analogues with phosphorus pentasulfide (Scheme 26). This conversion may also be achieved by using Lawesson's reagent (see p. 123). One of the best methods involves the reaction of iminoesters (**55**) or iminothiolates (**56**) with hydrogen sulfide. The starting compounds (**55**) and (**56**) are easily obtained by treatment of nitriles with alcohols or thiols (Scheme 27).

Alkyl and aryl dithioacids (**51**) are conveniently prepared by the action of carbon disulfide on a Grignard reagent (see Chapter 9, p. 148). THF is found to be an excellent solvent for this reaction. Chlorothionoformates (**57**) may be

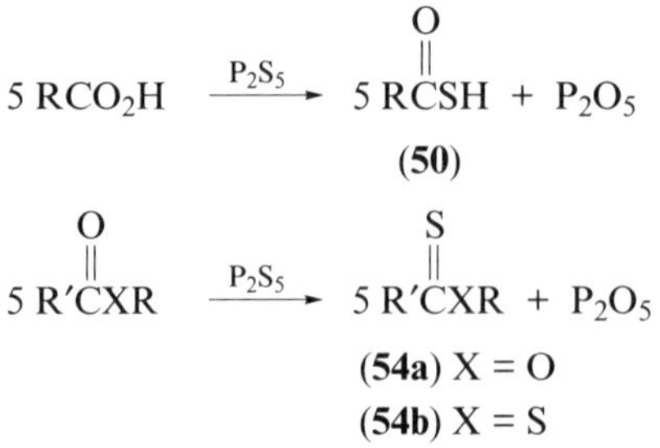

$$5\ RCO_2H \xrightarrow{P_2S_5} 5\ R\overset{\overset{\displaystyle O}{\|}}{C}SH + P_2O_5$$

(50)

$$5\ R'\overset{\overset{\displaystyle O}{\|}}{C}XR \xrightarrow{P_2S_5} 5\ R'\overset{\overset{\displaystyle S}{\|}}{C}XR + P_2O_5$$

(**54a**) X = O
(**54b**) X = S

Scheme 26

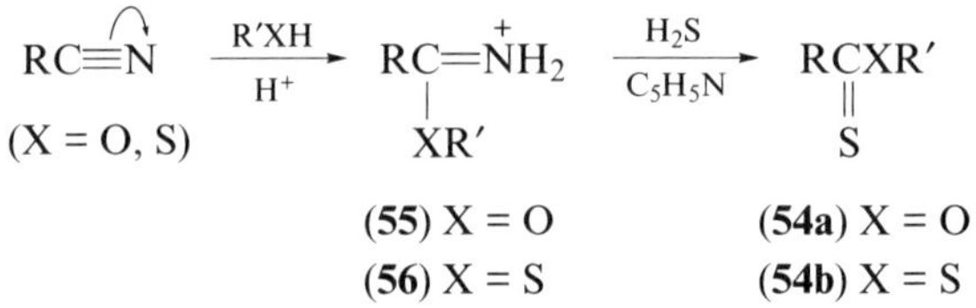

$$RC{\equiv}N \xrightarrow[H^+]{R'XH} R\underset{\underset{\displaystyle XR'}{|}}{C}{=}\overset{+}{N}H_2 \xrightarrow[C_5H_5N]{H_2S} R\underset{\underset{\displaystyle S}{\|}}{C}XR'$$

(X = O, S)

(**55**) X = O (**54a**) X = O
(**56**) X = S (**54b**) X = S

Scheme 27

similarly used to obtain the mono- and dithioesters (**54a**) and (**54b**) (Scheme 28). Chlorothionoformates (**57**) may also be used to prepare aromatic thioesters (**54a**) and (**54b**) (R = aryl) by a Friedel–Crafts-type reaction with a suitable aromatic substrate (Scheme 29).

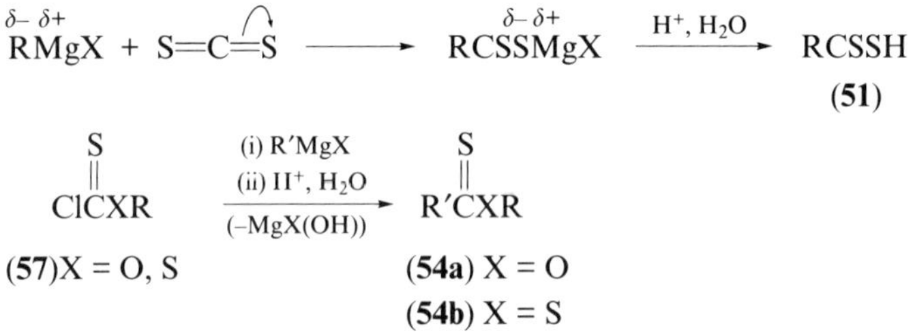

$$\overset{\delta-\ \delta+}{RMgX} + S{=}C{=}S \longrightarrow \overset{\delta-\ \delta+}{RCSSMgX} \xrightarrow{H^+,\ H_2O} RCSSH$$

(51)

$$Cl\overset{\overset{\displaystyle S}{\|}}{C}XR \xrightarrow[(-MgX(OH))]{(i)\ R'MgX \quad (ii)\ H^+,\ H_2O} R'\overset{\overset{\displaystyle S}{\|}}{C}XR$$

(**57**)X = O, S

(**54a**) X = O
(**54b**) X = S

Scheme 28

$$ArH + ClCSXR \xrightarrow[(-HCl)]{AlCl_3} ArCSXR$$

(**57**)

(**54a**) X = O
(**54b**) X = S

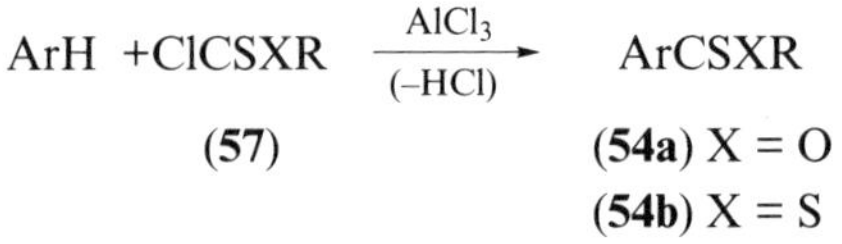

Scheme 29

Aryl thionoesters (**54a**) may be prepared by nucleophilic substitution reactions of aryl thiocarbonyl halides with alcohols or phenols; for instance, thiobenzoyl chloride (**58**) condenses with phenol to yield the diaryl ester (**54a**) (Scheme 30).

$$\underset{(\mathbf{58})}{PhCSCl} + PhOH \xrightarrow{\text{base}} \underset{(\mathbf{54a})}{PhCSOPh} + [HCl]$$

Scheme 30

In general, thiocarbonyl halides (**59**) function as thioacylation reagents with a variety of nucleophiles to yield the appropriate thio derivatives (**60**) (Scheme 31). For example, (**59**) on condensation with thiols, amines, potassium cyanide or potassium thiocyanide yields the corresponding thio compounds (**60**). Thiolocarboxylic acids (50) characteristically acylate alcohols and amines with desulfuration (Scheme 32). Dithiocarboxylic acid esters (**54b**) react with organolithium or Grignard reagents to give the dithioketals (**61**) after treatment with an alkyl halide (Scheme 33).

$$\underset{(\mathbf{59})}{RCSCl} + NuH \xrightarrow{\text{base}} \underset{(\mathbf{60})}{RCSNu} + [HCl]$$

Scheme 31

$$\underset{(\mathbf{50})}{RCOSH} + R'OH \longrightarrow RCOOR' + H_2S$$

$$\underset{(\mathbf{50})}{RCOSH} + R'NH_2 \longrightarrow RCONHR' + H_2S$$

Scheme 32

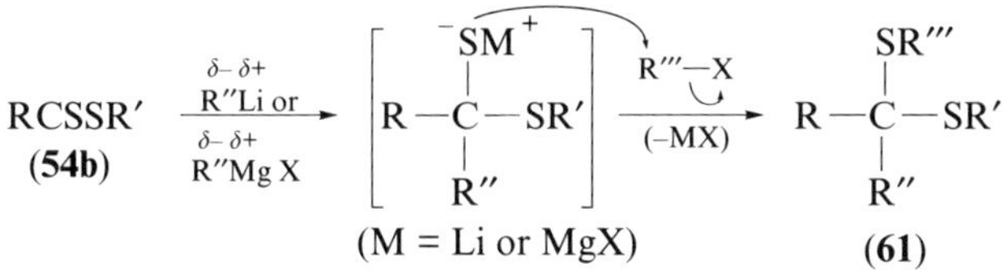

$$\underset{(\mathbf{54b})}{RCSSR'} \xrightarrow[\overset{\delta-\ \delta+}{R''MgX}]{\overset{\delta-\ \delta+}{R''Li}\text{ or}} \left[R{-}\underset{R''}{\overset{^{-}SM^{+}}{C}}{-}SR' \right] \xrightarrow[(-MX)]{R'''{-}X} \underset{(\mathbf{61})}{R{-}\underset{R''}{\overset{SR'''}{C}}{-}SR'}$$

(M = Li or MgX)

Scheme 33

Dithioesters like (**62**) which can exist in the enol form react with sulfur ylides like dimethylsulfoxonium methylide (**63**) to give ketenethiols (**64**). The mechanism of the reaction involves nucleophilic attack by the ylide on the enol form of (**62**) (Scheme 34). Similar compounds (**65**) may be synthesised by treatment of a suitable thioester (**62**) with an alkyl halide in the presence of a base (Scheme 35).

Dithioesters (**54b**) undergo the Clemmensen reduction with desulfuration, so that the thiono group is converted into the methylene group (Scheme 36).

Certain dithioesters possessing an electron-withdrawing group will participate in the Diels–Alder [4+2] cycloaddition reaction with dienes; thus, the dithioester (**54b**) yields the cycloadduct (**66**) (Scheme 37).

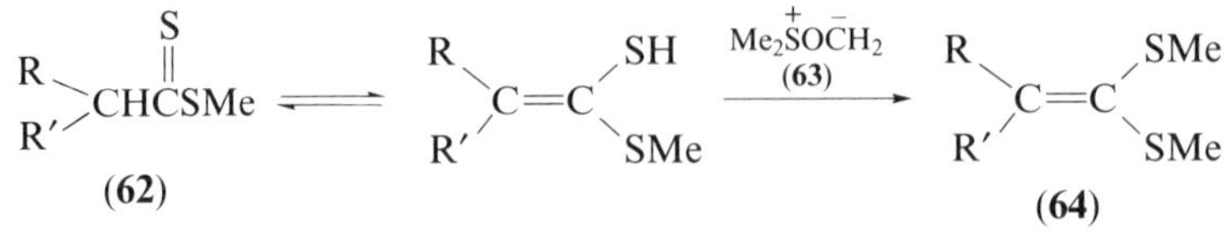

$$RR'CHC(=S)SMe \rightleftharpoons RR'C{=}C(SH)(SMe) \xrightarrow{Me_2\overset{+}{S}O\overset{-}{C}H_2\ (63)} RR'C{=}C(SMe)(SMe)$$

(62) (64)

Mechanism

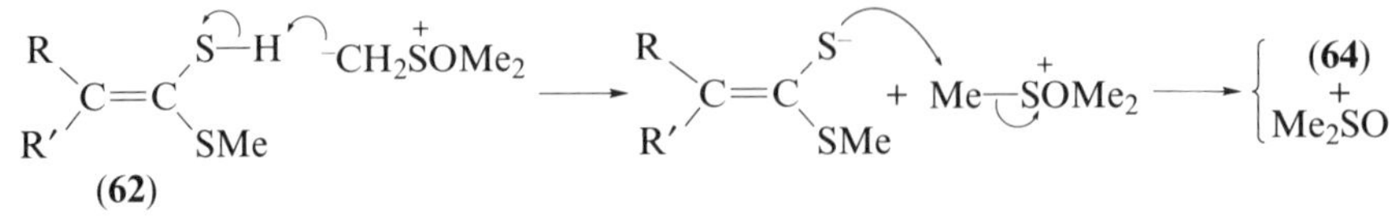

$$RR'C{=}C(S{-}H)(SMe)\ \ {}^{-}CH_2\overset{+}{S}OMe_2 \longrightarrow RR'C{=}C(S^-)(SMe) + Me{-}\overset{+}{S}OMe_2 \longrightarrow \{(64) + Me_2SO$$

(62)

Scheme 34

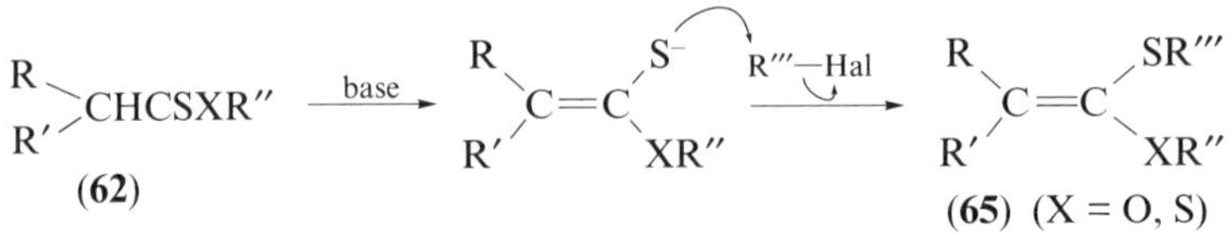

$$RR'CHCSXR'' \xrightarrow{base} RR'C{=}C(S^-)(XR'') \xrightarrow{R'''{-}Hal} RR'C{=}C(SR''')(XR'')$$

(62) (65) (X = O, S)

Scheme 35

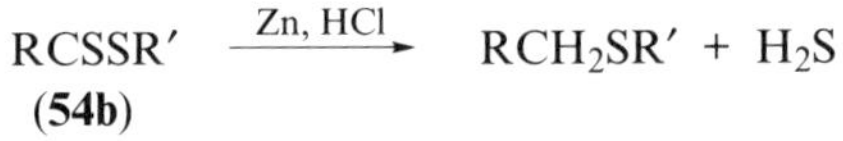

$$RCSSR' \xrightarrow{Zn,\ HCl} RCH_2SR' + H_2S$$

(54b)

Scheme 37

$$RC(=S)SR' + R''CH{=}CH{-}CH{=}CHR''' \longrightarrow (66)$$

(54b) (R = CN, CF_3) (66)

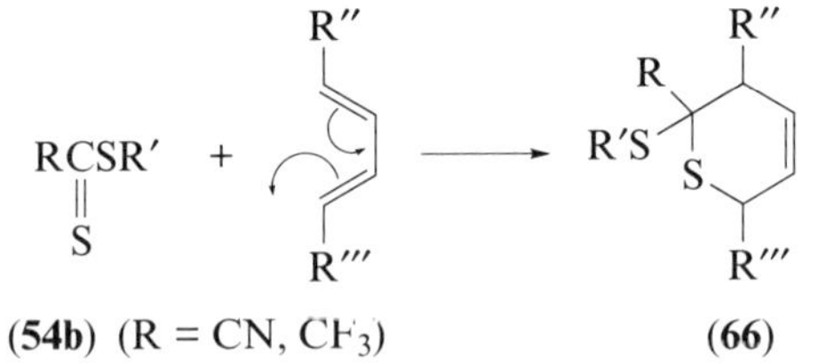

Scheme 37

Carbonic acid (**67**) can form three different types of thio derivatives, namely thionocarbonates (**68**), dithiocarbonates (**69**) and trithiocarbonates (**70**) (Figure 4). Only the trithioacid (**70**) is known to exist as the free acid, although all the acids can be isolated as their salts and esters. The mono-, di- and trithioesters (**71**)–(**73**) can be prepared by the reaction of thiophosgene with alkoxides, phenolates and thiols, respectively (Scheme 38).

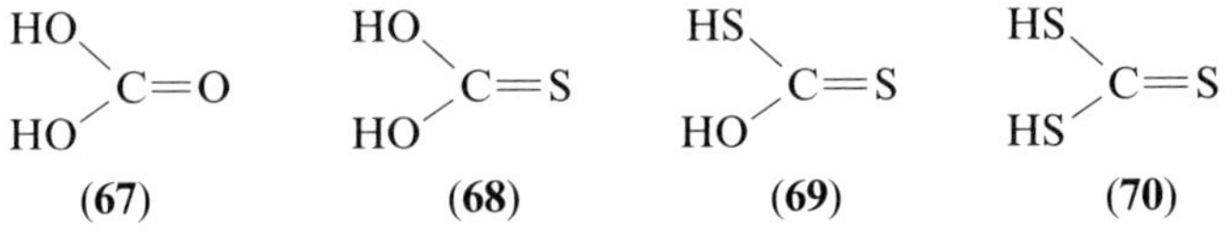

$(HO)_2C{=}O$ (**67**) $(HO)_2C{=}S$ (**68**) $(HS)(HO)C{=}S$ (**69**) $(HS)_2C{=}S$ (**70**)

Figure 4

$CSCl_2$ (thiophosgene) $\xrightarrow{RO^-\ (1\ equiv.)}$ $ClC(=S)OR$ $\xrightarrow{R'O^-\ (1\ equiv.)}$ $ROC(=S)OR'$ (**71**)

$ClC(=S)OR$ $\xrightarrow{R'S^-\ (1\ equiv.)}$ $ROC(=S)SR'$ (**72**)

$CSCl_2$ $\xrightarrow{R'S^-\ (1\ equiv.)}$ $ClC(=S)SR'$ $\xrightarrow{RO^-\ (1\ equiv.)}$ $ROC(=S)SR'$ (**72**)

$ClC(=S)SR'$ $\xrightarrow{RS^-\ (1\ equiv.)}$ $RSC(=S)SR'$ (**73**)

Scheme 38

Carbon disulfide is a valuable synthon (see Chapter 9, p. 147) which can be used for the synthesis of thiocarbonic acid derivatives. Thus, carbon disulfide reacts with ammonium polysulfide or hydrogen sulfide to give trithiocarbonic acid (**70**) or symmetrical esters (**73**) after reaction with an alkyl halide. With alkoxides or thiolates, carbon disulfide forms xanthates (**77**) or *S*-alkyl trithiocarbonates; the latter by further treatment with alkyl, acyl or diazonium halides affords the derivatives (**74**)–(**76**) (Scheme 39).

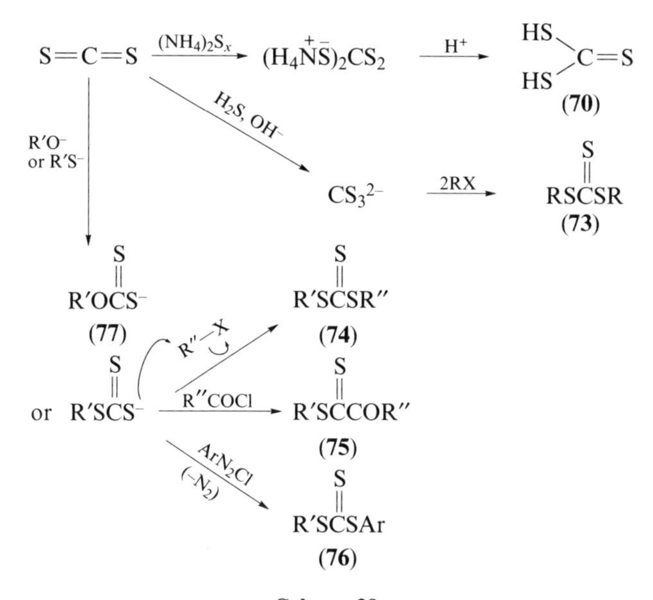

Scheme 39

The formation of xanthates (**77**) by reaction of an alcohol with carbon disulfide in the presence of alkali (the xanthate reaction) (Scheme 40) is of considerable industrial importance since it forms the basis of the manufacture of rayon and cellophane (see Introduction, p. 3, Chapter 2, p. 20 and Chapter 9, p. 147). The xanthate (**77**) on treatment with a soluble copper salt forms a yellow precipitate of copper xanthate; the name xanthate derives from the Greek *xanthos* meaning yellow. Cellulose is largely insoluble in solvents, but can be dissolved by treatment with carbon disulfide and alkali to give a solution of cellulose xanthate. The solution is then forced through a spinneret into a bath of sulfuric acid to re-form cellulose as a fine thread (rayon); alternatively, the solution may be extruded through a narrow slit into the acid, forming very thin sheets (cellophane). The various thiocarbonate esters undergo hydrolysis and aminolysis (nucleophilic substitutions). The reactions of trithiocarbonates (**73**) and xanthate esters (**72**) generally occur with preferential elimination of a thiol; however, the reactions may go in different ways depending on the nature of the reactants and the reaction conditions (Scheme 41).

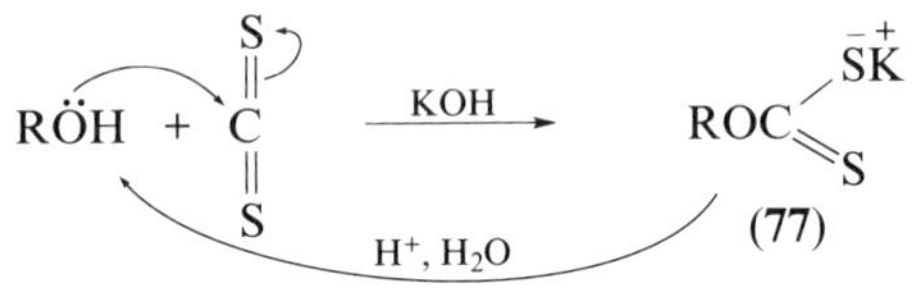

$$\mathrm{R\ddot{O}H} + \mathrm{S{=}C{=}S} \xrightarrow{\mathrm{KOH}} \mathrm{ROC({=}S)\bar{S}\overset{+}{K}}\ (\mathbf{77}) \qquad (\mathbf{77}) \xrightarrow{\mathrm{H^+,\ H_2O}} \mathrm{ROH} + \mathrm{CS_2}$$

Scheme 40

$$\mathrm{ROC({=}S)SR'}\ (\mathbf{72}) \xrightarrow{\mathrm{OH^-}} \mathrm{ROC({=}O)S^-} + \mathrm{R'SH}$$

$$\mathrm{ROC({=}S)SR'}\ (\mathbf{72}) \xrightarrow{\mathrm{H^+,\ H_2O}} \mathrm{ROH} + \mathrm{COS} + \mathrm{R'SH}$$

$$\mathrm{ROC({=}S)SR'}\ (\mathbf{72}) \xrightarrow{\mathrm{NH_3}} \mathrm{ROC({=}S)NH_2} + \mathrm{R'SH}$$

$$\mathrm{R'_2NC({=}S)SR} \xleftarrow[\mathrm{(-RSH)}]{\mathrm{R'_2NH}} \mathrm{(RS)_2CS}\ (\mathbf{73}) \xrightarrow{\mathrm{R'NH_2}} \mathrm{R'NHC({=}S)NHR'} + 2\mathrm{RSH}$$

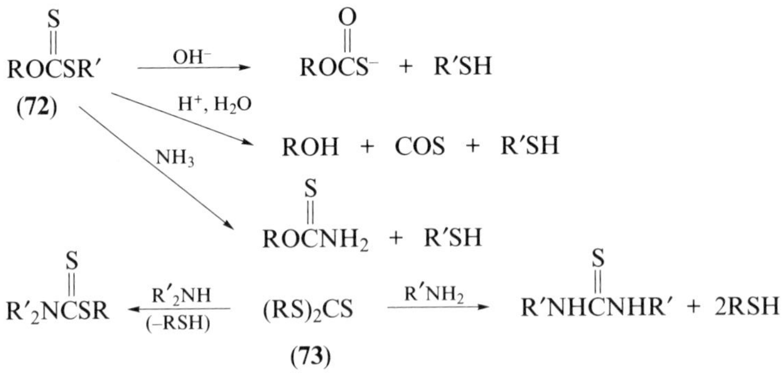

Scheme 41

The controlled oxidation using a peroxycarboxylic acid of trithiocarbonates (**78**) without α-hydrogens, and which therefore cannot form enethiols, yields the corresponding sulfines (**79**) (see p. 129). The use of other oxidants, e.g. concentrated nitric acid or an excess of the peroxycarboxylic acid, gives higher oxidation products like (**80**) and (**81**) (Scheme 42).

Cyclic thionocarbonates (**78**) by heating with trimethyl phosphite yield (*Z*)-alkenes (**84**) (the Corey–Winter reaction). The starting thionocarbonates (**78**) may be prepared by condensation of thiophosgene with vic-glycols (**82**) (X = O) or vic-thiols (**82**) (X = S) (Scheme 43). The Corey–Winter reaction may proceed via the carbenoid intermediate (**83**), which subsequently collapses to yield the (*Z*)-

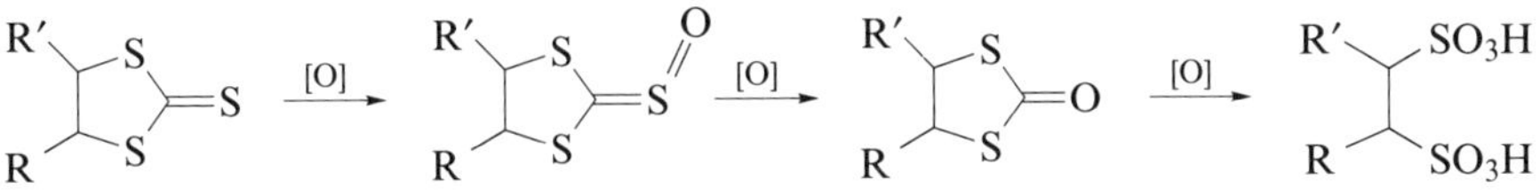

Scheme 42

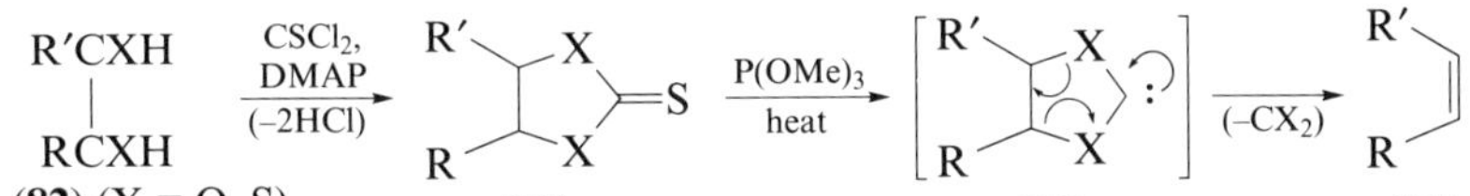

Scheme 43

alkene (**84**). The Corey–Winter reaction is a valuable route for the synthesis of unsaturated sugars and complex cycloalkenes. Alkenes may also be obtained by pyrolysis of methyl xanthates (**85**) containing at least one α-hydrogen atom (the Chugaev reaction) (Scheme 44). The methyl xanthates (**85**) are prepared by treatment of the appropriate alcohol with carbon disulfide and base followed by methyl iodide (Scheme 44) (see p. 135).

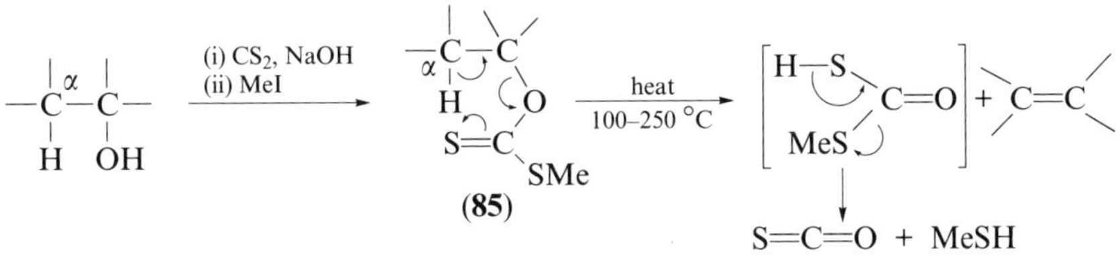

Scheme 44

Thioamides

The most widely used synthesis of thioamides (**86**) involves thionation of the corresponding amides by heating with phosphorus pentasulfide in xylene (Hofmann, 1878) (Scheme 45). In the preparation of primary thioamides (**87**) by this procedure, care must be taken to avoid the decomposition of the product into the nitrile and hydrogen sulfide—the reverse of the earliest preparative route to thioamides (Gay Lussac, 1815) (Scheme 46).

The reaction is often assisted by the addition of a tertiary amine (e.g. triethylamine) as a basic catalyst. The Willgerodt–Kindler reaction (1923), involving

$$\mathrm{RCONR'R''} \xrightarrow[\text{heat}]{P_2S_5,\ \text{xylene}} \mathrm{RCSNR'R''}\ (\mathbf{86})$$

Scheme 45

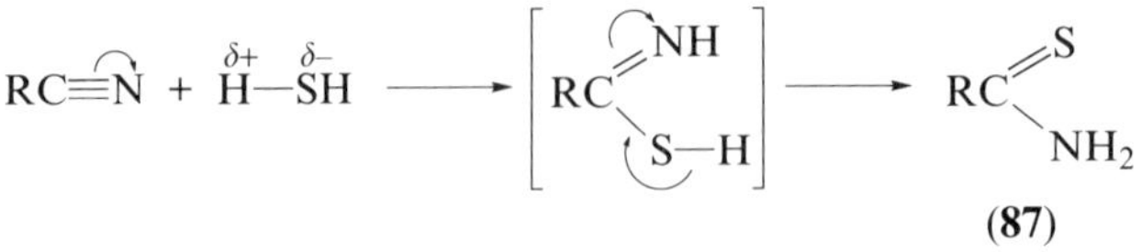

$$RC{\equiv}N + \overset{\delta+}{H}{-}\overset{\delta-}{S}H \longrightarrow \left[RC\begin{matrix} {=}NH \\ {-}S{-}H \end{matrix} \right] \longrightarrow RC\begin{matrix} {=}S \\ {-}NH_2 \end{matrix}$$

(87)

Scheme 46

heating a methyl ketone with a mixture of a secondary amine and sulfur, provides another useful route to thioamides (Scheme 47); a closely related method is the thiolation of enamines by sulfur at room temperature (Scheme 47). These two reactions are closely related since it is likely that the Willgerodt–Kindler reaction proceeds via enamine intermediates (see Chapter 2, p. 18). Tertiary aromatic thioamides (**88**) may be obtained by the Friedel–Crafts reaction using the appropriate thiocarbamoyl chlorides (**89**) (Scheme 48).

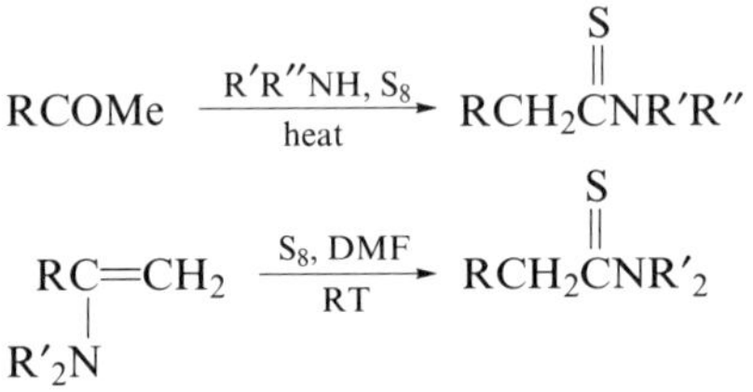

$$RCOMe \xrightarrow[\text{heat}]{R'R''NH,\ S_8} RCH_2\overset{S}{\overset{\|}{C}}NR'R''$$

$$R'_2N{-}C(R){=}CH_2 \xrightarrow[\text{RT}]{S_8,\ DMF} RCH_2\overset{S}{\overset{\|}{C}}NR'_2$$

Scheme 47

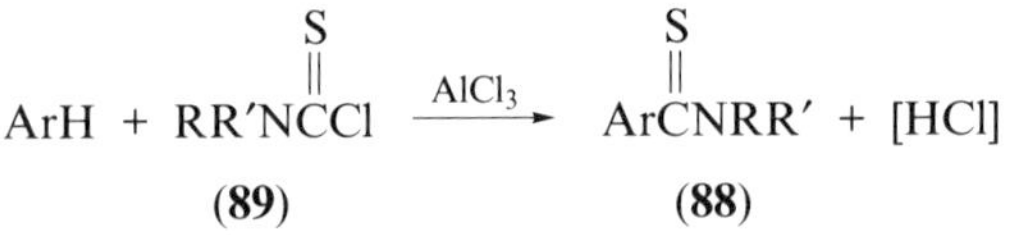

$$ArH + RR'N\overset{S}{\overset{\|}{C}}Cl \xrightarrow{AlCl_3} Ar\overset{S}{\overset{\|}{C}}NRR' + [HCl]$$

(89) **(88)**

Scheme 48

$$R\overset{S}{\overset{\|}{C}}NR'R'' \ \textbf{(86)} \xrightarrow{R'''NH_2} RC\begin{matrix} {-}NR'R'' \\ {=}NR''' \end{matrix} + H_2S \ \textbf{(90)}$$

$$\textbf{(86)} \xrightarrow{NH_2OH} RC\begin{matrix} {-}NR'R'' \\ {=}NOH \end{matrix} + H_2S \ \textbf{(91)}$$

$$\textbf{(86)} \xrightarrow{NH_2NH_2} RC\begin{matrix} {-}NR'R'' \\ {=}NNH_2 \end{matrix} + H_2S \ \textbf{(92)}$$

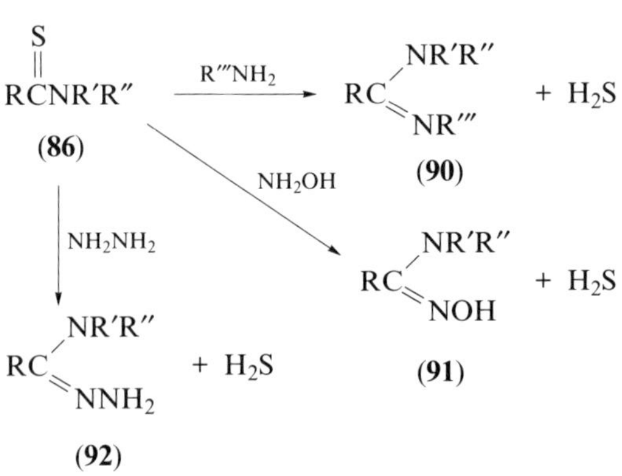

Scheme 49

Thioamides (**86**) are more resistant to hydrolysis than the corresponding amides. By reaction with amines, hydroxylamine and hydrazine they tend to eliminate hydrogen sulfide to yield the imino derivatives (**90**)–(**92**) (Scheme 49).

Thioamides (**86**) are readily reduced to amines (**93**) (Scheme 50). The reduction generally occurs more easily than with amides and can be achieved with a variety of reducing agents, e.g. metal–acid, sodium amalgam, lithium aluminium hydride or Raney nickel.

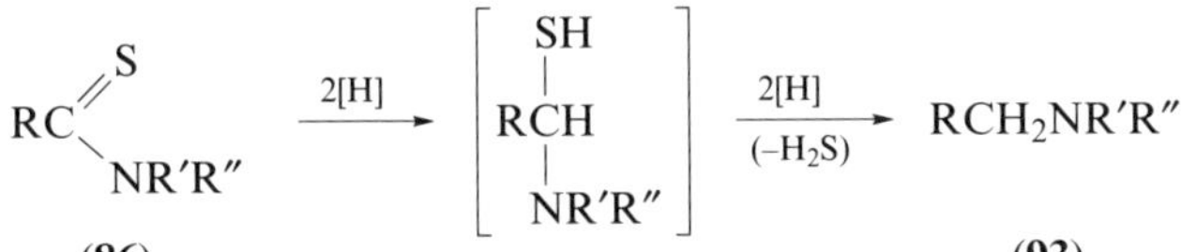

$$\underset{(\mathbf{86})}{RC(=S)NR'R''} \xrightarrow{2[H]} \left[RCH(SH)NR'R'' \right] \xrightarrow[(-H_2S)]{2[H]} \underset{(\mathbf{93})}{RCH_2NR'R''}$$

Scheme 50

Thioamides (**86**) can react with electrophiles either at the sulfur or the nitrogen atom; the former reaction occurs with acids to yield the *S*-protonated species and with alkyl halides to give imidothiolic esters (**94**) (Scheme 51). On the other hand, with primary or secondary thioamides, *N*-alkylation may occur, possibly via kinetically controlled *S*-alkylation, to give an intermediate product which subsequently rearranges under appropriate conditions (Scheme 52).

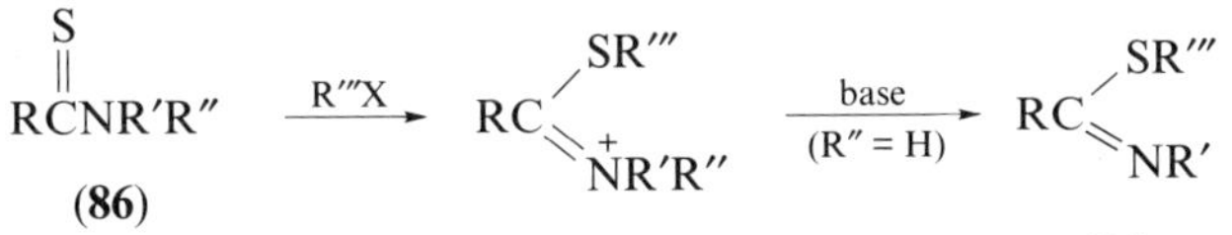

$$\underset{(\mathbf{86})}{RC(=S)NR'R''} \xrightarrow{R'''X} RC(SR''')=\overset{+}{N}R'R'' \xrightarrow[(R''=H)]{base} \underset{(\mathbf{94})}{RC(SR''')=NR'}$$

Scheme 51

$$RC(=S)NH_2 \xrightarrow[HCl]{R'Cl} RC(SR')=\overset{+}{N}H_2\ Cl^- \xrightarrow[(-HCl)]{K_2CO_3} RC(S-R')=NH$$

$$RC(SR')=\overset{+}{N}H_2\ Cl^- \xrightarrow{H^+,\ H_2O} RC(=S)NHR' \xleftarrow{heat} RC(S-R')=NH$$

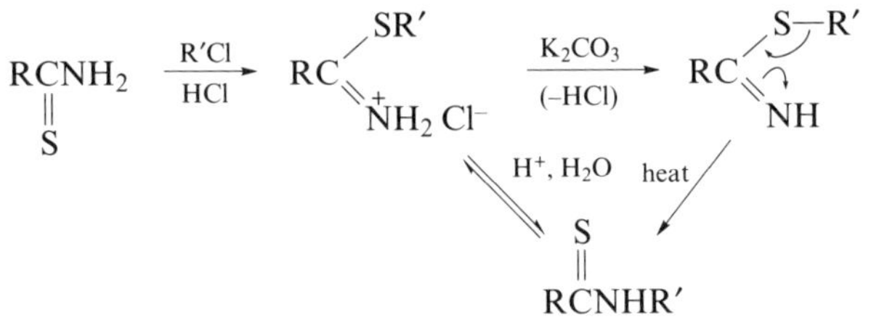

Scheme 52

Thioamides are useful synthons for the preparation of different types of nitrogen–sulfur heterocycles; for instance, thiazoles (**95**) are synthesised by condensation of a thioamide with an α-chlorocarbonyl compound (**96**) (Scheme 53). Vitamin B_1 (thiamine) (**97a**) (Figure 5) is a thiazole derivative and is a member of the vitamin B complex required for growth and the proper functioning of the nervous system. It is a cofactor in biological decarboxylation reactions and exists

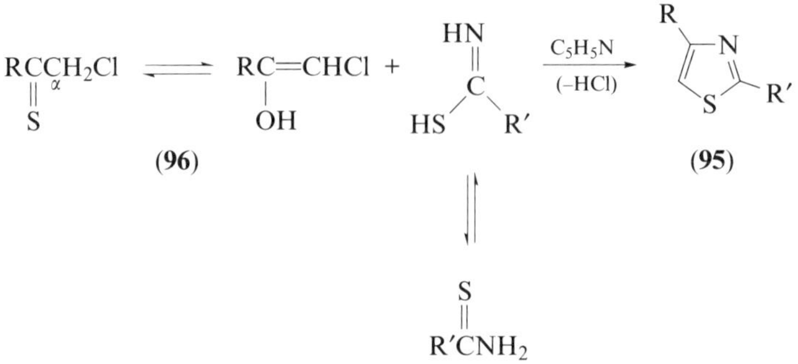

$$RC(=S)CH_2Cl \rightleftharpoons RC(OH)=CHCl + HN=C(SH)R' \xrightarrow[(-HCl)]{C_5H_5N} \text{(95)}$$

(96) (95)

$$HN=C(SH)R' \rightleftharpoons R'C(=S)NH_2$$

Scheme 53

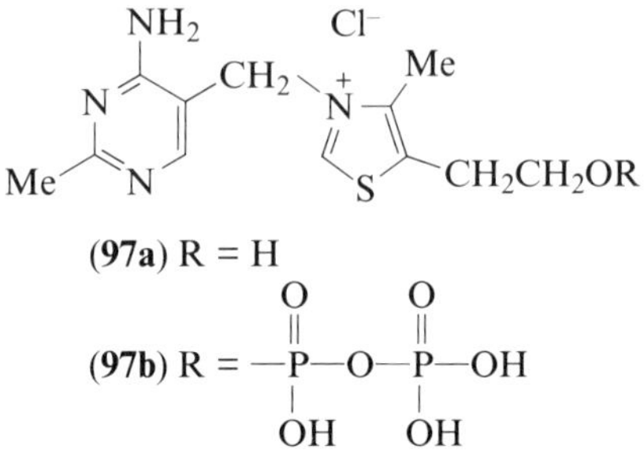

(97a) R = H

(97b) R = —P(=O)(OH)—O—P(=O)(OH)—OH

Figure 5

in vivo as the pyrophosphate (**97b**), which plays a similar role to the cyanide anion in benzoin condensations.

Thioureas

Thiourea (**98**) was first prepared in 1870 by heating ammonium thiocyanate (**99**) (Scheme 54). The reaction is analogous to the historic preparation of urea (Wöhler, 1828) which involved heating ammonium cyanate. Thioureas generally are stable crystalline solids which are useful in the synthesis of heterocyclic compounds. Symmetrical thioureas (**100**) may be obtained by the action of amines on carbon disulfide, and the procedure can be extended to the synthesis of cyclic thioureas (**101**) (Scheme 55). The reaction occurs via the intermediate (**102**) which on subsequent treatment with either ammonia or an amine yields the corre-

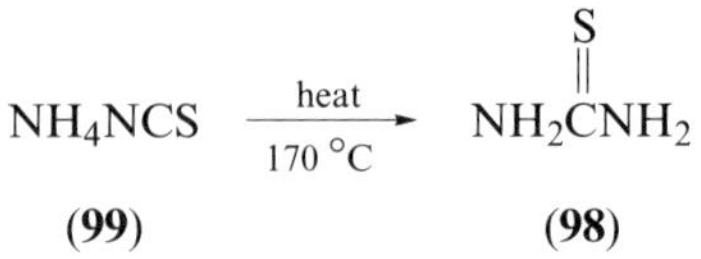

$$NH_4NCS \xrightarrow[170\,^{\circ}C]{heat} NH_2C(=S)NH_2$$

(99) (98)

Scheme 54

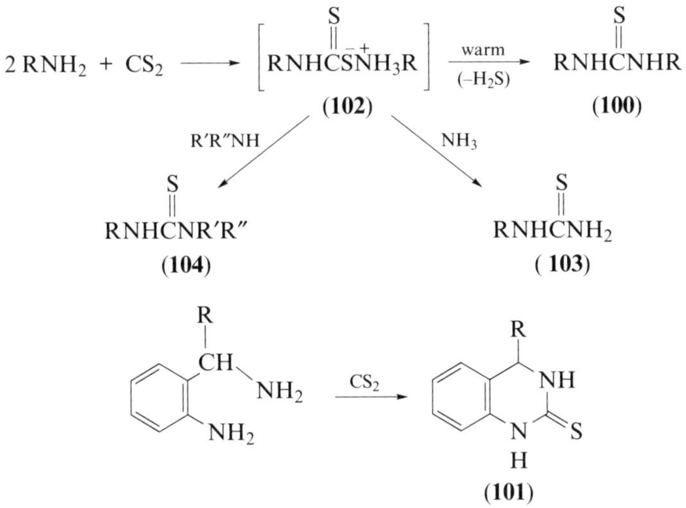

$$2\,RNH_2 + CS_2 \longrightarrow \left[RNH\overset{S}{\overset{\|}{C}}S^{-}\overset{+}{N}H_3R\right] \xrightarrow[(-H_2S)]{\text{warm}} RNH\overset{S}{\overset{\|}{C}}NHR$$

(102) (100)

(102) $\xrightarrow{R'R''NH}$ $RNH\overset{S}{\overset{\|}{C}}NR'R''$ (104)

(102) $\xrightarrow{NH_3}$ $RNH\overset{S}{\overset{\|}{C}}NH_2$ (103)

2-(RCH(NH₂))aniline $\xrightarrow{CS_2}$ (101)

Scheme 55

sponding thioureas (**103**) and (**104**) (Scheme 55). The reaction is assisted by certain catalysts, e.g. hydrogen peroxide, iodine, pyridine and sulfur.

The most general synthetic route to thioureas (**104**) is probably by treatment of an alkyl or aryl isothiocyanate (**105**) with ammonia or a primary or secondary amine (Scheme 56).

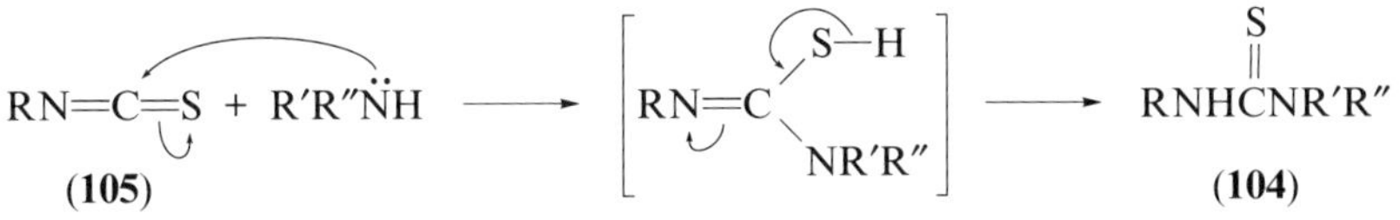

$$RN{=}C{=}S + R'R''\ddot{N}H \longrightarrow \left[RN{=}C\begin{matrix}S{-}H\\NR'R''\end{matrix}\right] \longrightarrow RNH\overset{S}{\overset{\|}{C}}NR'R''$$

(105) (104)

Scheme 56

Symmetrical thioureas (**100**) may be obtained by condensation of thiophosgene with amines (Scheme 57).

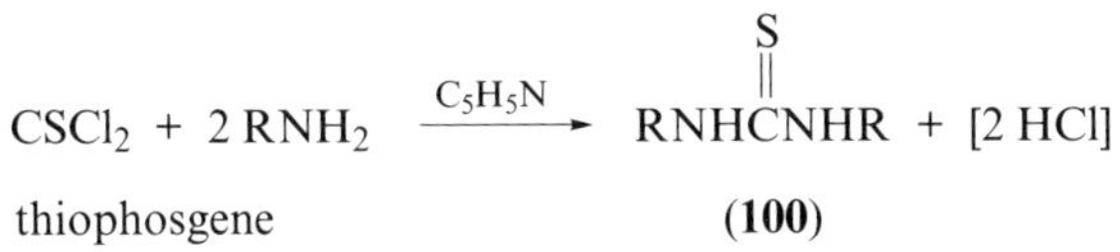

$$CSCl_2 + 2\,RNH_2 \xrightarrow{C_5H_5N} RNH\overset{S}{\overset{\|}{C}}NHR + [2\,HCl]$$

thiophosgene (100)

Scheme 57

REACTIONS

Thiourea (**98**) reacts with alkyl and benzyl halides to give the *S*-alkyl and *S*-benzyl derivatives; thus, on treatment with benzyl chloride, *S*-benzylisothiuronium chloride (**106**) is formed. This reagent is used in organic qualitative analysis for

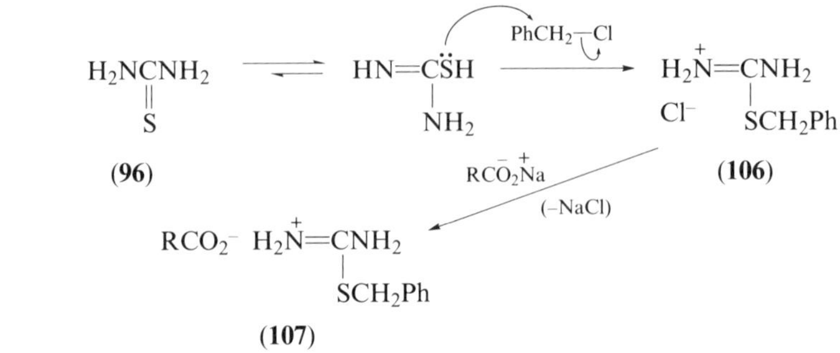

$H_2NC(=S)NH_2$ (96) ⇌ $HN{=}C(NH_2)\ddot{S}H$ —($PhCH_2$–Cl)→ $H_2\overset{+}{N}{=}C(NH_2)SCH_2Ph$ Cl^- (106) —($RCO_2^-\overset{+}{Na}$, (–NaCl))→ RCO_2^- $H_2\overset{+}{N}{=}C(NH_2)SCH_2Ph$ (107)

Scheme 58

the characterisation of carboxylic and sulfonic acids, with which it yields crystalline isothiuronium salts (**107**) (Scheme 58).

Thioureas generally react with alkyl halides to give *S*-alkylisothiuronium salts. On the other hand, by reaction with acyl halides, either the *N*- or *S*-acylthioureas may be formed. The nature of the product appears especially dependent on the reaction temperature. *S*-Acylthioureas (**108**) on heating rearrange to the *N*-acyl derivatives (**109**) (Scheme 59).

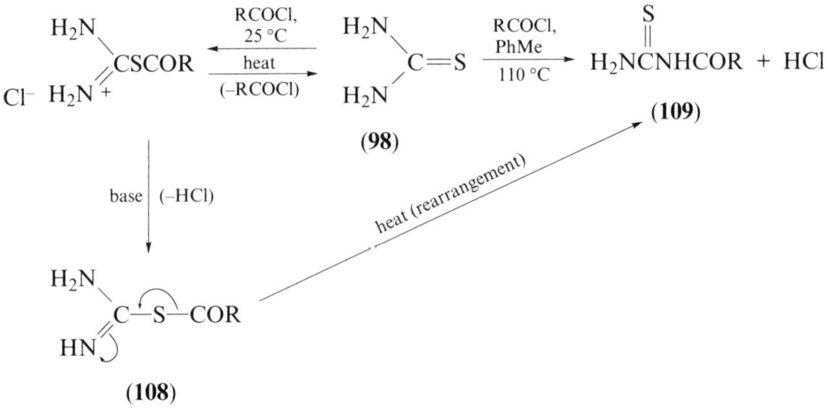

$(H_2N)_2C{=}S$ (98) —(RCOCl, 25 °C)→ Cl^- $H_2N(H_2\overset{+}{N}{=})CSCOR$; reverse: heat (–RCOCl)

Cl^- $H_2N(H_2\overset{+}{N}{=})CSCOR$ —(base (–HCl))→ $H_2N(HN{=})C{-}S{-}COR$ (108) —(heat (rearrangement))→ $H_2NC(=S)NHCOR$ (109)

$(H_2N)_2C{=}S$ (98) —(RCOCl, PhMe, 110 °C)→ $H_2NC(=S)NHCOR$ + HCl (109)

Scheme 59

Isothiocyanates

Isothiocyanates (**105**) are very reactive molecules, but rather less so than isocyanates. Some are physiologically active and occur as natural products; for example, allyl isothiocyanate (**105**) ($R = CH_2CH{=}CH_2$) is a major component of mustard oil and horseradish root. Isothiocyanates can be prepared by thermal rearrangement of the analogous thiocyanates (**111**); the latter are obtained by treatment of alkyl halides or tosylates with potassium thiocyanate (**110**) (Scheme 60).

$$\overset{+}{K}\overset{-}{S}CN + RX \longrightarrow RSCN \xrightarrow{\text{heat}} RNCS$$

(**110**) (X = halogen or OTs) (**111**) (**105**)

Scheme 60

Other preparative routes to isothiocyanates involve reaction of amines with carbon disulfide and alkali and subsequent treatment of the dithiocarbamate (**112**) with phosgene or ethyl chloroformate. Alternatively, the dithiocarbamate may be reacted with mercury(II) chloride or lead tetranitrate (Scheme 61). Amines can also be reacted with thiophosgene or *N*,*N*-diethyl thiocarbamoyl chloride (**113**) (Scheme 61).

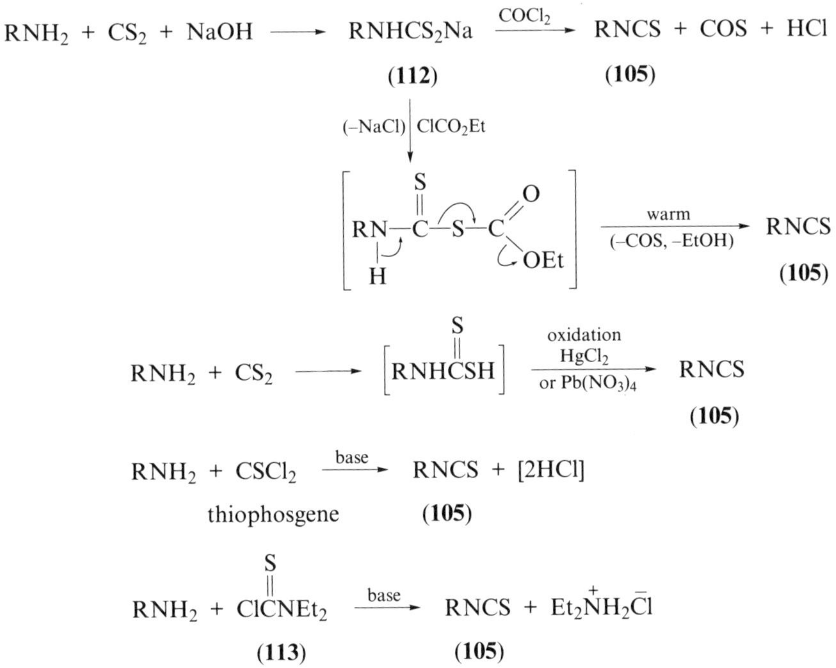

$$RNH_2 + CS_2 + NaOH \longrightarrow RNHCS_2Na \xrightarrow{COCl_2} RNCS + COS + HCl$$

(**112**) (**105**)

(**112**) $\xrightarrow[(-NaCl)]{ClCO_2Et}$ $[RNH{-}C(=S){-}S{-}C(=O)OEt]$ $\xrightarrow[(-COS,\ -EtOH)]{\text{warm}}$ RNCS (**105**)

$$RNH_2 + CS_2 \longrightarrow [RNHC(=S)SH] \xrightarrow[\text{or } Pb(NO_3)_4]{\text{oxidation, } HgCl_2} RNCS$$

(**105**)

$$RNH_2 + CSCl_2 \xrightarrow{\text{base}} RNCS + [2HCl]$$

thiophosgene (**105**)

$$RNH_2 + ClC(=S)NEt_2 \xrightarrow{\text{base}} RNCS + Et_2\overset{+}{N}H_2\overset{-}{Cl}$$

(**113**) (**105**)

Scheme 61

REACTIONS

Isothiocyanates (**105**) react with nucleophilic reagents such as amines, alcohols, thiols and acids to yield thioureas (**104**), thiono- or dithiocarbamates (**114**) (X = O or S), or amides (**115**) (Scheme 62).

Isothiocyanates (**105**) are valuable in heterocyclic synthesis; thus, they will condense with thiol acids (**116**) (X = S) or amino acids (**116**) (X = NH) to form heterocycles like (**117**) and (**118**) (Scheme 63).

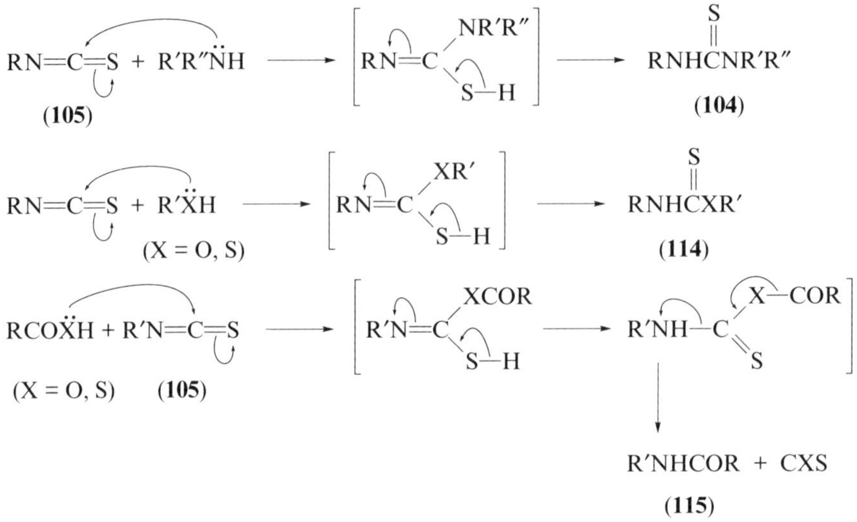

$$RN{=}C{=}S + R'R''\ddot{N}H \longrightarrow [RN{=}C(NR'R'')(S{-}H)] \longrightarrow RNHC(=S)NR'R''$$

(105) (104)

$$RN{=}C{=}S + R'\ddot{X}H \longrightarrow [RN{=}C(XR')(S{-}H)] \longrightarrow RNHC(=S)XR'$$

(X = O, S) (114)

$$RCO\ddot{X}H + R'N{=}C{=}S \longrightarrow [R'N{=}C(XCOR)(S{-}H)] \longrightarrow [R'NH{-}C(=S){-}X{-}COR] \longrightarrow R'NHCOR + CXS$$

(X = O, S) (105) (115)

Scheme 62

$$RN{=}C{=}S + H\ddot{X}CH_2CO_2H \longrightarrow [RNHC(=S){-}X{-}CH_2C(=O)OH] \xrightarrow{(-H_2O)}$$

(105) (116) (X = S or NH)

(117) X = S
(118) X = NH

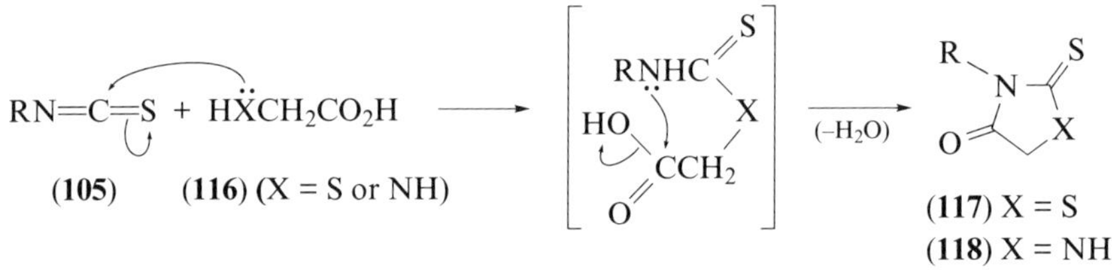

Scheme 63

BIOLOGICAL ACTIVITY

Compounds containing the thiono group display a wide spectrum of biological activity (see Chapter 10, p. 230).[6] For example, α-naphthylthiourea (**119**) is a rodenticide, chlorothiamid (**120**) is a herbicide, thiobenzamides (**121**) (X, Y and Z are electron-donating groups, e.g. Me or OMe) are antibacterial and antitubercular drugs, and several isothiocyanates like (**122**) and (**123**) are anthelmintics (Figure 6).

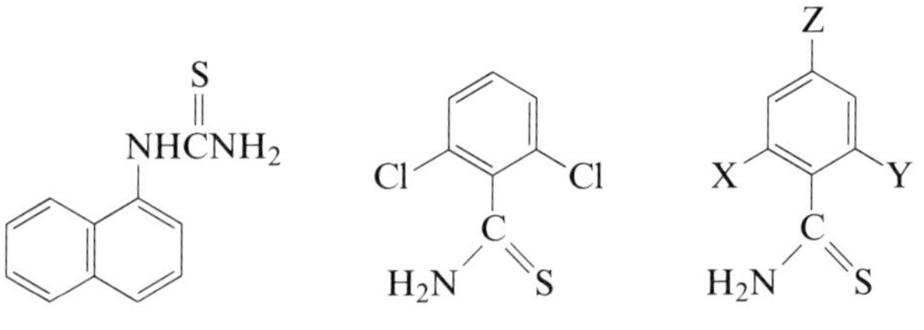

S NHCNH₂ — (119)

Cl Cl, C, H_2N S — (120)

Z, X Y, C, H_2N S — (121)

Figure 6 *(continued)*

Figure 6 *(continued)*

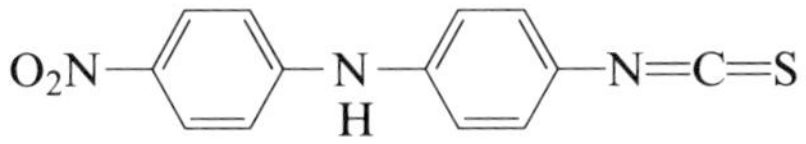

O_2N—C_6H_4—NH—C_6H_4—N=C=S

(122)

S=C=N—C_6H_4—N=C=S

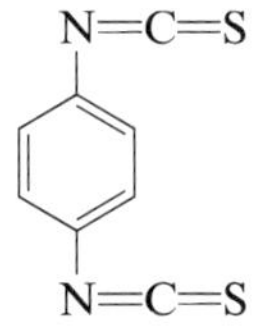

(123)

Figure 6

References

1 R. Meyer, in *Organosulfur Chemistry* (Ed. M. J. Janssen), Interscience, New York, 1967, Chap. 13, p. 219.
2 A. Ohno, in *Organic Chemistry of Sulfur* (Ed. S. Oae), Plenum Press, New York, 1977, Chap. 5, p. 189.
3 R. Mayer, in *Sulfur in Organic and Inorganic Chemistry* (Ed. A. Senning), Vol. 3, Dekker, New York, 1972, Chap. 27, p. 325.
4 F. Duus, in *Comprehensive Organic Chemistry* (Eds D. H. R. Barton and W. O. Ollis), Vol. 3, Pergamon Press, Oxford, 1979, p. 373.
5 M. P. Cara and M. I. Levinson, *Tetrahedron*, **41**, 5061 (1985).
6 J. R. Cashman, G. G. Skellern and Y. Segall, in *Sulfur-Containing Drugs and Related Organic Compounds—Chemistry, Biochemistry and Toxiology*, (Ed. L. A. Damani), Vol. 1, Part B, Ellis Horwood, Chichester, 1989, Chaps 2–4.

9 MISCELLANEOUS ORGANOSULFUR COMPOUNDS

Carbon disulfide

Carbon disulfide (**1**) is one of the earliest-known organic compounds, being first prepared by heating carbon and sulfur together in the absence of air (Lampadius, 1796).[1] Carbon disulfide is currently manufactured by the vapour phase reaction of methane with sulfur in the presence of a catalyst, e.g. charcoal, silica, aluminium oxide or magnesium oxide (Scheme 1).

$$CH_4 + 4S \xrightleftharpoons{700\,°C} \underset{(\mathbf{1})}{CS_2} + 2\,H_2S$$

$$CH_4 + 2S \rightleftharpoons \underset{(\mathbf{1})}{CS_2} + 2\,H_2$$

Scheme 1

Carbon disulfide is an important industrial solvent for the extraction of oils and waxes. In organic chemistry, it is widely used as a solvent for Friedel–Crafts reactions. Carbon disulfide functions as an electrophilic reagent. It is more susceptible than carbon dioxide to nucleophilic attack as the energy required to convert C=S to C—S (188 kJ mol^{-1}) is much less than required for the analogous conversion of C=O to C—O (305 kJ mol^{-1}). Carbon disulfide (**1**) thus undergoes nucleophilic additions with alcohols and phenols to yield the corresponding xanthates (**2**) (the xanthate reaction) (Scheme 2)

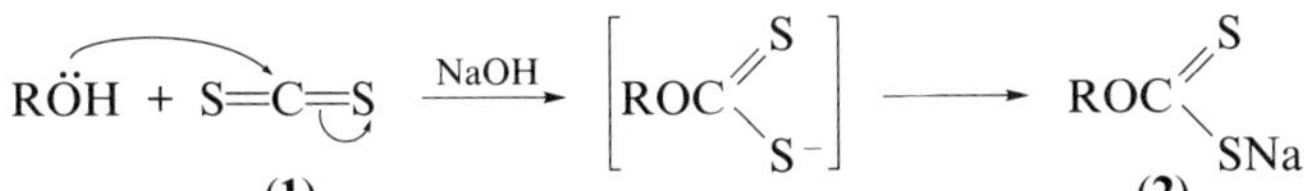

$$R\ddot{O}H + \underset{(\mathbf{1})}{S{=}C{=}S} \xrightarrow{NaOH} \left[ROC(=S)S^-\right] \longrightarrow \underset{(\mathbf{2})}{ROC(=S)SNa}$$

(R = alkyl or aryl)

Scheme 2

The xanthate reaction is of industrial importance since it is employed in the manufacture of cellophane and rayon (see Chapter 8, p. 135). Similar nucleophilic addition reactions occur between carbon disulfide (**1**) and primary and secondary amines to give the corresponding dithiocarbamates (**3**) and (**4**) (Scheme 3).[2]

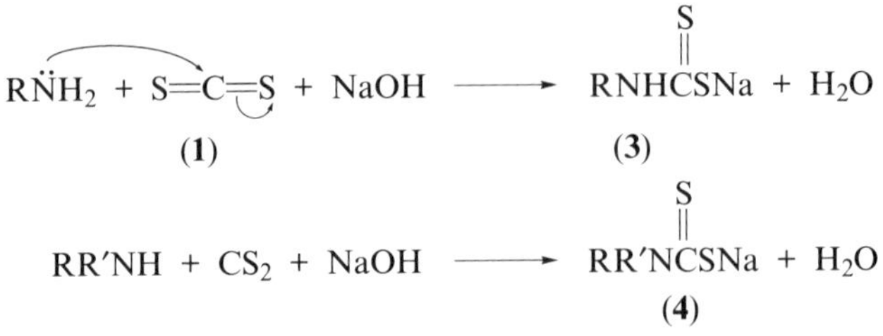

$$R\ddot{N}H_2 + S{=}C{=}S + NaOH \longrightarrow RNHC({=}S)SNa + H_2O$$

(**1**) (**3**)

$$RR'NH + CS_2 + NaOH \longrightarrow RR'NC({=}S)SNa + H_2O$$

(**4**)

Scheme 3

Several dithiocarbamates are important commercial fungicides,[3] e.g. maneb, which is manufactured from carbon disulfide and ethylenediamine (**5**). The intermediate disodium dithiocarbamate (**6**) is finally treated with a solution of a soluble manganese salt to yield maneb (**7**) (Scheme 4). Maneb is a very widely used agricultural fungicide (see Chapter 11, p. 234).[3]

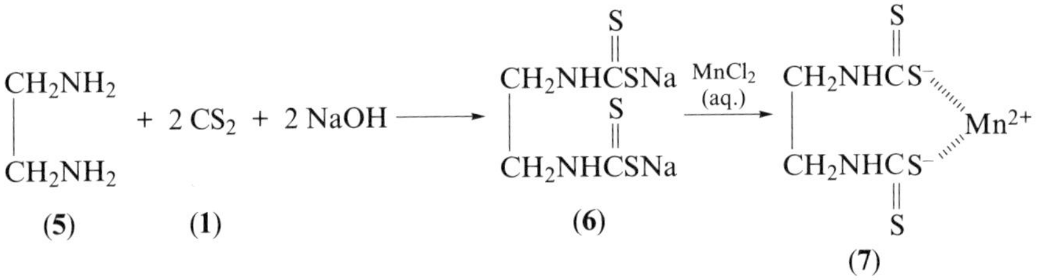

$$\begin{matrix} CH_2NH_2 \\ | \\ CH_2NH_2 \end{matrix} + 2\,CS_2 + 2\,NaOH \longrightarrow \begin{matrix} CH_2NHC({=}S)SNa \\ | \\ CH_2NHC({=}S)SNa \end{matrix} \xrightarrow[\text{(aq.)}]{MnCl_2} \begin{matrix} CH_2NHC({=}S)S^- \\ | \\ CH_2NHC({=}S)S^- \end{matrix} Mn^{2+}$$

(**5**) (**1**) (**6**) (**7**)

Scheme 4

Carbon disulfide (**1**) reacts with aniline and potassium hydroxide in boiling ethanol to yield *sym*-diphenylthiourea (**8**) (Scheme 5), which is used as a rubber vulcanisation accelerator (see Chapter 11, p. 219).

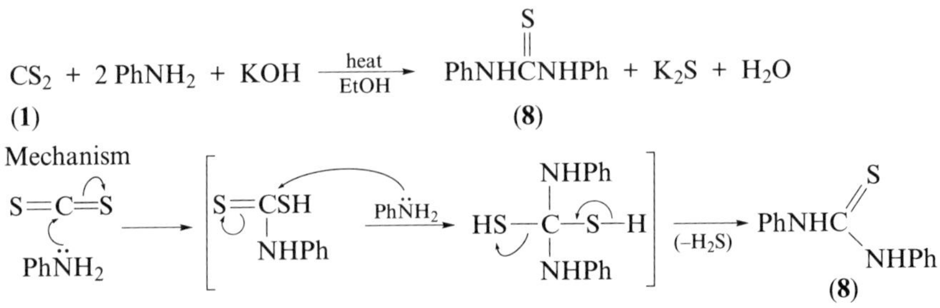

$$CS_2 + 2\,PhNH_2 + KOH \xrightarrow[\text{EtOH}]{\text{heat}} PhNHC({=}S)NHPh + K_2S + H_2O$$

(**1**) (**8**)

Mechanism

$$S{=}C{=}S + Ph\ddot{N}H_2 \longrightarrow \left[S{=}C(SH)NHPh \xrightarrow{Ph\ddot{N}H_2} HS{-}C(NHPh)_2{-}S{-}H \right] \xrightarrow{(-H_2S)} PhNHC({=}S)NHPh$$

(**8**)

Scheme 5

With organometallic reagents, carbon disulfide reacts to yield dithioacids (**9**) (see Chapter 8, p. 131) (Scheme 6).

Scheme 6

Carbon disulfide (**1**) reacts with compounds containing reactive methyl or methylene groups.[1] For instance, with acetophenone (**10**) or acetonitrile (**11**), the reactions may be used to obtain dithioacids (**12**), dithioesters (**13**) or ketene dithioacetals (**14**) (Scheme 7). Aromatic compounds containing reactive methyl groups, e.g. 2,4-dinitrotoluene (**15**), also react with carbon disulfide to yield the corresponding dithioacids (Scheme 8).

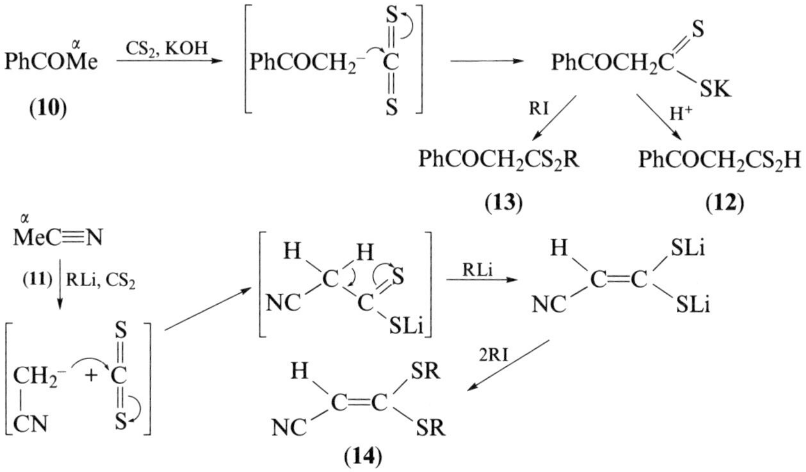

Scheme 7

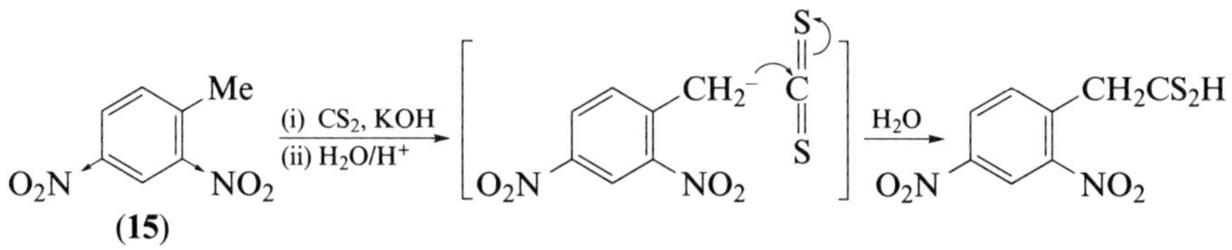

Scheme 8

With phenolic ethers (**16**) and certain pyrroles (**17**), carbon disulfide (**1**) will insert into the nucleophilic carbon–hydrogen bond by a Friedel–Crafts-type reaction (Scheme 9).

Carbon disulfide is a useful synthon for the preparation of thiocarbonic acid derivatives (see Chapter 8, p. 135). Carbon disulfide will react with alkynes (**18**) under forcing conditions; the reaction is facilitated by the presence of electron-

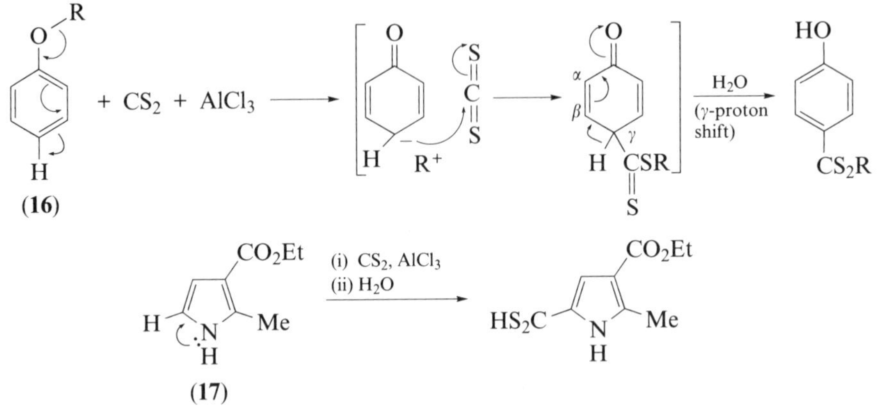

R, O, H + CS_2 + $AlCl_3$ → [O, H, R^+, S=C=S → O, α, β, γ, H, CSR, S] → H_2O (γ-proton shift) → HO, CS_2R

(16)

CO_2Et, H, N, H, Me — (i) CS_2, $AlCl_3$ (ii) H_2O → CO_2Et, HS_2C, N, H, Me

(17)

Scheme 9

withdrawing groups. The reaction involves the addition of carbon disulfide across the carbon–carbon triple bond to give the heterocyclic carbene intermediate (**19**), which can be further reacted with, for example, benzaldehyde and carbon disulfide to give the compounds (**20**) and (**21**) (Scheme 10). Carbon disulfide (**1**) also reacts with 1,3-dipolar compounds, here behaving as a dipolarophile like other heterocumulenes; thus, carbon disulfide reacts with azomethine imines (**22**) to give 1,3,4-thiadiazolidin-5-thiones (**23**) (Scheme 11).

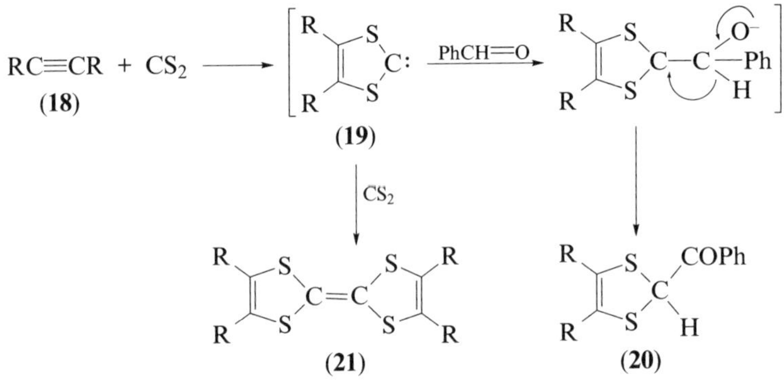

RC≡CR + CS_2 → [R, S, C:, S, R] —PhCH=O→ [R, S, C–C–Ph, O⁻, H, S, R]

(18) (19)

CS_2

R, S, C=C, S, R (21)

R, S, C, COPh, H, S, R (20)

Scheme 10

ArCH=$\overset{+}{N}$–$\overset{-}{N}$Me ↔ Ar$\overset{+}{C}$HN$\overset{-}{N}$Me, Me — S=C=S → S, 5, 1S, N–Me, 4, Ar, 2, N 3, H, Me

(22) (23)

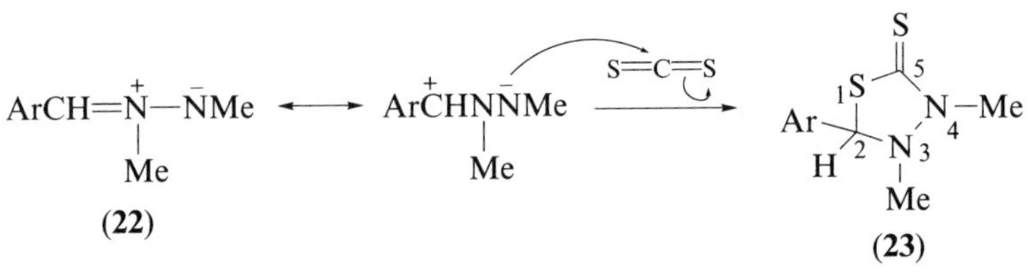

Scheme 11

Carbon disulfide (**1**) reacts with compounds that possess acidic methylene α-hydrogens, like 1,3-dithian (**24**), sulfoxides and sulfones, in the presence of a strong base. Thus, 1,3-dithian (**24**) may be converted into either the dithioate ester (**25**) or the dithioacetal (**26**) via the intermediate (**27**) or (**28**), depending on the amount of base used (Scheme 12). Sulfoxides ($RSOCH_2R'$) and sulfones (RSO_2CH_2R') react similarly with carbon disulfide, butyllithium and an alkyl halide ($R''X$) to yield either (**29**) or (**30**) (Scheme 13).

Scheme 12

Scheme 13

Carbon disulfide (**1**) reacts with enamines, amides and nitrogen heterocyclic compounds; for example, aziridines (**31**) give products like (**32**) and (**33**) (Scheme 14). The relative amounts of the products (**32**) and (**33**) will be governed by the stereoelectronic nature of the groups R and R′.

Carbon disulfide (**1**) may be chlorinated by the action of chlorine in the presence of iodine catalyst (Scheme 15). The reaction probably involves initial addition of chlorine to the C=S bond followed by loss of sulfur dichloride to yield trichloromethanesulfenyl chloride (**34**) (Scheme 15). Trichloromethanesulfenyl chloride (**34**) is a synthetic intermediate in the commercial manufacture of several valuable agricultural fungicides containing the $NSCCl_3$ moiety (see Chapter 11, p. 235). An important example is the foliage fungicide captan (**35**), which is prepared from butadiene and maleic anhydride (**36**) via tetrahydrophthalimide (**37**)

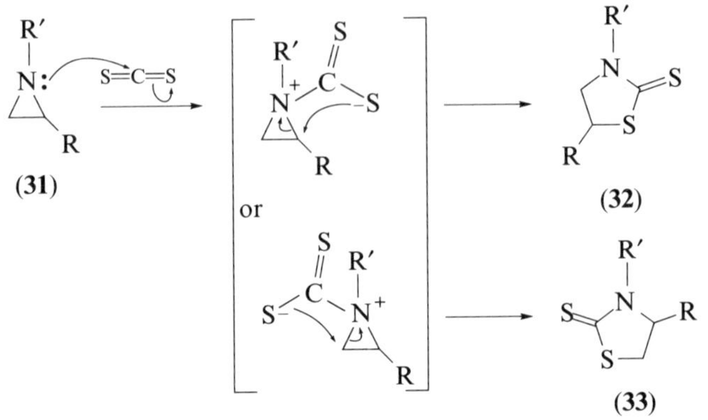

(31) (32) (33)

Scheme 14

$$CS_2 + 3Cl_2 \xrightarrow{I_2,\ \text{catalyst}} Cl_3CSCl + SCl_2$$

(1) (34)

Mechanism

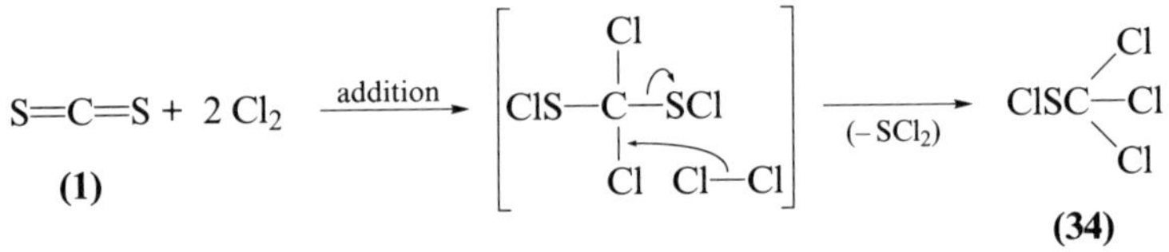

$$S{=}C{=}S + 2\,Cl_2 \xrightarrow{\text{addition}} [ClS{-}CCl_2{-}SCl] \xrightarrow{(-SCl_2)} ClSCCl_3$$

(1) (34)

Scheme 15

butadiene (36) Diels–Alder reaction (cycloaddition) $(NH_4)_2CO_3$ heat (37) acidic hydrogen NaOH $(-H_2O)$ $Cl_3CS{-}Cl$ (34) (– NaCl) (35)

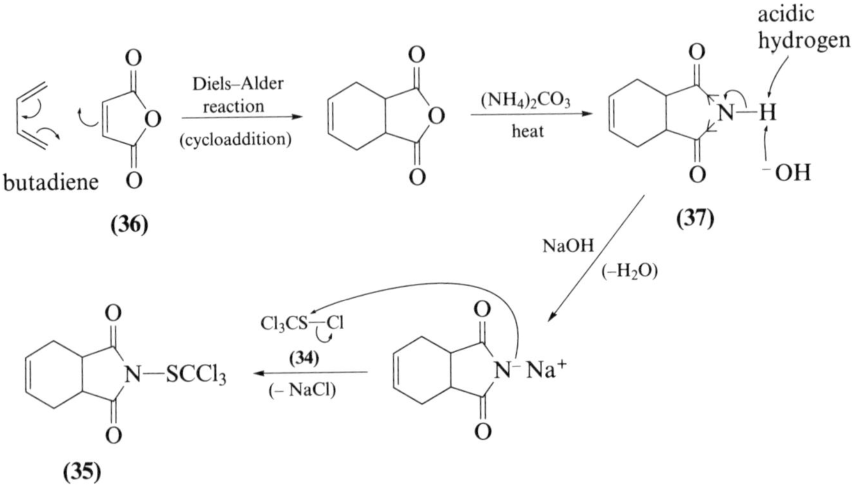

Scheme 16

(Scheme 16).[3] Tetrahydrophthalimide (**37**) contains an acidic imino hydrogen atom and forms a sodium salt; the latter undergoes a nucleophilic substitution reaction with trichloromethanesulfenyl chloride (**34**), yielding captan (**35**) (Scheme 16).

Dithiocarbamates

These are easily prepared by the reaction of amines with carbon disulfide (**1**) in the presence of alkali (Scheme 17).[2] The synthesis of dithiocarbamates (**4**) was first reported by Debus in 1850. Dithiocarbamates (**4**) form metal chelates, and sodium dimethyl dithiocarbamate is used in quantitative inorganic analysis for the estimation of metals, e.g. copper and zinc. Dithiocarbamates are also employed as vulcanisation accelerators and antioxidants in the rubber industry, and as agricultural fungicides.[3] The parent dithiocarbamic acids are unstable, decomposing to thiocyanic acid and hydrogen sulfide; however, the salts and esters are stable compounds. Dithiocarbamates (**4**) are oxidised by mild oxidants to the thiuram disulfides (**38**) (Scheme 17).

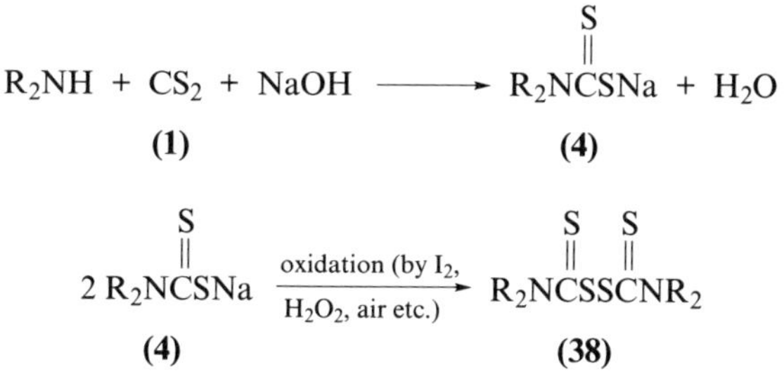

$$R_2NH + \underset{(\mathbf{1})}{CS_2} + NaOH \longrightarrow \underset{(\mathbf{4})}{R_2NC(=S)SNa} + H_2O$$

$$2\,\underset{(\mathbf{4})}{R_2NC(=S)SNa} \xrightarrow[H_2O_2,\ \text{air etc.})]{\text{oxidation (by } I_2,} \underset{(\mathbf{38})}{R_2NC(=S)SSC(=S)NR_2}$$

Scheme 17

Thiuram disulfide (**38**) contains two thiocarbamoyl groups attached to sulfur. Thiram (**38**) (R = Me), initially manufactured as a vulcanisation accelerator (see Chapter 11, p. 219), was discovered to be strongly fungicidal by Tisdale and Williams (1934) and is used as a seed dressing to combat fungi causing damping-off diseases (see Chapter 11, p. 234). Thiuram disulfides (**38**) react with a variety of reagents capable of supplying hydrogen, e.g. amines and hydrogen sulfide (Scheme 18).

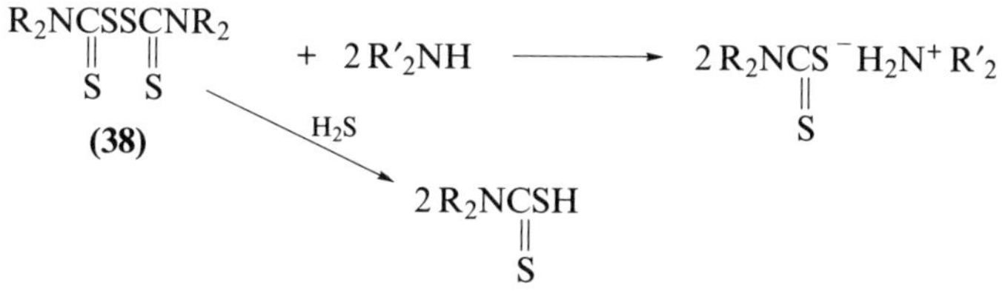

$$\underset{(\mathbf{38})}{R_2NC(=S)SSC(=S)NR_2} + 2\,R'_2NH \longrightarrow 2\,R_2NC(=S)S^-\ H_2N^+R'_2$$

$$\underset{(\mathbf{38})}{R_2NC(=S)SSC(=S)NR_2} \xrightarrow{H_2S} 2\,R_2NC(=S)SH$$

Scheme 18

Tetramethylthiuram disulfide (thiram) (**38**) (R = Me) forms insoluble chelates, with certain metals, e.g. copper and zinc. This reaction is extensively used for the detection and quantitative analysis of heavy metals. Thiram forms both 1:1 and 1:2 metal chelates (Scheme 19).

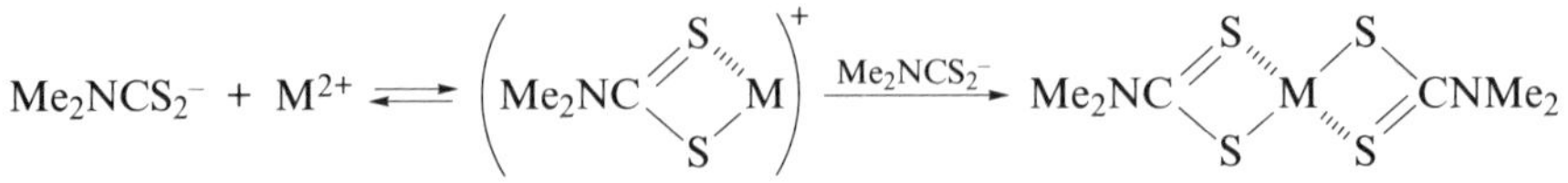

$$Me_2NCS_2^- + M^{2+} \rightleftharpoons \left(Me_2NC\langle S_2 \rangle M \right)^+ \xrightarrow{Me_2NCS_2^-} Me_2NC\langle S_2 \rangle M \langle S_2 \rangle CNMe_2$$

1:1 Complex 1:2 Complex

Scheme 19

The fungicidal properties of dithiocarbamates, like thiram, are probably associated with their ability to chelate with essential trace metals such as copper and zinc. Additionally, the 1:1 metal chelates are themselves fungicidal with the ability to penetrate lipid barriers in the fungal cell, and are probably the ultimate toxicants in these fungicides.[3] Alkyl esters of dithiocarbamic acids are readily prepared by treatment of sodium dithiocarbamate (**4**) with the appropriate alkyl halide (Scheme 20).

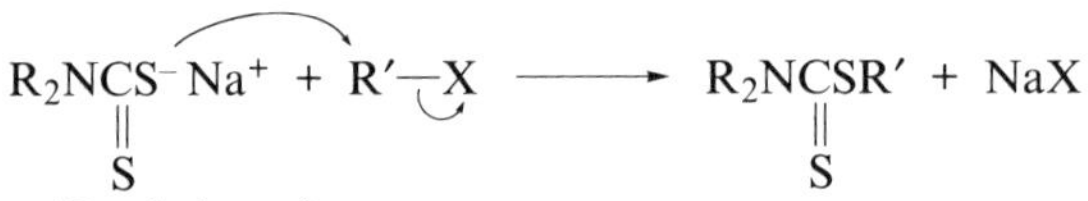

$$R_2NC(=S)S^- Na^+ + R'{-}X \longrightarrow R_2NC(=S)SR' + NaX$$

(X = halogen)
(**4**)

Scheme 20

The reactions of diamines with carbon disulfide in the presence of alkali afford a useful synthetic route to the ethylenebisdithiocarbamates (**6**) (see p. 148). Several of these compounds, such as the manganese salt maneb (**7**), are important fungicides, especially for the control of potato and tomato blight.[3]

Alkyl thiocyanates and isothiocyanates

Alkyl thiocyanates (**39**) are prepared by heating alkyl halides or tosylates with sodium or potassium thiocyanate (Scheme 21). The reaction may be performed in acetone, ethanol or water, but goes best in polar aprotic solvents like DMF or DMSO. The thiocyanate ion is an ambidentate nucleophile owing to resonance (Figure 1).[4] In principle, therefore, the reaction shown in Scheme 21 may yield

$$RX + \overset{+}{M}\overset{-}{SCN} \xrightarrow{\text{heat}} RSCN + \overset{+}{M}\overset{-}{X}$$

(**39**)

(X = halogen or OTs, M = Na or K)

Scheme 21

$$S^{-}\!-\!C\equiv N \longleftrightarrow S=C=N^{-}$$

Figure 1

either the alkyl thiocyanate or isothiocyanate; experiments have shown that as the alkyl group R is changed from primary to secondary to tertiary, there is an increasing tendency to form the corresponding alkyl isothiocyanate.

Alkyl thiocyanates (**39**) are reasonably stable, volatile oils, with a slight garlic-like odour. They are readily oxidised to either sulfonyl cyanates (**40**) or sulfonic acids (**41**), and by reduction they yield thiols (Scheme 22).

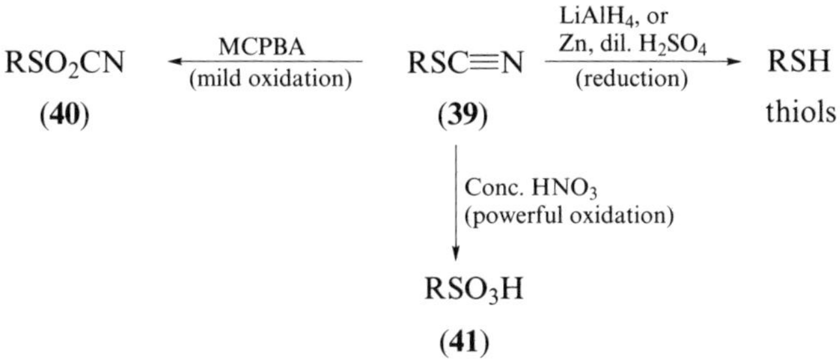

$$\underset{(\mathbf{40})}{RSO_2CN} \xleftarrow[\text{(mild oxidation)}]{\text{MCPBA}} \underset{(\mathbf{39})}{RSC\equiv N} \xrightarrow[\text{(reduction)}]{\text{LiAlH}_4\text{, or Zn, dil. }H_2SO_4} \underset{\text{thiols}}{RSH}$$

$$\underset{(\mathbf{39})}{RSC\equiv N} \xrightarrow[\text{(powerful oxidation)}]{\text{Conc. }HNO_3} \underset{(\mathbf{41})}{RSO_3H}$$

Scheme 22

Thiocyanogen (**42**) can be obtained by the reaction of bromine and lead thiocyanate (Scheme 23). Thiocyanogen (**42**) is a gas which functions as a pseudohalogen and will add to carbon–carbon double bonds (thiocyanation).

$$Pb(SCN)_2 + Br_2 \xrightarrow{Et_2O} \underset{(\mathbf{42})}{N\equiv CSSC\equiv N} + PbBr_2$$

Scheme 23

When alkyl thiocyanates (**39**) are heated they isomerize to the corresponding isothiocyanates (**43**) (Scheme 24).[5] Isothiocyanates (**43**) are better obtained by heating a primary amine with carbon disulfide (**1**) and mercury(II) chloride (Hofmann mustard oil reaction, 1868) (Scheme 25).

Aliphatic and aromatic isothiocyanates (**43**) may also be conveniently prepared by heating dithiocarbamates (**3**) with ethyl chloroformate (**44**) (Scheme 26). The intermediate (**45**) is stable in dilute acid but immediately decomposes in hot dilute potassium hydroxide, probably via a cyclic transition state in which the base

$$\underset{(\mathbf{39})}{RSC\equiv N} \xrightarrow{\text{heat, 180 °C}} \underset{(\mathbf{43})}{RN=C=S}$$

Scheme 24

$$\underset{(1)}{R\ddot{N}H_2 + CS_2} \xrightarrow{\text{heat}} [RNHC(=S)SH] \xrightarrow{HgCl_2} \underset{(43)}{RNCS} + HgS + 2HCl$$

Scheme 25

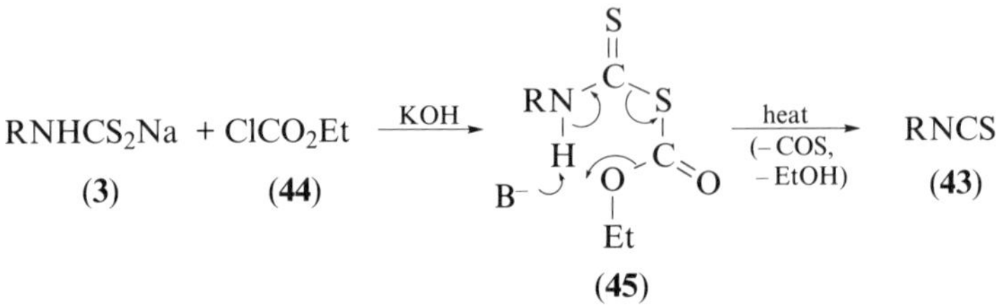

$$\underset{(3)}{RNHCS_2Na} + \underset{(44)}{ClCO_2Et} \xrightarrow{KOH} \underset{(45)}{[\text{cyclic intermediate}]} \xrightarrow[(-COS,\ -EtOH)]{\text{heat}} \underset{(43)}{RNCS}$$

Scheme 26

$$RNHCS_2NH_4 + \underset{(46)}{COCl_2} \longrightarrow \underset{(43)}{RNCS} + COS + HCl + NH_4Cl$$

Mechanism

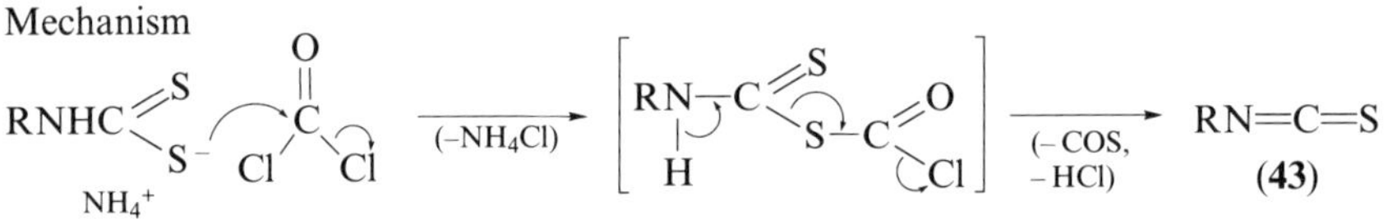

$$RNHC(=S)S^- \; NH_4^+ + ClC(=O)Cl \xrightarrow{(-NH_4Cl)} \left[RNHC(=S)S{-}C(=O)Cl \right] \xrightarrow{(-COS,\ -HCl)} \underset{(43)}{RN{=}C{=}S}$$

Scheme 27

(B^-) facilitates removal of the imidic proton (Scheme 26). An analogous synthesis of isothiocyanates involves the action of phosgene (**46**) on ammonium dithiocarbamates; the probable mechanism of the reaction is shown in Scheme 27.

Alkyl isothiocyanates are lachrymatory, vesicatory liquids with a powerful mustard odour; they are hydrolysed and reduced to primary amines (Scheme 28). Some alkyl isothiocyanates and thiocyanates occur in nature; for instance, compounds (**47**) and (**48**) have been isolated from radish and penny cress, respectively (Figure 2).

Several alkyl thiocyanates have insecticidal properties;[4] an example is isobornyl thiocyanate (**49**) prepared from isoborneol (**50**) by sequential treatment with chloroacetyl chloride and sodium thiocyanate. In this reaction, the more reactive acyl chlorine atom reacts faster than the alkyl chlorine and is consequently preferentially substituted (Scheme 29).

$$\underset{(43)}{RNCS} + 2H_2O \xrightarrow[\text{(acid hydrolysis)}]{HCl} RNH_2 + CO_2 + H_2S$$

$$\underset{(43)}{RNCS} \xrightarrow[\text{4[H] (reduction)}]{\text{Zn, dil. } H_2SO_4} RNH_2 + H_2C{=}S$$

Scheme 28

$MeS(=O)CH{=}CHCH_2CH_2NCS$ (47) $H_2C{=}CHCH_2NCS$ (48)

Figure 2

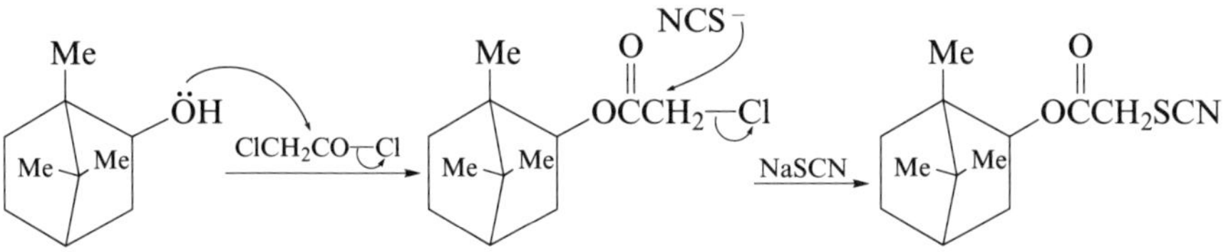

(50) $\xrightarrow{ClCH_2CO-Cl}$ ...$OC(=O)CH_2-Cl$ + NCS^- $\xrightarrow{NaSCN}$ ...$OC(=O)CH_2SCN$ (49)

Scheme 29

Sulfonyl thiocyanates

Sulfonyl thiocyanates[6] (**51**) may be prepared by the action of thiocyanogen (**42**) on the appropriate sodium sulfinates (**52**) (Scheme 30). Sulfonyl thiocyanates (**51**) will undergo free radical addition to alkenes and alkynes to give the adducts (**53**) and (**54**) (Scheme 30).

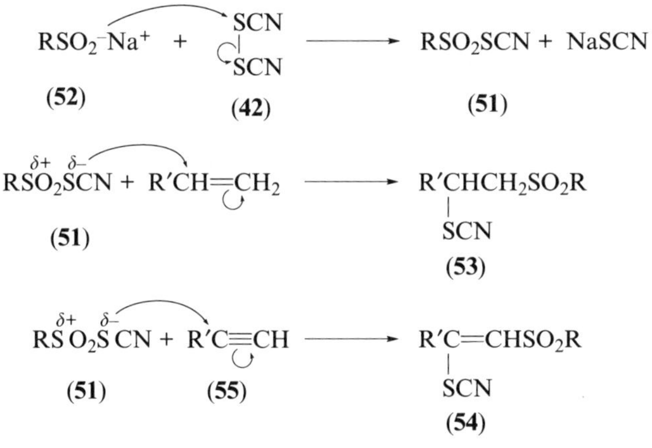

$$RSO_2^-Na^+ + NCS{-}SCN \longrightarrow RSO_2SCN + NaSCN$$
(**52**) (**42**) (**51**)

$$RSO_2^{\delta+}SCN^{\delta-} + R'CH{=}CH_2 \longrightarrow R'CH(SCN)CH_2SO_2R$$
(**51**) (**53**)

$$RSO_2^{\delta+}SCN^{\delta-} + R'C{\equiv}CH \longrightarrow R'C(SCN){=}CHSO_2R$$
(**51**) (**55**) (**54**)

Scheme 30

Chlorosulfonyl isocyanate[6] (CSI) (**57**) is a very reactive compound which was first prepared from cyanogen chloride (**56**) and sulfur trioxide (Graf, 1956) (Scheme 31). CSI (**57**) undergoes [2 + 2] cycloaddition reactions with various alkenes; initial adducts like (**58**) and (**59**) may be hydrolysed to the β-lactams (**60**) and (**61**). The cycloaddition reaction also occurs with enamines (**62**) to give products like (**63**) and (**64**) (Scheme 32).

$$\underset{(\mathbf{56})}{CNCl} + SO_3 \longrightarrow \underset{(\mathbf{57})}{ClSO_2NCO}$$

Scheme 31

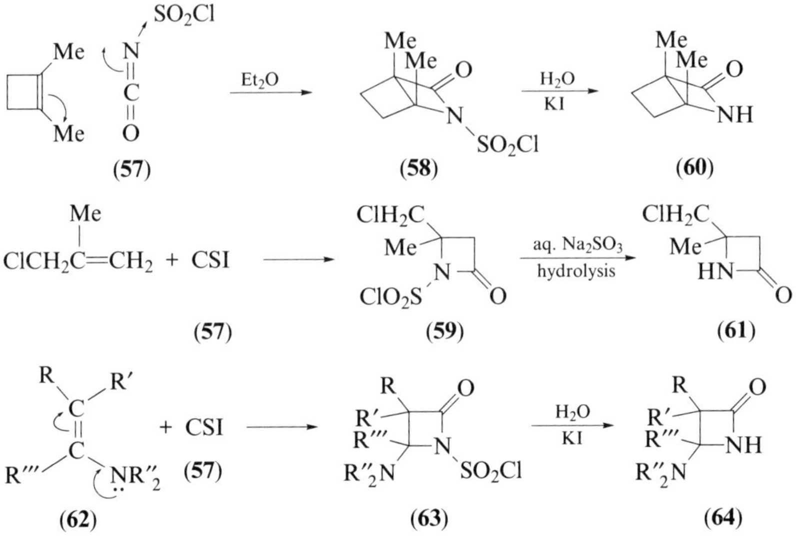

Scheme 32

The high reactivity of CSI is probably associated with the powerful electron-withdrawing character of the chlorosulfonyl moiety, which activates the adjacent isocyanate group with respect to nucleophilic addition. The [2+2] cycloadditions with alkenes may therefore be depicted as in Scheme 33.

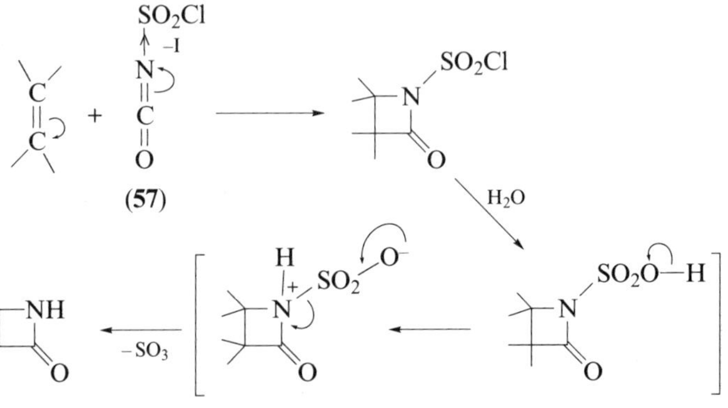

Scheme 33

The facile hydrolysis of the *N*-sulfonyl chlorides (**58**) and (**59**) to the β-lactams (**60**) and (**61**) may involve the initial formation of the *N*-sulfonic acid which then collapses to give the lactam (Scheme 33). Surprisingly, attempts to convert the *N*-sulfonyl chlorides to the corresponding sulfonamides by treatment with amines are largely unsuccessful, and the major product is always the β-lactam, probably formed as a result of the transformations depicted in Scheme 33. Unsaturated *N*-chlorosulfonyl β-lactams like (**65**) are often unstable and may rearrange to give a product corresponding to a formal 1,4-cycloaddition. For instance, the reaction of CSI (**57**) with 1,3-cyclohexadiene (**66**) at RT affords the [2 + 2] cycloadduct (**65**) which on hydrolysis gives the corresponding lactam (**67**). However, by treatment with boiling chloroform, the product (**68**) is isolated; this corresponds to the 1,4-cycloadduct of cyclohexadiene (**66**) and CSI (Scheme 34). The compound (**68**) on hydrolysis yields the lactam (**69**) which on reduction with lithium aluminium hydride affords the 2-azabicyclo[2.2.2]octene (**70**) (Scheme 34).

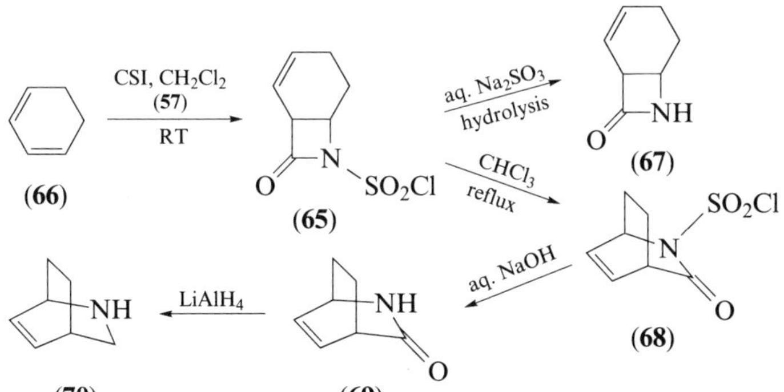

CSI, CH_2Cl_2 (**57**) RT; aq. Na_2SO_3 hydrolysis; $CHCl_3$ reflux; aq. NaOH; $LiAlH_4$

(**66**) (**65**) (**67**) (**68**) (**69**) (**70**)

Scheme 34

RC≡CR′ alkynes; CSI RT

(**72**) (**73**) (**74**) (**71**)

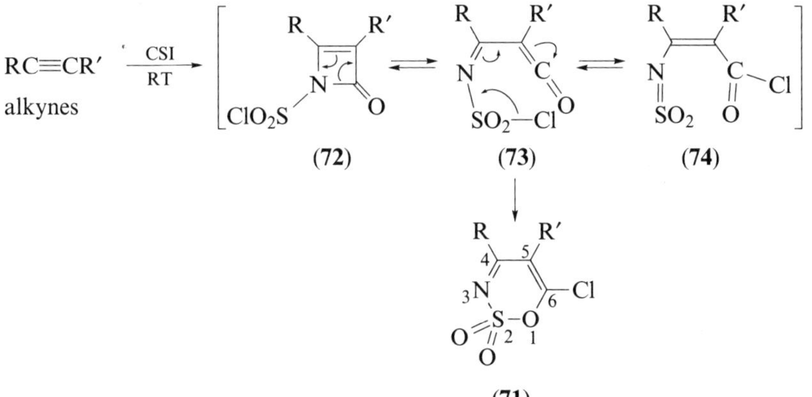

Scheme 35

CSI (**57**) also adds to alkynes at RT to yield 1,2,3-oxathiazine-2,2-dioxides (**71**) (Scheme 35). The initial adduct (**72**) probably undergoes electrophilic ring-opening to yield (**73**), then a sigmatropic 1,5-chloride shift to form (**74**) and finally electrophilic ring closure to give (**71**) (Scheme 35).

CSI reacts with carboxylic acids to form nitriles via the initial adducts (**75**) which lose carbon dioxide to give the *N*-chlorosulfonylcarboxamides (**76**); which with DMF afford good yields of the nitriles (**77**) (Scheme 36). This provides a valuable procedure for the conversion of carboxylic acids into the nitriles. In addition, thiophene (**78**), as well as other reactive heterocycles, reacts with CSI to yield the *N*-chlorosulfonylcarboxamide (**79**) which on treatment with DMF affords the nitrile (**80**) (Scheme 37).

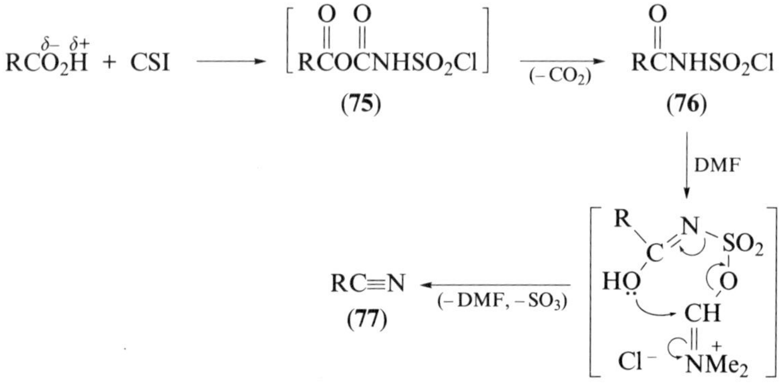

$$\overset{\delta-\ \delta+}{RCO_2H} + CSI \longrightarrow [RC(O)OC(O)NHSO_2Cl]\ (\mathbf{75}) \xrightarrow{(-CO_2)} RC(O)NHSO_2Cl\ (\mathbf{76})$$

$$(\mathbf{76}) \xrightarrow{DMF} [\text{intermediate: R, C, N, }SO_2\text{, HO, O, CH, }Cl^-\text{, }\overset{+}{N}Me_2] \xrightarrow{(-DMF,\ -SO_3)} RC{\equiv}N\ (\mathbf{77})$$

Scheme 36

$$\text{thiophene}\ (\mathbf{78}) \xrightarrow[Et_2O,\ 0\ °C]{CSI\ (equiv.)} \text{2-thienyl–}CONHSO_2Cl\ (\mathbf{79}) \xrightarrow{DMF} \text{2-thienyl–}CN\ (\mathbf{80})$$

Scheme 37

$$RCH{=}NR'\ (\mathbf{81}) + 2\,CSI \xrightarrow[RT]{CH_2Cl_2} (\mathbf{82})$$

(**82**): six-membered ring with N–R', C–R, two C=O, and two N–SO_2Cl groups (ClO_2S, SO_2Cl)

Mechanism

$$RCH{=}\ddot{N}R' + ClO_2S{-}N{=}C{=}O \longrightarrow [R\overset{+}{C}HN(R')C(=O)N^{-}SO_2Cl] + ClO_2S\ddot{N}{=}C{=}O \longrightarrow (\mathbf{82})$$

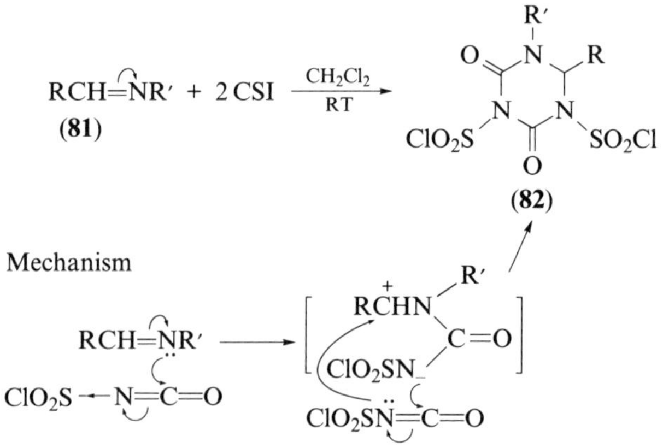

Scheme 38

CSI (**57**) will also react with carbon–nitrogen double bonds, e.g. in anils (**81**), to yield cycloadducts. Here, the [2 + 2] cycloadducts are not isolated, but the six-membered triazinediones (**82**) are formed by addition of two equivalents of CSI, as indicated in Scheme 38.

CSI also reacts with diaryl azines (**83**) to afford the bicyclic triazolotriazolones (**84**), which are probably formed by 1,3-dipolar cycloaddition of CSI (two equivalents) to the azine (Scheme 39). Phenyl isothiocyanate undergoes a similar 1,3-dipolar cycloaddition reaction with a diaryl azine (**83**) to give the adduct (**85**) (Scheme 40). The main difference is that phenyl isothiocyanate is much less reactive than CSI, and consequently the cycloaddition requires more drastic conditions, namely prolonged boiling in xylene.

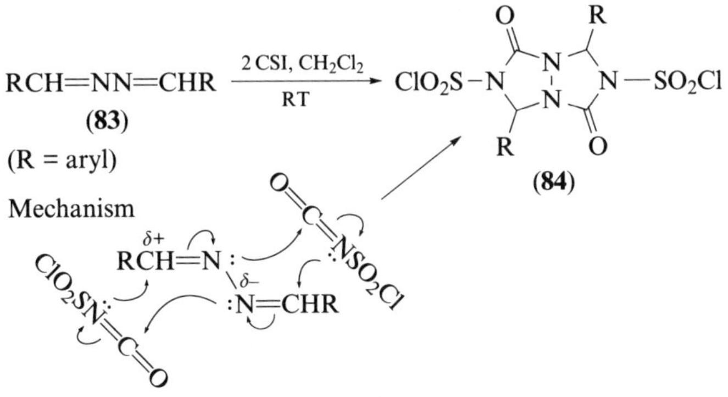

RCH=NN=CHR (**83**) (R = aryl) —[2 CSI, CH_2Cl_2, RT]→ ClO_2S–N ... N–SO_2Cl (**84**)

Mechanism

Scheme 39

RCH=NN=CHR (**83**) + 2 PhNCS —[xylene, reflux]→ Ph–N ... N–Ph (**85**)

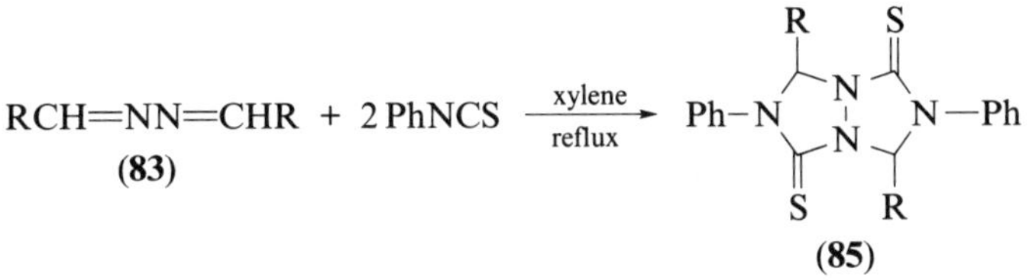

Scheme 40

Sulfamic acid and derivatives[7]

Sulfamic acid (**86**) can be manufactured by reaction of ammonia and sulfur trioxide followed by hydrolysis of the initially formed ammonium sulfamate (**87**) (Scheme 41). Sulfamic acid (**86**) has also been prepared by reaction of urea with

$$2\,NH_3 + SO_3 \longrightarrow NH_2\overset{-}{S}O_3\overset{+}{N}H_4\ (\mathbf{87}) \xrightarrow{H^+,\ H_2O} NH_2SO_3H\ (\mathbf{86})$$

Scheme 41

fuming sulfuric acid (oleum). Ammonium sulfamate (**87**) is manufactured from ammonia and sulfur trioxide, or by oxidation of ammonium thiosulfate. *N*-methylsulfamic acid is obtained by the action of fuming sulfuric acid on *N*-methylurea. Sodium *N*-cyclohexylsulfamate (cyclamate) (**88**), a non-calorific artificial sweetener, is manufactured by heating cyclohexylamine with sulfamic acid (**86**) in xylene (Scheme 42) (see Chapter 11, p. 240).

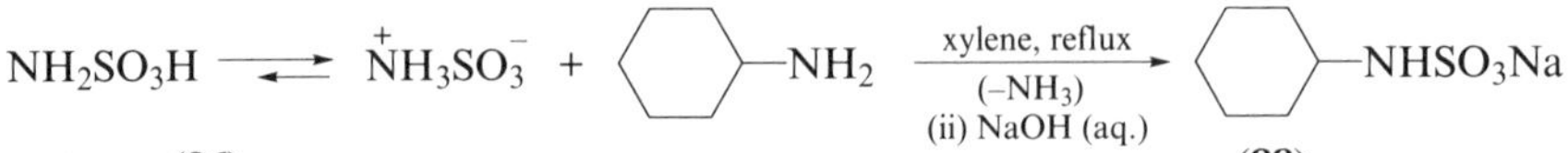

$NH_2SO_3H \rightleftharpoons \overset{+}{N}H_3SO_3^-$ + C_6H_{11}–NH_2 $\xrightarrow[\text{(-NH}_3\text{)}\ \text{(ii) NaOH (aq.)}]{\text{xylene, reflux}}$ C_6H_{11}–$NHSO_3Na$

(**86**) (**88**)

Scheme 42

Many sulfamic acid derivatives are sweet and much work has been carried out on structure–taste relationships. For example, research showed that sodium *exo*-2-norbornylsulfamate (**89**) was some five times sweeter than sodium cyclamate (**88**), although the corresponding *endo*-isomer (**90**) was tasteless (Figure 3).

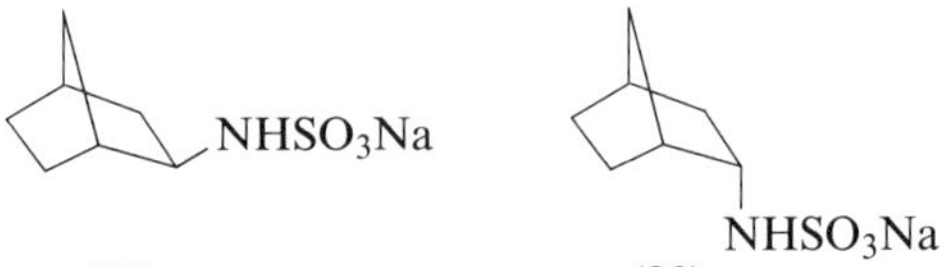

$NHSO_3Na$ (**89**) $NHSO_3Na$ (**90**)

Figure 3

Simple *meta*-substituted phenylsulfamates (**91**) are sweet, but not the corresponding *ortho*-isomers or *para*-isomers. The free arylsulfamic acids (**92**) were first prepared by Kanetani (1980) by acidification of the corresponding ammonium salts (Figure 4).[7] Phenylsulfamic acid (**92**) (Ar = Ph) was obtained by this procedure; this compound has for long been postulated as an intermediate in the 'baking process' for the sulfonation of aniline to sulfanilic acid (see Chapter 7, p. 99).

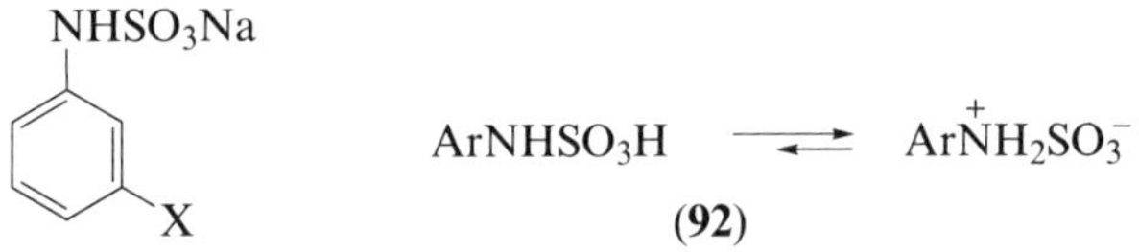

$NHSO_3Na$, X

(**91**) (X = F, Cl, Br, Me, CN)

$ArNHSO_3H \rightleftharpoons Ar\overset{+}{N}H_2SO_3^-$

(**92**)

Figure 4

Many heterocyclic sulfamates have been synthesised, e.g. the compounds (**93**)–(**95**) (Figure 5). Compound (**93**) is sweet but not the corresponding sulfone nor the oxacompound (**94**) or the morpholino derivative (**95**); in contrast, all the carbon analogues of these heterocycles are sweet.

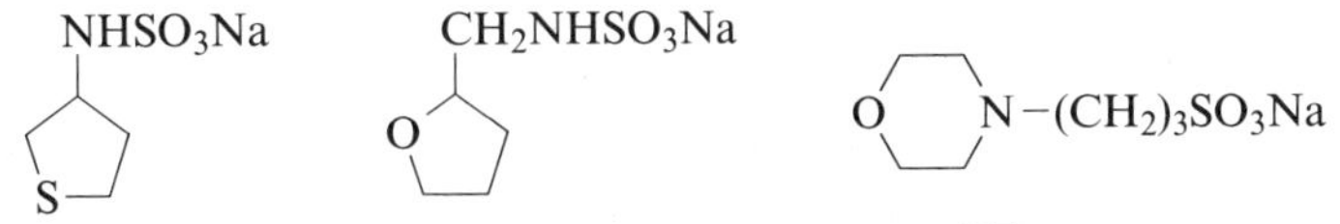

(93) (94) (95)

Figure 5

Monobactams

3-Acylamino-2-oxoazetidine-1-sulfonates, or monobactams (**96**), are monocyclic β-lactam antibiotics which are active against Gram-negative bacteria.[7] Monobactams (**96**) may be prepared by either sulfonation of an azetidinone (**97**) with a sulfur trioxide complex or by cyclisation of an acylsulfamate (**98**) (Figure 6).

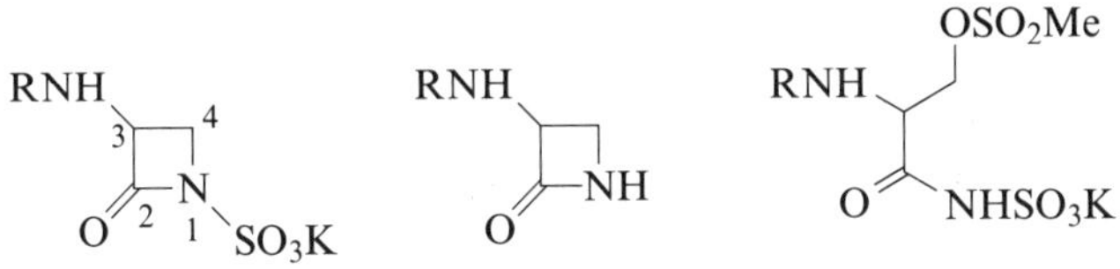

(96) (97) (98)

Figure 6

The antibacterial activity of monobactams has provided a great stimulus for synthetic work, and a large number of different compounds have been prepared.[7] Sulfamic acid (**86**) can be used for sulfamation of fatty acid monoglycerides and other compounds. Reaction of aniline with hot sulfamic acid (170 °C) afforded a mixture of sulfonic acids, namely orthanilic acid (**99**), sulfanilic acid (**100**) and the 2,4-disulfonic acid (**101**) (Scheme 43).

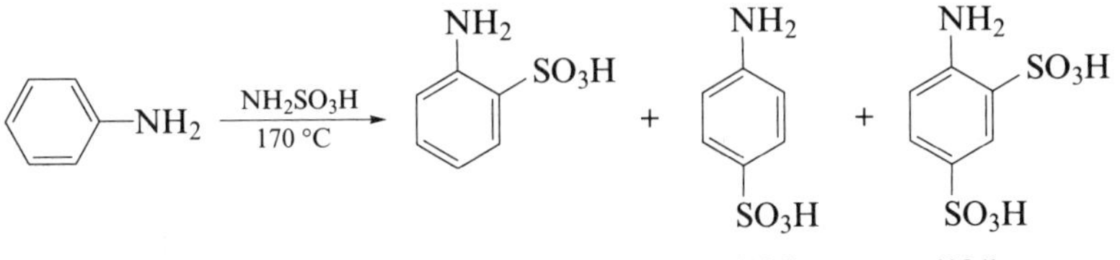

NH_2SO_3H, 170 °C

(99) (100) (101)

Scheme 43

Sulfamoyl derivatives

The sulfamoyl esters (**102**) and (**103**) are prepared by condensation of the corresponding sulfamoyl halides (**104**) and (**105**) with the appropriate alcohol, alkoxide, phenol or phenoxide; a suitable base, e.g. pyridine, is needed for the reaction with alcohols (Scheme 44). The yields of the sulfamoyl esters are often poor but can be substantially increased by the use of phase transfer catalysts, and this discovery has greatly extended the range of sulfamate esters that can be efficiently synthesised. CSI (**57**) reacts with phenols to give aryl sulfamates (**106**) after hydrolysis of the aryloxysulfonyl isocyanate (**107**) (Scheme 45).

$$RR'NSO_2Cl + R''OH \xrightarrow{\text{base}} RR'NSO_2OR'' + [HCl]$$

(**104**) (**102**)

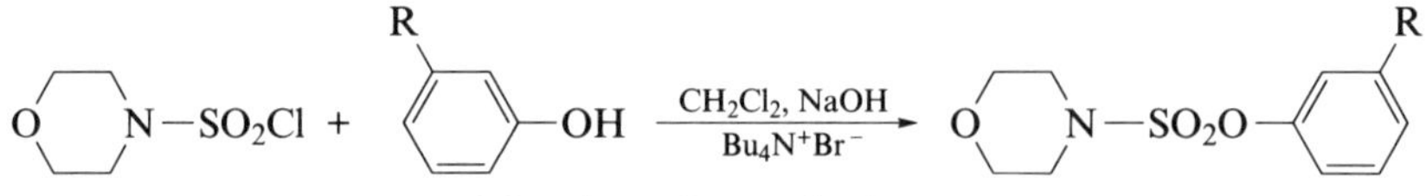

(**105**) (phase transfer catalyst) (**103**)

CH_2Cl_2, NaOH, $Bu_4N^+Br^-$

Scheme 44

$$ArOH + Cl{-}SO_2N{=}C{=}O \longrightarrow ArOSO_2NCO \xrightarrow[(-CO_2)]{H_2O} ArOSO_2NH_2$$

(**57**) (**107**) (**106**)

Scheme 45

The intermediate sulfonyl isocyanate (**107**) reacts with the ω-haloalcohols (**108**) to give *N*-carboxylsulfamates (**109**) which readily cyclise on treatment with triethylamine or sodium hydride to yield 2-oxazolidones (**110**) (Scheme 46). However, when (**109**) (Ar = Ph, X = Br) was heated with sodium hydride in boiling THF, the 2-oxazolidone (**110**) was not isolated, but rather the phenoxysulfonylpiperidazine (**111**) (Scheme 46). The formation of the piperazine derivative (**111**) is believed to involve the initial formation of the 2-oxazolidone (**110**) which

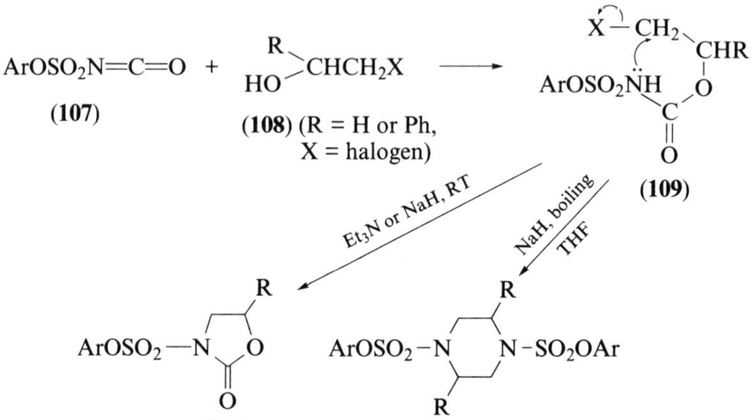

$ArOSO_2N{=}C{=}O$ + R(HO)$CHCH_2X$ ⟶ (**109**)

(**107**) (**108**) (R = H or Ph, X = halogen) (**109**)

Et_3N or NaH, RT → (**110**); NaH, boiling THF → (**111**)

(**110**) (**111**)

Scheme 46

then suffers nucleophilic attack by the iodide anion with ring opening by carbon–oxygen bond cleavage to give the intermediate (**112**) which eliminates carbon dioxide to form (**113**). Compound (**113**) is in equilibrium with the aziridine (**114**) and dimerises to the piperazine (**111**) (Scheme 47).

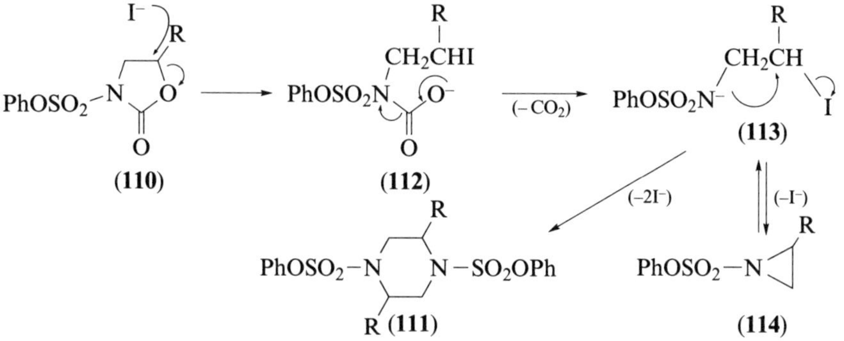

Scheme 47

Aryl chlorosulfonates (**115**) react with amines to give the sulfamoyl chlorides (**104**), formed by the amine attacking the electrophilic sulfur atom with subsequent loss of the phenol via sulfur–oxygen bond cleavage. The sulfamoyl chloride (**104**) then reacts with a second mole of the amine to yield the sulfamide (**116**) (Scheme 48).

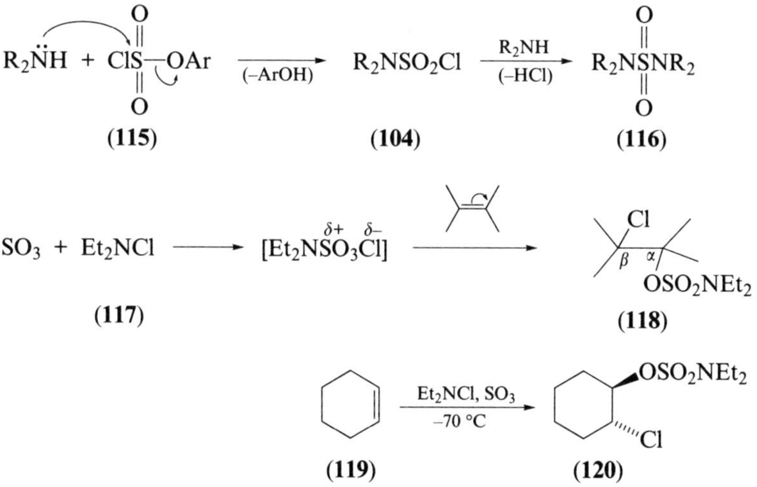

$$R_2NH + ClSO_2OAr \xrightarrow[(-ArOH)]{} R_2NSO_2Cl \xrightarrow[(-HCl)]{R_2NH} R_2NSO_2NR_2$$

(**115**) (**104**) (**116**)

$$SO_3 + Et_2NCl \longrightarrow [Et_2\overset{\delta+}{N}SO_3\overset{\delta-}{Cl}] \longrightarrow$$ (**118**)

(**117**)

(**119**) $\xrightarrow[-70\ °C]{Et_2NCl,\ SO_3}$ (**120**)

Scheme 48

A large number of sulfamate esters can be prepared by addition of an activated electrophile across an alkene carbon–carbon double bond. The activated electrophiles are formed by addition of sulfur trioxide to dialkylchloramines like

(**117**) at –70 °C, and on subsequent reaction with alkenes *trans* addition occurs to give *β*-chloroalkyl sulfamate esters (**118**) (Scheme 48). For instance, cyclohexane (**119**) reacts with *N,N*-diethylchloramine–sulfur trioxide to form the *trans*-2-chlorosulfamate (**120**) (Scheme 48).

Alkynes also react with activated electrophiles; thus, diphenylethyne (**121**) with *N,N*-diethylchloramine–sulfur trioxide affords a mixture of the *cis*-sulfamate (**122**) and *trans*-sulfamate (**123**) in a 2:1 ratio (Scheme 49).

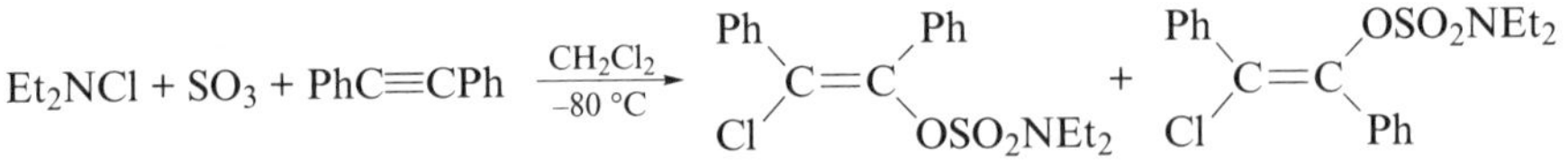

$$\underset{(\mathbf{117})}{Et_2NCl} + SO_3 + \underset{(\mathbf{121})}{PhC{\equiv}CPh} \xrightarrow[-80\ ^\circ C]{CH_2Cl_2} \underset{(\mathbf{122})}{Ph(Cl)C{=}C(Ph)OSO_2NEt_2} + \underset{(\mathbf{123})}{Ph(Cl)C{=}C(OSO_2NEt_2)Ph}$$

Scheme 49

A wide variety of sulfamate esters have been synthesised and screened as herbicides, pharmaceutical agents and sweeteners (see Chapter 11, p. 240). Sulfamates containing a primary amino group (**124**) are conveniently prepared by condensation of the appropriate alcohol with sulfamoyl chloride (**125**) in DMF in the presence of sodium hydride (Scheme 50). An illustrative example is provided by the conversion of substituted *β*-phenylethanols (**126**) to the corresponding sulfamates (**127**) (Scheme 50). Compounds of type (**127**) exhibit anticonvulsant and carbonic anhydrase activity and may be useful in the treatment of epilepsy and glaucoma. Sulfamoyl chlorides (**125**) may be prepared by treatment of amines or amine hydrochlorides with sulfuryl chloride (**128**). An analogous reaction also occurs with dialkyl sulfonamides (**129**) (Scheme 51).

$$ROH + \underset{(\mathbf{125})}{NH_2SO_2Cl} \xrightarrow{NaH,\ DMF} \underset{(\mathbf{124})}{ROSO_2NH_2} + [HCl]$$

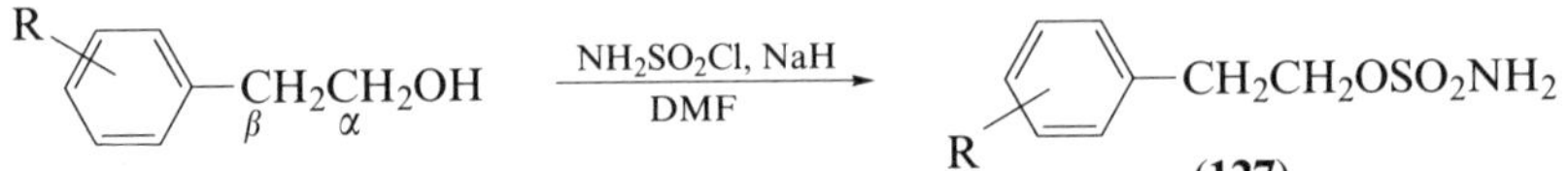

$$\underset{(\mathbf{126})}{R{-}C_6H_4{-}\underset{\beta}{C}H_2\underset{\alpha}{C}H_2OH} \xrightarrow[DMF]{NH_2SO_2Cl,\ NaH} \underset{(\mathbf{127})}{R{-}C_6H_4{-}CH_2CH_2OSO_2NH_2}$$

(R = alkyl, alkoxy, CF_3, NO_2)

Scheme 50

$$RR'NH + \underset{(\mathbf{128})}{SO_2Cl_2} \xrightarrow{NEt_3} \underset{(\mathbf{125})}{RR'N\ SO_2Cl} + [HCl]$$

$$\underset{(\mathbf{129})}{RNHSO_2NHR} + \underset{(\mathbf{128})}{SO_2Cl_2} \xrightarrow{NEt_3} \underset{(\mathbf{125})}{2RNHSO_2Cl}$$

Scheme 51

The mechanism of this last reaction may be postulated as shown in Scheme 52. Here, the initial nucleophilic attack by the sulfonamide anion (**129**) is followed by protonation of the intermediate and attack by the chloride anion to yield two moles of the sulfamoyl chloride (**125**). A wide range of sulfamoyl derivatives can be prepared by nucleophilic displacement of the chlorine atom in sulfamoyl chlorides (**125**). Examples include condensations with ureas, alcohols, compounds containing acidic hydrogens and nitrogen heterocycles to give the corresponding sulfamoyl derivatives (**130**)–(**133**) (Scheme 53).

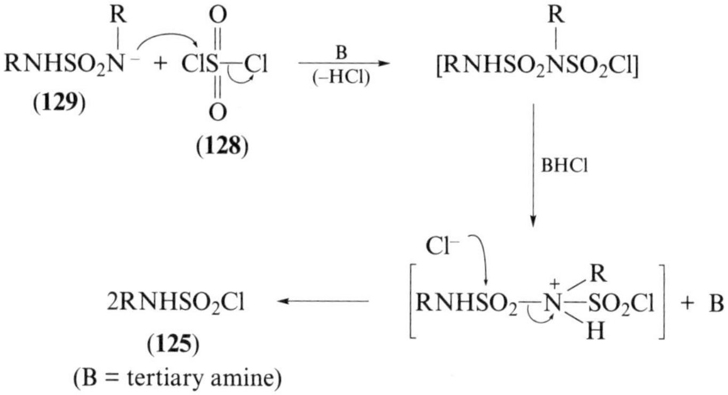

$$RNHSO_2N(R)^- \; (\mathbf{129}) + ClSO_2Cl \; (\mathbf{128}) \xrightarrow[(-HCl)]{B} [RNHSO_2N(R)SO_2Cl] \xrightarrow{BHCl} \left[RNHSO_2\text{-}\overset{+}{N}(R)(H)\text{-}SO_2Cl\right] Cl^- + B \longrightarrow 2RNHSO_2Cl \; (\mathbf{125})$$

(B = tertiary amine)

Scheme 52

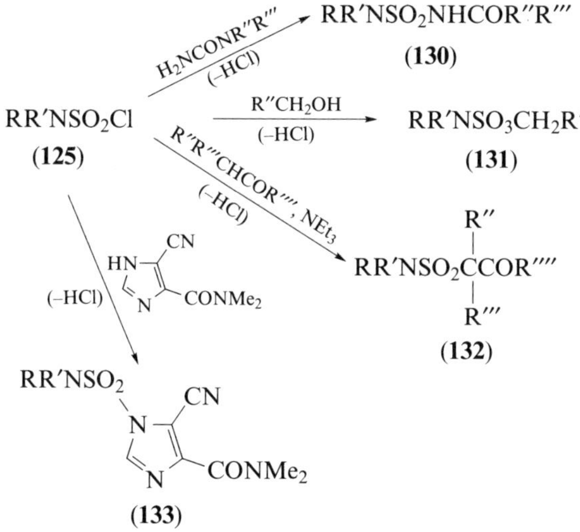

$$RR'NSO_2Cl \; (\mathbf{125}) \xrightarrow[(-HCl)]{H_2NCONR''R'''} RR'NSO_2NHCOR''R''' \; (\mathbf{130})$$

$$RR'NSO_2Cl \; (\mathbf{125}) \xrightarrow[(-HCl)]{R''CH_2OH} RR'NSO_3CH_2R'' \; (\mathbf{131})$$

$$RR'NSO_2Cl \; (\mathbf{125}) \xrightarrow[(-HCl)]{R''R'''CHCOR'''',\ NEt_3} RR'NSO_2C(R'')(R''')COR'''' \; (\mathbf{132})$$

$$RR'NSO_2Cl \; (\mathbf{125}) \xrightarrow[(-HCl)]{\text{4-cyano-5-(CONMe}_2\text{)imidazole (HN)}} \text{1-(RR'NSO}_2\text{)-4-CN-5-CONMe}_2\text{-imidazole} \; (\mathbf{133})$$

Scheme 53

Sulfamoyl chloride (**125**) can also be used in cyclisation reactions; for example, the imidazole (**134**) may be converted to the thiadiazine derivative (**135**) (Scheme 54).

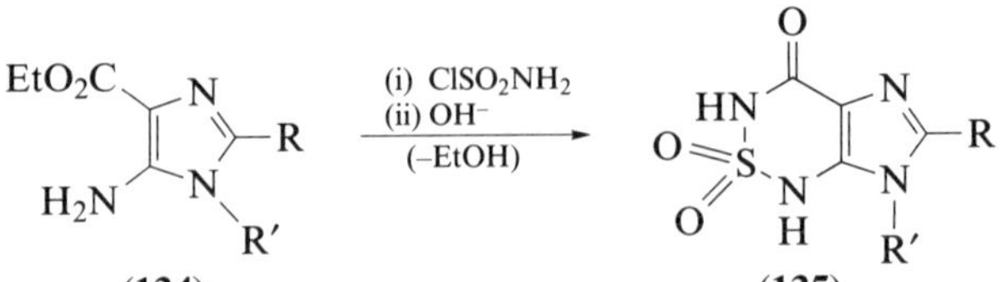

(i) $ClSO_2NH_2$
(ii) OH^-
(–EtOH)

(134) (135)

Scheme 54

Sulfamides

Sulfamide (**136**) can be prepared by reaction of sulfuryl chloride (**128**) with excess ammonia in dichloromethane at –50 °C, followed by extraction of the product with acetonitrile (Scheme 55). By using primary amines in the condensation, the method can be extended to obtain *N,N′*-disubstituted sulfamides (**137**) (Scheme 55). A valuable route to di- and trisubstituted sulfamides involves the reaction of catechol sulfate (**138**) with aniline. Aniline reacts slowly with (**138**) to give the sulfamoyl ester (**139**). The unreacted aniline then promotes base-catalysed elimination of catechol to form the transient, highly reactive *N*-phenylsulfonylimine (**140**) which is trapped by aniline to yield *N,N′*-diphenylsulfamide (**141**). By reaction of the ester (**139**) with an alkylamine, the procedure can be used to obtain trisubstituted sulfamides (**142**) (Scheme 56). In general, the amination of the 2-hydroxyphenyl sulfamate esters using alkylamines is an efficient route for the synthesis of trialkylsulfamides (**143**) (Scheme 57).

$$\underset{(\mathbf{128})}{SO_2Cl_2} + 2\,NH_3 \xrightarrow{NEt_3} \underset{(\mathbf{136})}{H_2NSO_2NH_2} + [2\,HCl]$$

$$\underset{(\mathbf{128})}{SO_2Cl_2} + 2\,RNH_2 \xrightarrow{NEt_3} \underset{(\mathbf{137})}{RNHSO_2NHR} + [2\,HCl]$$

Scheme 55

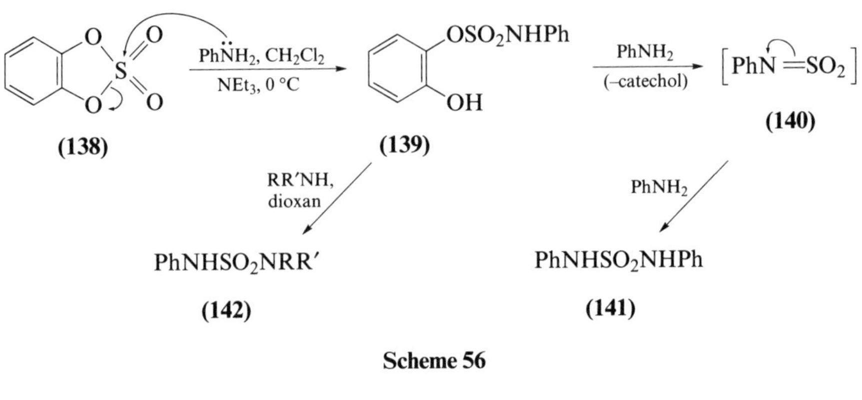

$PhNH_2$, CH_2Cl_2, NEt_3, 0 °C

(138) (139) OSO_2NHPh, OH

$PhNH_2$ (–catechol) $[PhN{=}SO_2]$ (140)

RR′NH, dioxan → $PhNHSO_2NRR'$ (142)

$PhNH_2$ → $PhNHSO_2NHPh$ (141)

Scheme 56

(138) $\xrightarrow{R\ddot{N}H_2}$ SO_2—N(H)(R) $H\ddot{N}R'R''$

$RNHSO_2NR'R''$ $\xleftarrow{R'R''\ddot{N}H}$ $[RN{=}SO_2]$ +

(143)

Scheme 57

CSI (**57**) (see p. 157) is a useful reagent for the synthesis of sulfamides. CSI (**57**) by sequential treatment with 2-chloroethanol and a primary or secondary amine affords the corresponding chloroethoxycarbonylsulfamide (**144**). In the reaction with the amine, the sulfonyl chloride is more reactive than the alkyl chloride and is therefore selectively substituted to yield (**144**) (Scheme 58).

$ClSO_2N{=}C{=}O$ + $H\ddot{O}CH_2CH_2Cl$ ⟶ $Cl{-}SO_2NHC(O)OCH_2CH_2Cl$

(**57**) 2-chloroethanol

$RR'\ddot{N}H$, NEt_3 (– HCl)

$RR'NSO_2NHC(O)OCH_2CH_2Cl$

(**144**)

Scheme 58

CSI (**57**) can also be used to obtain *N*,*N*-disubstituted sulfamides (**145**) by treatment with pentachlorophenol. This reaction gives the sulfamide (**146**) via the unstable carbamic acid intermediate (**147**). On heating with a secondary amine, (**146**) affords a good yield of the corresponding *N*,*N*-dialkylsulfamide (**145**). The final step involves elimination of the pentachlorophenoxy anion, which is an excellent leaving group (Scheme 59).

Sulfamide (**136**) may be used in the synthesis of nitriles (**148**) by a one-pot reaction with acid chlorides involving elimination of sulfamic acid (**86**) (Scheme 60). The reaction is believed to involve the initial formation of the *N*-acylsulfamide followed by enolisation and subsequent elimination of sulfamic acid (**86**) (Scheme 60).

Several sulfamoylamidine derivatives have shown[7] activity as histamine H_2 receptor antagonists and are valuable as gastric secretion inhibitors (see Chapter 11, p. 232), e.g. 3-(4-thiazolemethylthio)propionamidines (**149**), prepared by Lewis acid catalysed reaction of the propionitrile (**150**) with sulfamide (**136**) (Scheme 61).

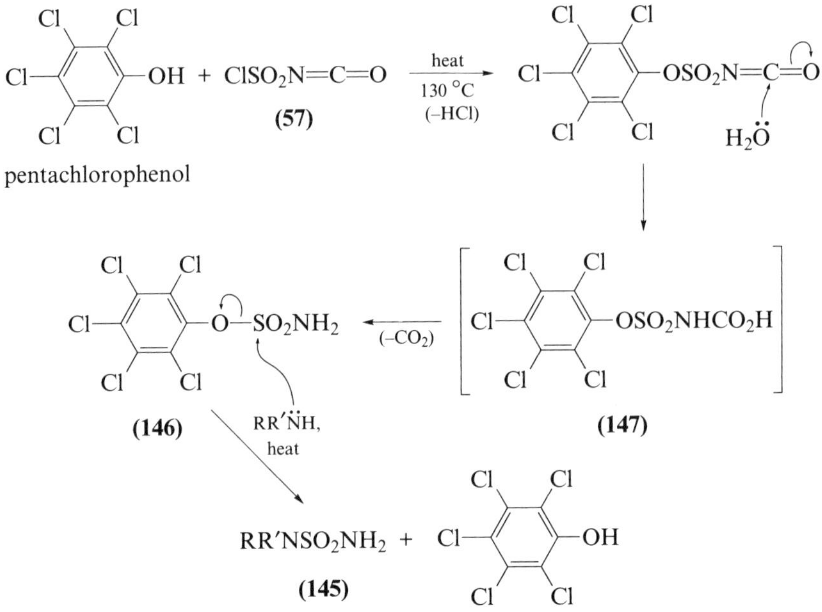

Scheme 59

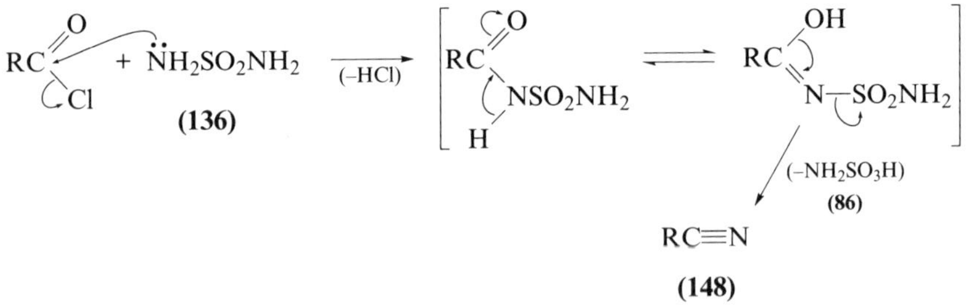

Scheme 60

The formation of (**149**) involves nucleophilic addition of sulfamide (**136**) across the carbon–nitrogen triple bond of (**150**), followed by tautomerisation of the initial iminoproduct (Scheme 61). A series of analogous *N*-sulfamoylamidines in which the length of the chain between the thiazole nucleus and the amidine moiety was varied also inhibited gastric juice secretion.[7]

N,*N*′-Dialkylsulfamides (**137**), react with alkaline sodium hypochlorite to yield dialkyldiazenes (**151**); in non-aqueous media, the intermediate thiadiaziridine, 1,1-dioxides (**152**) were isolated (Scheme 62).

N,*N*′-Diphenylsulfamide (**153**) on heating rearranges to the *N*-phenylsulfonamide (**154**); on the other hand, when (**153**) is heated in aniline, amine exchange occurs to yield the *ortho*-isomer (**155**) (Scheme 63).

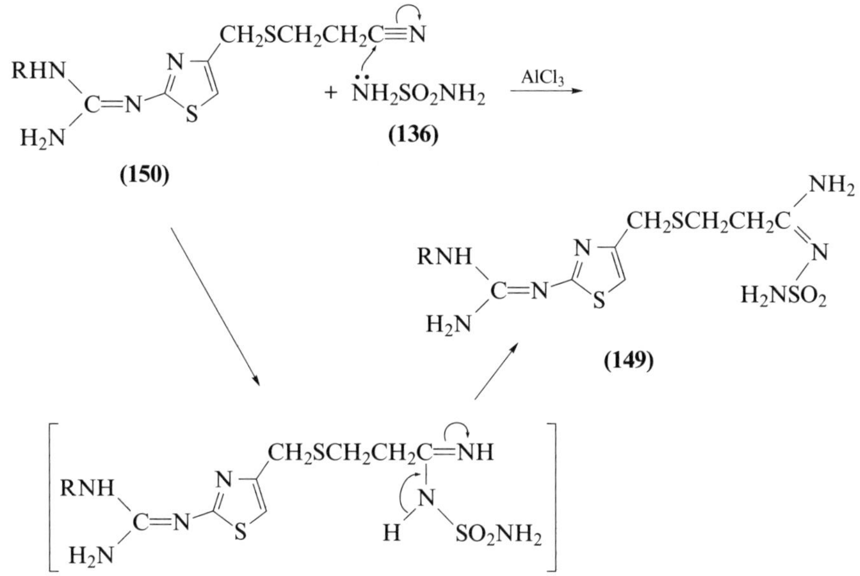

RHN, H_2N C=N (thiazole) $CH_2SCH_2CH_2C{\equiv}N$ + $\ddot{N}H_2SO_2NH_2$ $\xrightarrow{AlCl_3}$

(150) **(136)**

RNH, H_2N C=N (thiazole) $CH_2SCH_2CH_2C(NH_2)=N-SO_2NH_2$

(149)

[RNH, H_2N C=N (thiazole) $CH_2SCH_2CH_2C(=NH)-N(H)SO_2NH_2$]

Scheme 61

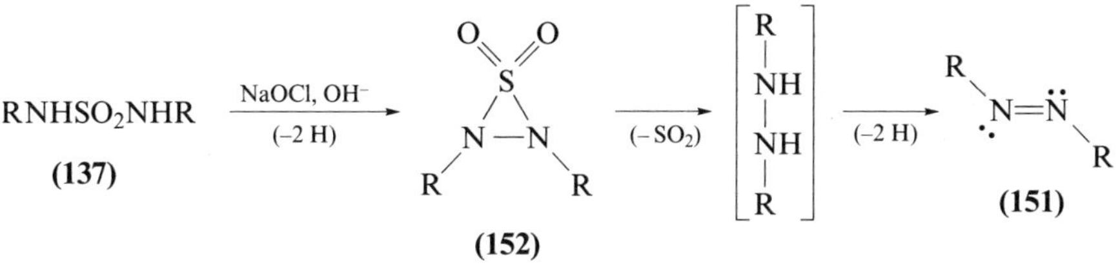

$RNHSO_2NHR$ $\xrightarrow[(-2\ H)]{NaOCl,\ OH^-}$ (152) $\xrightarrow{(-SO_2)}$ [RNHNHR] $\xrightarrow{(-2\ H)}$ $R\ddot{N}=\ddot{N}R$

(137) **(152)** **(151)**

Scheme 62

$PhNHSO_2NHPh$ $\xrightarrow{heat}$ $PhNHSO_2-C_6H_4-NH_2$

(153) **(154)**

heat in $PhNH_2$

$C_6H_5NHSO_2-C_6H_4-NH_2$ (ortho)

(155)

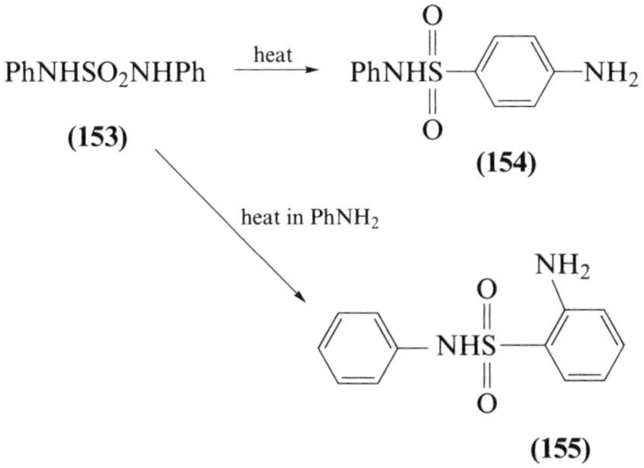

Scheme 63

Sultones and sultams[8]

Sultones are heterocyclic compounds containing the OSO_2 group and are internal esters of the corresponding hydroxysulfonic acids. Sultones are thus the sulfur analogues of lactones. Sultams are the sulfur analogues of lactams, i.e. they are internal esters of the corresponding aminosulfonic acids. The γ-sultone (**156**) and δ-Sultone (**157**) can be prepared by a number of routes, such as by cyclisation of the appropriate hydroxysulfonic acid or halosulfonic acid or a derivative (Scheme 64).

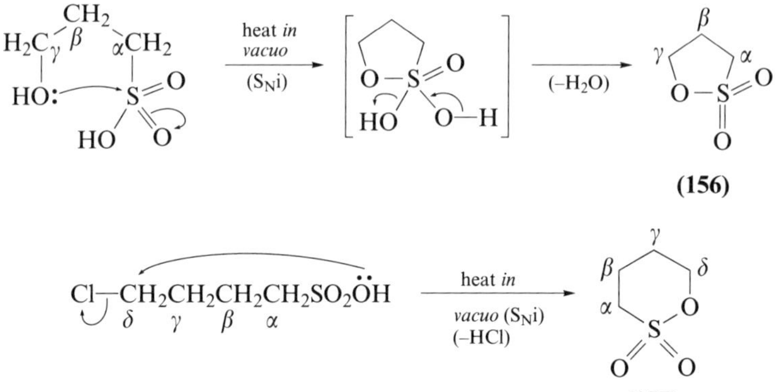

heat *in vacuo* (S_Ni) (–H_2O) (**156**)

heat *in vacuo* (S_Ni) (–HCl) (**157**)

Scheme 64

The cyclisation of aromatic hydroxysulfonic acid derivatives has been widely used to obtain aromatic sultones like (**158**) and (**159**). The reaction involves sulfur–oxygen bond formation and is effected by a wide variety of reagents, e.g. H_2SO_4, PCl_5, $POCl_3$ and $ClSO_3H$ (Scheme 65).[8]

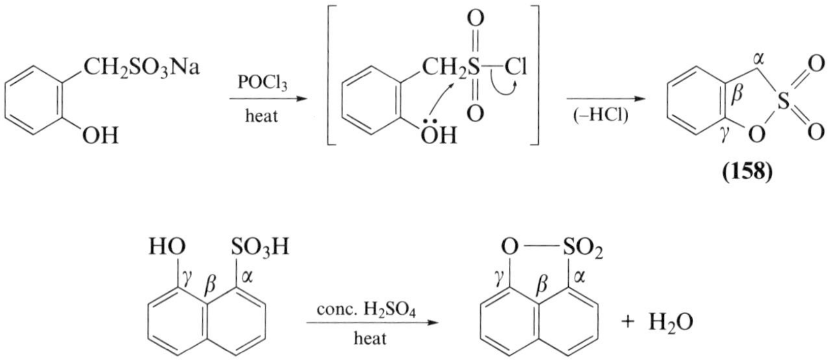

$POCl_3$ heat (–HCl) (**158**)

conc. H_2SO_4 heat + H_2O (**159**)

Scheme 65

Aliphatic γ-sultones and δ-sultones may be synthesised by the sulfonation of alkenes by reaction with the sulfur trioxide–dioxan complex or gaseous sulfur trioxide (Scheme 66).

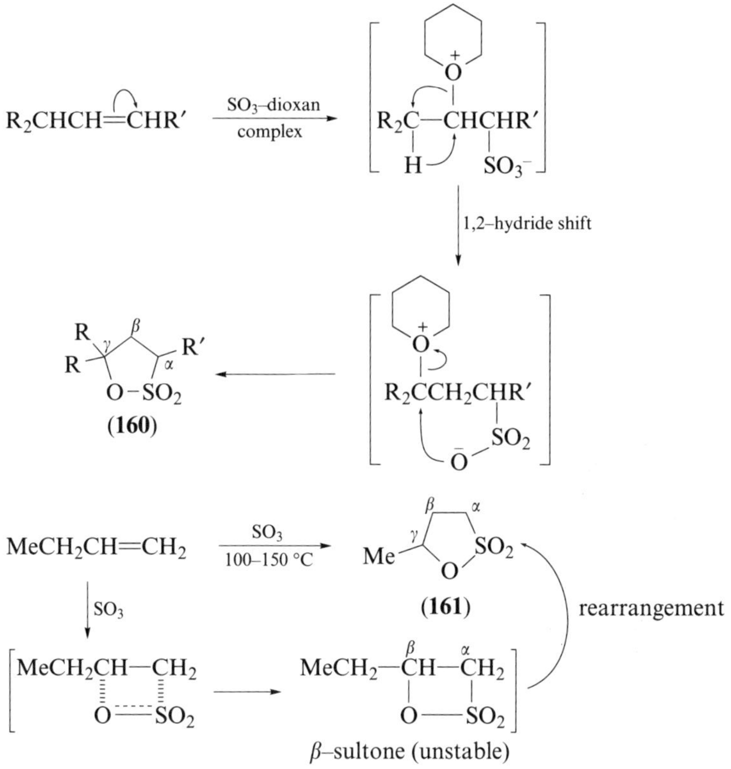

$R_2CHCH{=}CHR'$ → (SO_3–dioxan complex) → $[R_2C(H)–CH(O^+\text{-dioxan})CHR'(SO_3^-)]$ → (1,2-hydride shift) → $[R_2C(O^+\text{-dioxan})CH_2CHR'(SO_2O^-)]$ → (160)

$MeCH_2CH{=}CH_2$ → (SO_3, 100–150 °C) → (161)

$MeCH_2CH{=}CH_2$ → (SO_3) → $[MeCH_2CH–CH_2 \cdots O{=}SO_2]$ → $[MeCH_2–CH–CH_2, O–SO_2]$ β–sultone (unstable) → rearrangement → (161)

Scheme 66

The formation of the sultone (**160**) probably involves addition of the complex across the alkene double bond, a 1,2-hydride shift and an intramolecular nucleophilic substitution reaction. The sultone (**161**) is formed by addition of sulfur trioxide to give the unstable β-sultone which rearranges to the more stable γ-isomer (**161**). Another useful route to sultones is by metallation of alkanesulfonate esters; for example, butane-1,3-dimethylsulfonate (**162**), prepared from butane-1,3-diol, yields the δ-sultone, namely 6-methyl-1,2-oxathiin-2,2-dioxide (**163**) (Scheme 67).

Another example is the conversion of cyclohexene (**164**) to the bicyclic sultone (**165**) (Scheme 67). This allows the synthesis of sultones with partially determined stereochemistry. Both reactions depend on the high leaving group properties of the methanesulfoxy anion. Unsaturated δ-sultones like (**166**) can be obtained by 1,4-cycloaddition of sulfenes (generated *in situ*) to α,β-unsaturated carbonyl compounds (Scheme 68).

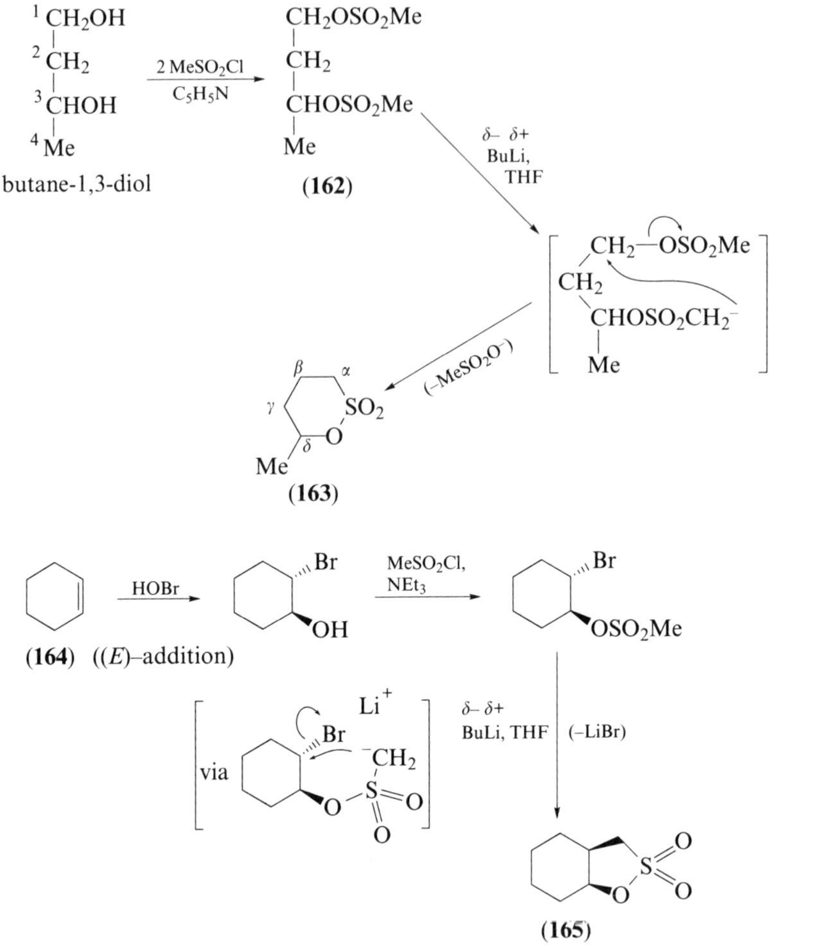

$^{1}CH_2OH-{}^{2}CH_2-{}^{3}CHOH-{}^{4}Me$ (butane-1,3-diol) $\xrightarrow[C_5H_5N]{2\,MeSO_2Cl}$ $CH_2OSO_2Me-CH_2-CHOSO_2Me-Me$ (**162**) $\xrightarrow[THF]{\delta-\ \delta+\ BuLi,}$ [CH_2-OSO_2Me, CH_2, $CHOSO_2CH_2^-$, Me] $\xrightarrow{(-MeSO_2O^-)}$ (**163**) (α, β, γ, δ; SO_2, O, Me)

(**164**) ((*E*)-addition) $\xrightarrow{HOBr}$ (Br, OH) $\xrightarrow{MeSO_2Cl,\ NEt_3}$ (Br, OSO_2Me) $\xrightarrow[(-LiBr)]{\delta-\ \delta+\ BuLi,\ THF}$ (**165**)

[via (Li^+, Br, $^-CH_2$, O–S(=O)=O)]

Scheme 67

sulfene (NEt_2; $\delta-$ CH_2=SO_2 $\delta+$; O) $\xrightarrow{MeSO_2Cl,\ NEt_3}$ (**166**) (NEt_2; α, β, γ, δ; O, SO_2)

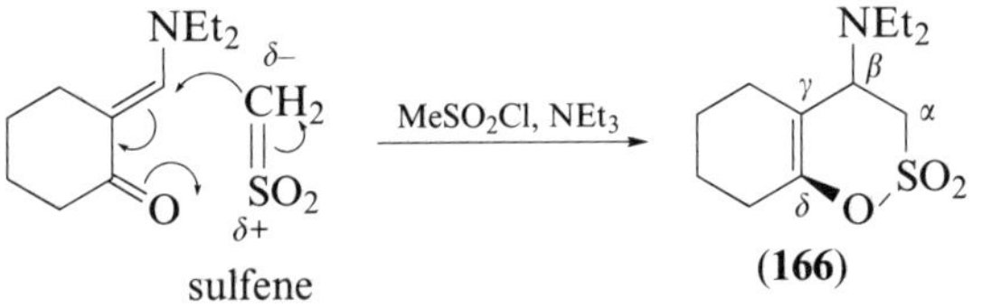

Scheme 68

Aromatic δ-sultones (**167**) may be prepared from alkanesulfonyl salicylaldehydes (**168**) by intramolecular carbanion condensation (Scheme 69).

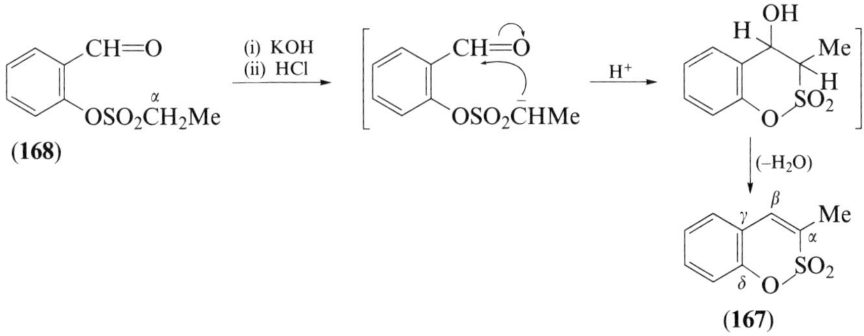

(168) → (i) KOH, (ii) HCl → [intermediate] → H^+ → [cyclic intermediate] → $(-H_2O)$ → (167)

Scheme 69

REACTIONS OF SULTONES

The most important reactions of sultones involve nucleophilic attack at carbon, consequently, they function as sulfoalkylating agents.[8] Thus, β-sultones, γ-sultones, and δ-sultones react with a wide range of nucleophilic reagents e.g. alcohols, amines, organic acid salts and organometallic reagents: a few of these nucleophilic ring-opening reactions involving carbon–oxygen bond cleavage are shown (Scheme 70).

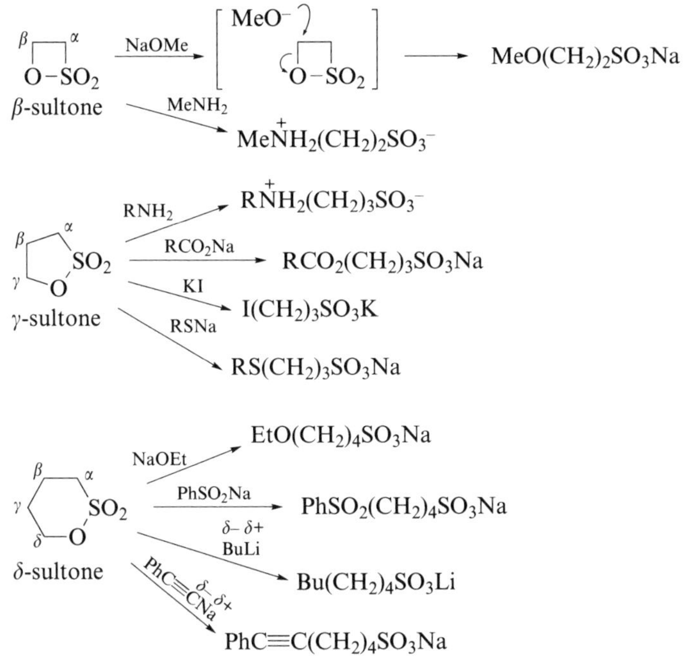

β-sultone → NaOMe → $MeO(CH_2)_2SO_3Na$

β-sultone → $MeNH_2$ → $Me\overset{+}{N}H_2(CH_2)_2SO_3^-$

γ-sultone → RNH_2 → $R\overset{+}{N}H_2(CH_2)_3SO_3^-$

γ-sultone → RCO_2Na → $RCO_2(CH_2)_3SO_3Na$

γ-sultone → KI → $I(CH_2)_3SO_3K$

γ-sultone → RSNa → $RS(CH_2)_3SO_3Na$

δ-sultone → NaOEt → $EtO(CH_2)_4SO_3Na$

δ-sultone → $PhSO_2Na$ → $PhSO_2(CH_2)_4SO_3Na$

δ-sultone → $\overset{\delta-}{Bu}\overset{\delta+}{Li}$ → $Bu(CH_2)_4SO_3Li$

δ-sultone → $PhC{\equiv}\overset{\delta-}{C}\overset{\delta+}{Na}$ → $PhC{\equiv}C(CH_2)_4SO_3Na$

Scheme 70

Aromatic sultones, unlike the aliphatic analogues, react with nucleophiles at the electrophilic sulfur atom. As an illustration, naphthalene-1,8-sultone (**169**) reacts with sodium hydroxide, ammonia or a Grignard reagent to give the products (**170**)–(**172**) (Scheme 71). The mechanism of these reactions involves initial nucleophilic attack at the sulfur atom with opening of the sultone ring system (Scheme 71).

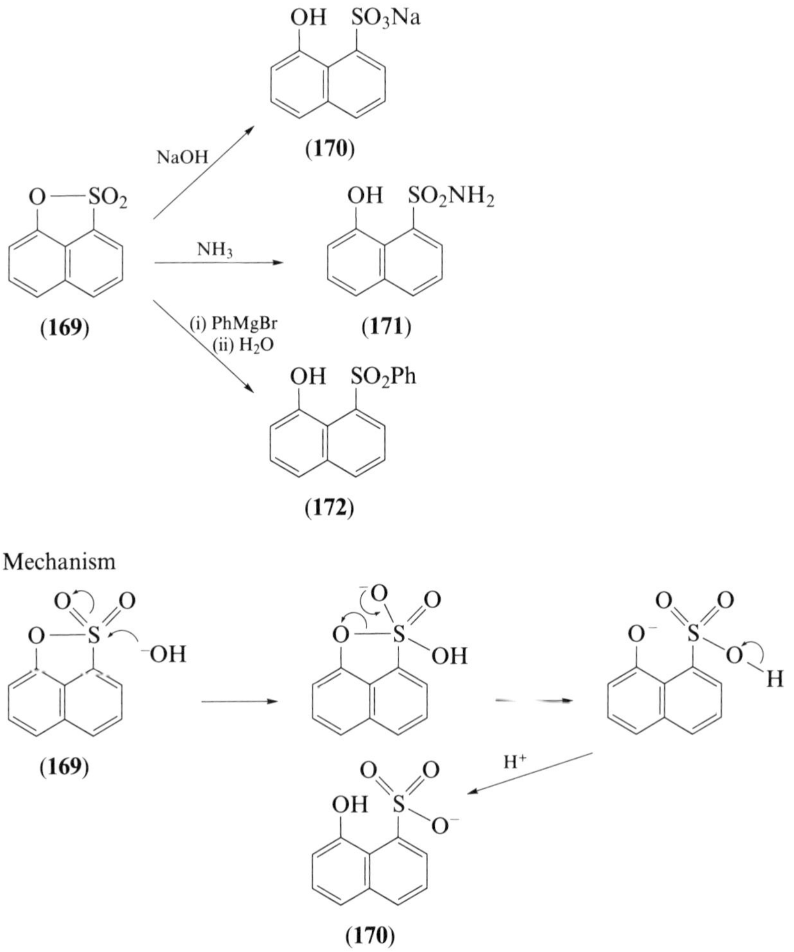

Mechanism

Scheme 71

Sultams may be prepared by cyclisation of the appropriate aminosulfonic acids and derivatives; the ring closure involves nitrogen–sulfur bond formation. The method can be applied to the synthesis of β-sultams, γ-sultams and δ-sultams; examples are the syntheses of (**173**) and (**174**) (Scheme 72). Modifications of the procedure include aminolysis of the corresponding sultones, and oxidation of

2-aminothiols to the sulfonyl chlorides followed by base-catalysed cyclisation. The latter provides a useful synthetic route to β-sultams. Scheme 73 shows examples of these methods in the synthesis of sultams **(175)**–**(178)**.

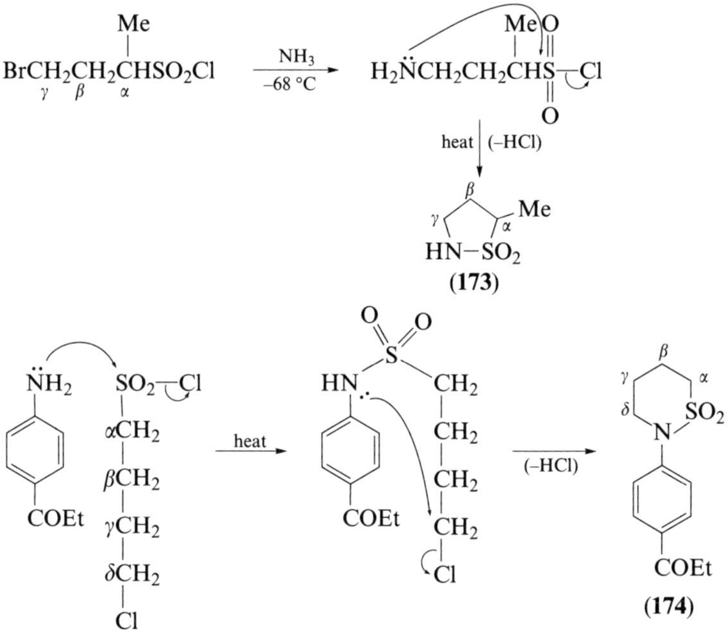

$BrCH_2CH_2CH(Me)SO_2Cl$ (γ β α) $\xrightarrow[-68\ ^\circ C]{NH_3}$ $H_2\ddot{N}CH_2CH_2CH(Me)S(O)_2$–Cl

heat (–HCl)

β, γ, α, Me, HN–SO_2

(173)

$\ddot{N}H_2$, COEt, SO_2–Cl, αCH_2, βCH_2, γCH_2, δCH_2, Cl

heat

O, O, S, HN, CH_2, CH_2, CH_2, CH_2, Cl, COEt

(–HCl)

β, γ, α, δ, SO_2, N, COEt

(174)

Scheme 72

$(CH_2)_n$, SO_2, O (n = 3, 4)

(i) $ArNH_2$
(ii) $POCl_3$, warm

$(CH_2)_n$, SO_2, N, Ar

(175)

Me, Me, O, SO_2 + $H_2\ddot{N}$–C_6H_4–Me

heat (–H_2O)

Me, β, γ, α, δ, Me, N, SO_2, Me

via

Me, Me, OH, $SO_2\ddot{N}H$–C_6H_4–Me

(–H_2O)

(176)

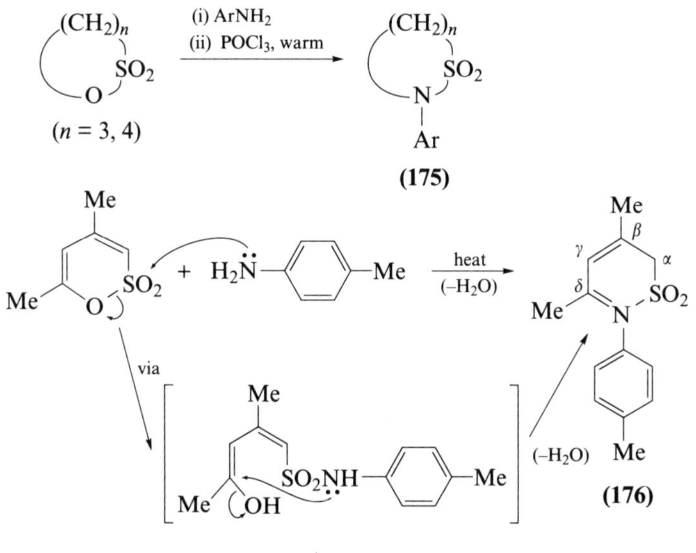

Scheme 73 *(continued)*

Scheme 73 *(continued)*

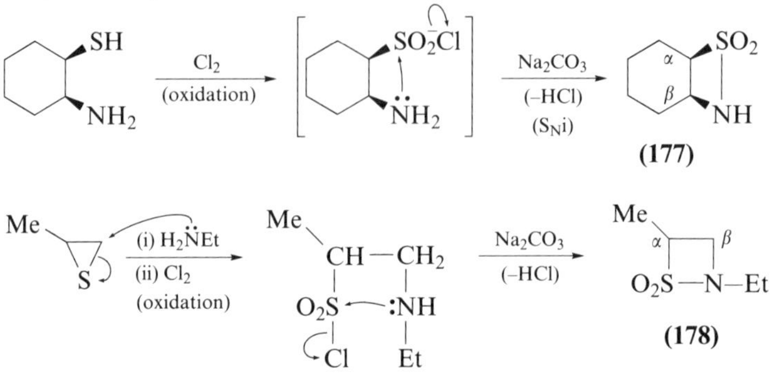

(177)

(178)

Scheme 73

Sultams can also be prepared by decomposition of the appropriate halo- and hydroxyalkanesulfonamides; the cyclisations involve carbon–nitrogen bond formation, as illustrated by the formation of the sultam (**179**). In this reaction, the sulfonyl chlorine atom is more reactive towards nucleophilic attack than the alkyl chlorine and consequently reacts first (Scheme 74).

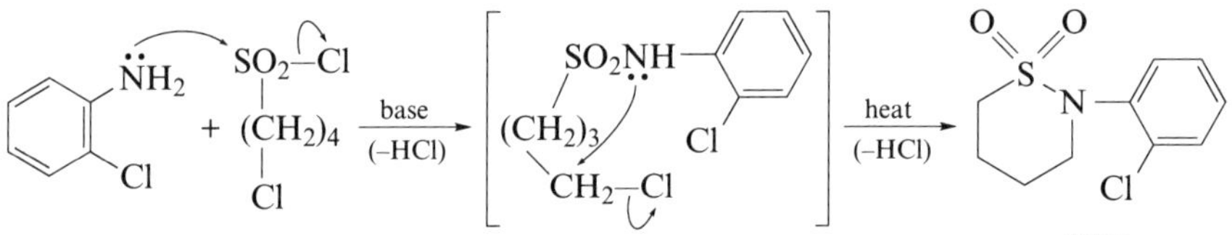

(179)

Scheme 74

N-(2-bromoalkyl) alkanesulfonamides (**180**) by metallation yield δ-sultams (**181**) via the resonance-stabilised carbanion (Scheme 75).

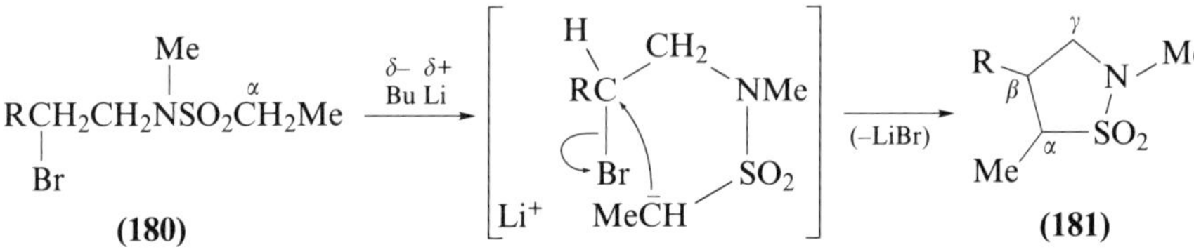

(180)

resonance-stabilised carbanion

(181)

Scheme 75

Sultams may also be obtained by flash vacuum pyrolysis (FVP) of sulfonyl azides; the reaction occurs via the sulfonyl nitrene (**182**) which contains an electron-deficient nitrogen atom and consequently rearranges by proton transfer to give the sultam (**183**) (Scheme 76).

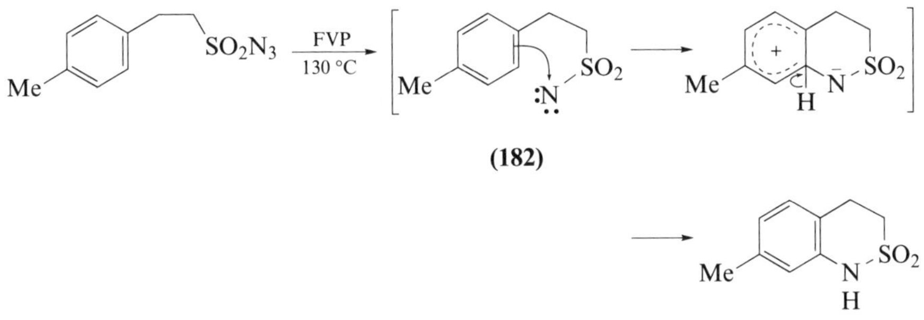

Me–C6H4–CH2CH2–SO_2N_3 —FVP, 130 °C→ [**(182)**] → **(183)**

Scheme 76

Sulfenes (see Chapter 7, p. 114) generated from sulfonyl chlorides by treatment with a tertiary amine may be used to prepare β-sultams. Thus, the sulfonyl chloride (**184**) may be converted to the sultam (**185**) via the sulfene (**186**) by reaction with an activated alkene (Scheme 77)

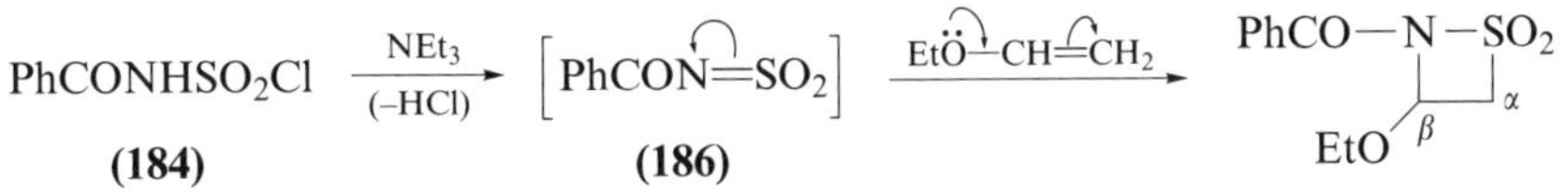

$PhCONHSO_2Cl$ (**184**) —NEt_3, (–HCl)→ [$PhCON{=}SO_2$] (**186**) —$EtO{-}CH{=}CH_2$→ (**185**)

Scheme 77

REACTIONS OF SULTAMS[8]

Sultams, e.g. (**187**)–(**189**), suffer hydrolysis by treatment with acid or alkali. The reaction opens the sultam ring (Scheme 78). With acids, the hydrolysis involves initial protonation of the nitrogen atom followed by nucleophilic attack by X^-. In alkaline hydrolysis, the reaction involves nucleophilic attack by OH^- and subsequent cleavage of the nitrogen–sulfur bond (Scheme 78). The order of reactivity of sultams towards hydrolysis is $\beta > \gamma > \delta$. This is to be expected since it parallels the relative instability of the ring systems, where the four-membered ring is more reactive than the five- and six-membered rings.

Sultams like (**190**) undergo aminolysis with amines by nitrogen–sulfur bond cleavage (similar to hydrolysis), and suitably functionalised sultams (**191**) may also suffer carbon–nitrogen bond cleavage by a bimolecular (E2) elimination reaction (Scheme 79).

β-Sultams (**187**) may be alkylated by treatment with activated alkyl halides—RCH_2Br reacts rapidly under phase transfer conditions—and they can also be acylated in the 2-position (Scheme 80). δ-Sultams (**192**) are conveniently alkylated and acylated via the lithium salts. Similar reactions occur with aromatic sultams, e.g. naphthalene-1,8-sultam (**189**). These also undergo typical aromatic

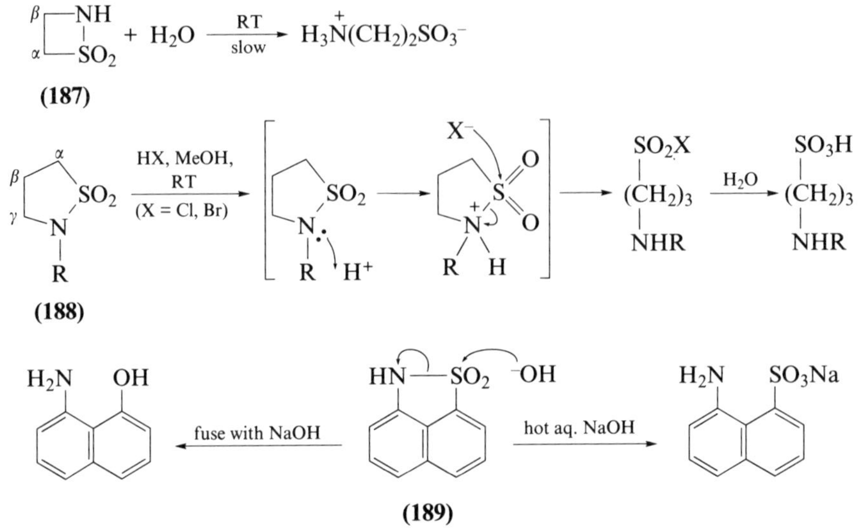

(187) + H_2O —RT, slow→ $H_3\overset{+}{N}(CH_2)_2SO_3^-$

(188) —HX, MeOH, RT (X = Cl, Br)→ [ ] → $SO_2X(CH_2)_3NHR$ —H_2O→ $SO_3H(CH_2)_3NHR$

(189): fuse with NaOH → 8-amino-1-naphthol (H_2N, OH); hot aq. NaOH → H_2N, SO_3Na

Scheme 78

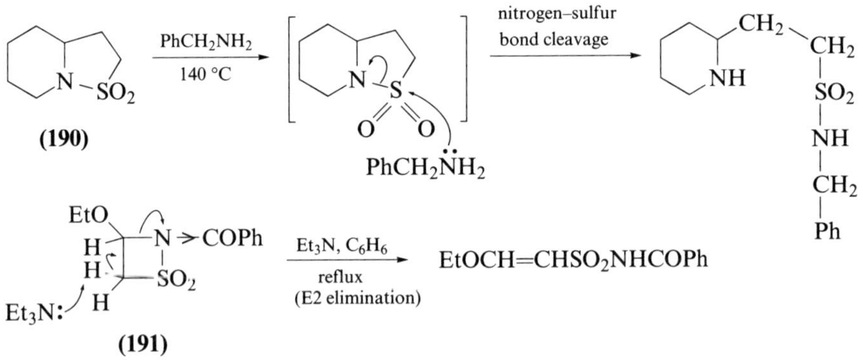

(190) —$PhCH_2NH_2$, 140 °C→ [ ] —nitrogen–sulfur bond cleavage→ $CH_2CH_2SO_2NHCH_2Ph$ piperidine (NH)

(191) —Et_3N, C_6H_6, reflux (E2 elimination)→ $EtOCH{=}CHSO_2NHCOPh$

Scheme 79

(187) —RCH_2–Br (–Br^-)→ N–CH_2R

—RCOCl→ N–COR

—MeN=C=O→ [N–C(–O–H)=NMe] —rearrangement→ N–CONHMe

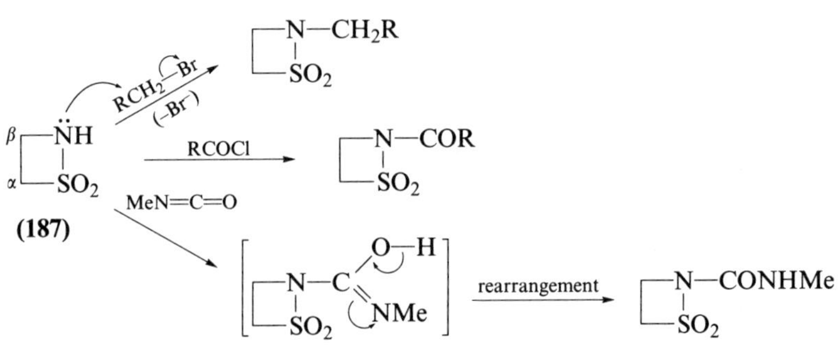

Scheme 80 *(continued)*

Scheme 80 *(continued)*

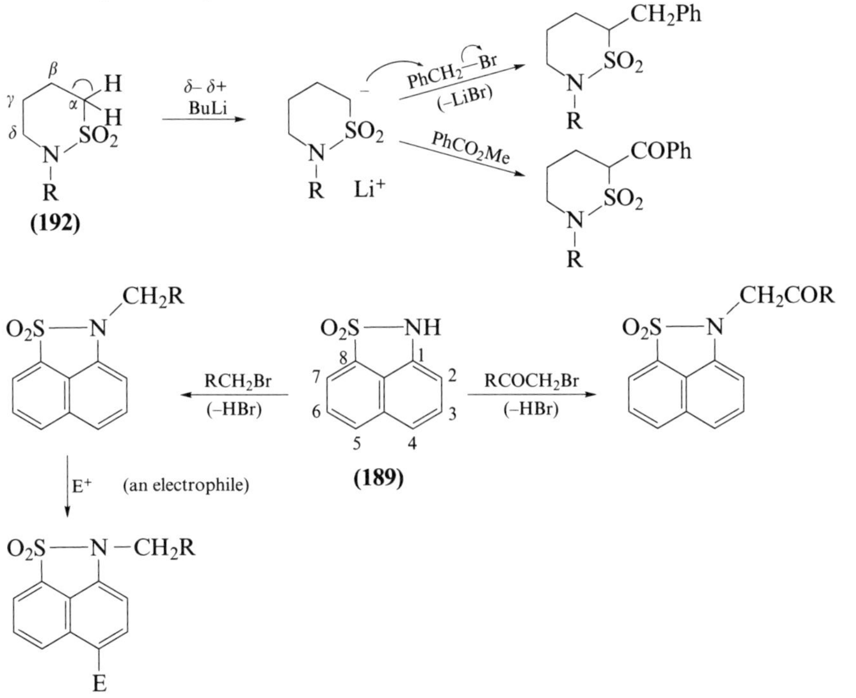

Scheme 80

electrophilic substitution reactions such as nitration and halogenation in the 4-position (Scheme 80).

References

1 A. D. Dunn and W.-D. Rudolf, *Carbon Dioxide in Organic Chemistry*, Ellis Horwood, Chichester, 1989.
2 G. D. Thorn and R. A. Ludwig, *The Dithiocarbamates*, Elsevier, Amsterdam, 1962.
3 R. J. Cremlyn, *Agrochemicals: Preparation and Mode of Action*, Wiley, Chichester, 1991.
4 R. Guy, in *The Chemistry of Cyanates and their Thio Derivatives* (Ed. S. Patai), Part. 2, Wiley, Chichester, 1977, Chap. 18, p. 819.
5 L. Drobnica, P. Kristian and J. Augustin, in *The Chemistry of Cyanates and their Thio Derivatives* (Ed. S. Patai), Part 2, Wiley, Chichester, 1977, p. 1003.
6 D. N. Dhar and K. S. M. Murthy, *Synthesis*, 434 (1988).
7 G. A. Benson and W. J. Spillane, in *The Chemistry of Sulfonic Acids, Esters and their Derivatives* (Eds S. Patai and Z. Rappoport), Wiley, Chichester, 1991 Chap. 22, p. 947.
8 A. J. Buglass and J. G. Tillett, in *The Chemistry of Sulfonic Acids, Esters and their Derivatives* (Eds S. Patai and Z. Rappoport), Wiley, Chichester, 1991, Chap. 19, p. 789.

10 THE UTILITY OF ORGANOSULFUR COMPOUNDS IN ORGANIC SYNTHESIS

A number of synthetic applications of different classes of organosulfur compounds have already been mentioned in previous chapters. The purpose of this chapter is to highlight the major organosulfur reagents, giving their important uses in synthetic organic chemistry.

Sulfur or sulfonium ylides[1,2a,2b]

Ylides are dipolar compounds containing positive and negative charges on adjacent atoms; the most important examples are phosphorus (**1a**) and sulfur (**2a**) ylides (Figure 1), in which a carbanion is directly attached to a positively charged phosphorus or sulfur atom. The structures (**1a**) and (**2a**) are stabilised by resonance with the non-polar contributing forms (**1b**) and (**2b**); ylides are basic compounds containing the negative charge on the carbon atom. Both types of ylides are generally prepared from either phosphonium (**3**) or sulfonium (**4**) salts (see Chapter 6, p. 81), which in turn derive from phosphines or sulfides by treatment with alkyl halides (Scheme 1).

$$R_3\overset{+}{P}-\overset{-}{C}R'_2 \longleftrightarrow R_3P{=}CR'_2$$

(**1a**) (**1b**)

$$R_2\overset{+}{S}-\overset{-}{C}R'_2 \longleftrightarrow R_2S{=}CR'_2$$

(**2a**) (**2b**)

Figure 1

Some specific examples of the formation of sulfur ylides by the deprotonation of sulfonium salts are shown in Scheme 2. The abstraction of a proton from the corresponding sulfonium salt as shown in Schemes 1 and 2, is the most common procedure for the generation of sulfur or sulfonium ylides; however, the most direct synthesis of sulfur ylides (**2a**) involves addition of a sulfide to a carbene (**5**),

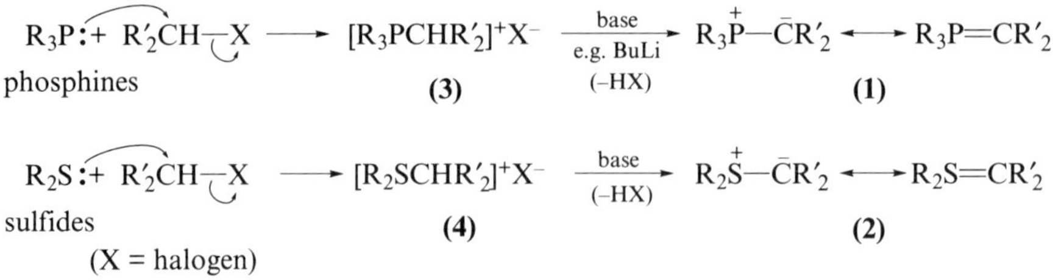

$$R_3P: + R'_2CH{-}X \longrightarrow [R_3PCHR'_2]^+X^- \xrightarrow[\text{e.g. BuLi}\ (-HX)]{\text{base}} R_3\overset{+}{P}{-}\overset{-}{C}R'_2 \longleftrightarrow R_3P{=}CR'_2$$

phosphines (3) (1)

$$R_2S: + R'_2CH{-}X \longrightarrow [R_2SCHR'_2]^+X^- \xrightarrow[(-HX)]{\text{base}} R_2\overset{+}{S}{-}\overset{-}{C}R'_2 \longleftrightarrow R_2S{=}CR'_2$$

sulfides (4) (2)

(X = halogen)

Scheme 1

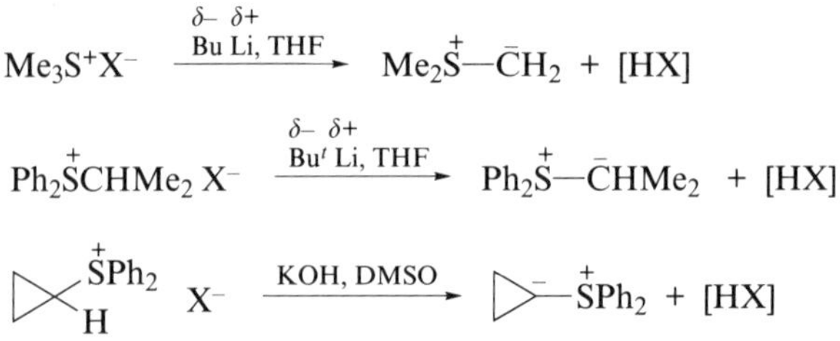

$$Me_3S^+X^- \xrightarrow{\overset{\delta-\ \delta+}{\text{Bu Li, THF}}} Me_2\overset{+}{S}{-}\overset{-}{C}H_2 + [HX]$$

$$Ph_2\overset{+}{S}CHMe_2\,X^- \xrightarrow{\overset{\delta-\ \delta+}{\text{Bu}^t\text{ Li, THF}}} Ph_2\overset{+}{S}{-}\overset{-}{C}HMe_2 + [HX]$$

$$\text{cyclopropyl}(H)\overset{+}{S}Ph_2\ X^- \xrightarrow{\text{KOH, DMSO}} \text{cyclopropyl}^-{-}\overset{+}{S}Ph_2 + [HX]$$

Scheme 2

$$R_2S: + :CR'_2 \rightleftharpoons R_2\overset{+}{S}{-}\overset{-}{C}R'_2$$

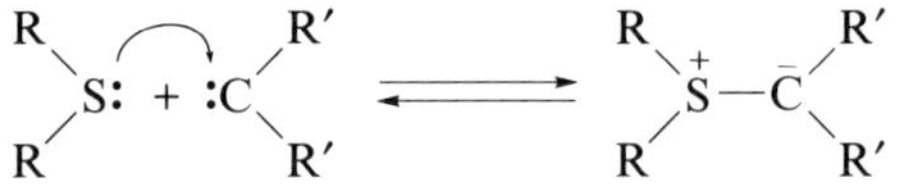

(5)

Scheme 3

(6) $\xrightarrow{:CCl_2}$ (9) [S⁺–⁻CCl₂, H] $\rightleftharpoons$ [S⁺–CHCl₂ ylide] $\longrightarrow$ (7) [4-CHCl₂] + (8) [2-CHCl₂]

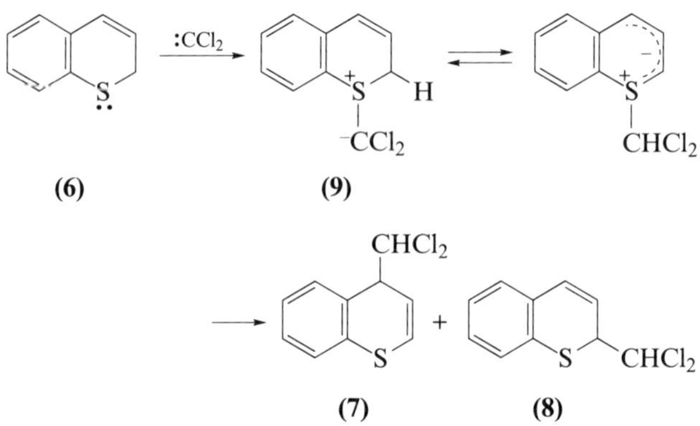

Scheme 4

an electron-deficient molecule (Scheme 3). For example, the generation of dichlorocarbene, from chloroform and base, in the presence of 2*H*-1-benzothiopyran (**6**) yields the insertion products (**7**) and (**8**), explainable on the basis of the formation of the ylide intermediate (**9**) (Scheme 4).

Diazoalkanes, e.g. ethyl diazoacetate, may also be employed as carbene precursors; the novel conversion of the cephalosporin (**10**) to the penicillin nucleus (**11**) (Scheme 5) illustrates the potential of this method.

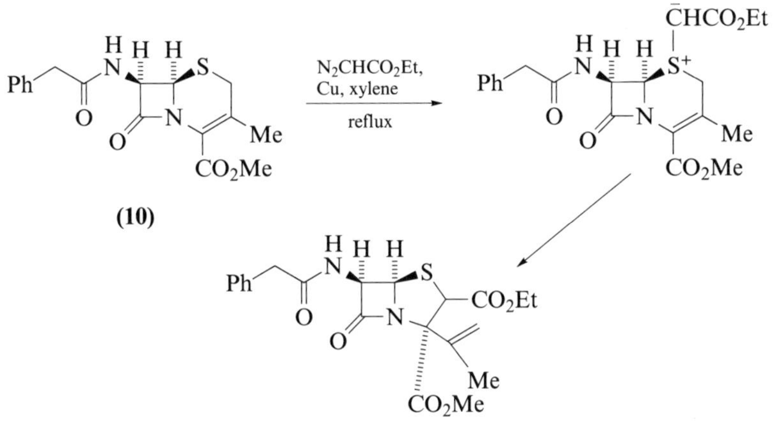

N_2CHCO_2Et, Cu, xylene, reflux

(**10**)

(**11**)
53%

Scheme 5

Stabilised ylides can be obtained by this procedure; thus, diazo compounds like (**12**) and (**13**) when irradiated with UV light or warmed with copper powder form the corresponding sulfur ylides (**14**) and (**15**) via the intermediate carbenes which are formed *in situ* (Scheme 6).

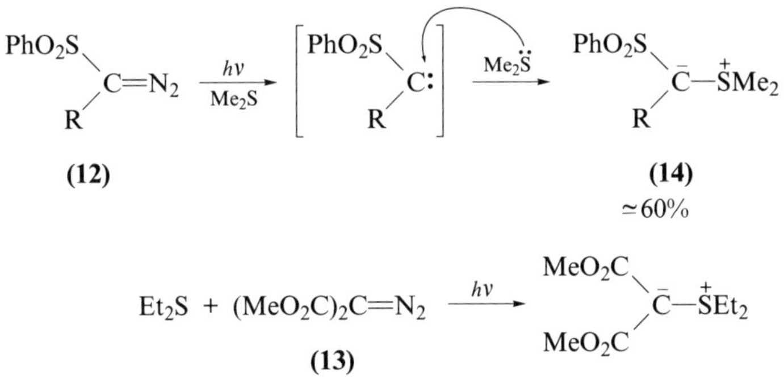

$$PhO_2S(R)C{=}N_2 \xrightarrow[Me_2S]{h\nu} [PhO_2S(R)C:] \xrightarrow{Me_2\ddot{S}} PhO_2S(R)\bar{C}{-}\overset{+}{S}Me_2$$

(**12**) (**14**) ≃60%

$$Et_2S + (MeO_2C)_2C{=}N_2 \xrightarrow{h\nu} (MeO_2C)_2\bar{C}{-}\overset{+}{S}Et_2$$

(**13**) [via $(MeO_2C)_2C$:] (**15**)

Scheme 6

Sulfur ylides containing only alkyl, vinyl or aryl groups are very unstable and must generally be generated and used at low temperatures (approximately –70 °C). On the other hand, ylides containing carbonyl, cyano, nitro or sulfonyl groups are relatively stable and may be isolated and stored at room temperature without appreciable decomposition.

In the structure of ylides, the ability of sulfur to stabilise an adjacent negative charge is an interesting phenomenon which probably involves the delocalisation of electron density into the 'low-lying' 3*d*-orbitals of sulfur. Essentially, the question revolves around the relative importance of the resonance forms (**2a**) and (**2b**) (Scheme 1) in the ground state structure of sulfur ylides. The evidence tends to indicate that sulfur ylides do not contain appreciable double-bond character, so that the structure (**2a**) is probably the major resonance contributor, which would be in agreement with the preferred geometry of carbon and sulfur.[1]

REACTIONS[1,2a,2b]

Phosphorus ylides are very important because of their use in the well-known Wittig reaction (1954) for the synthesis of alkenes. In the Wittig reaction, a phosphorus ylide (**1**) reacts with an aldehyde or ketone to yield the corresponding alkene (**16**) (Scheme 7). The reaction involves nucleophilic attack by the ylide (**1**) on the electrophilic carbonyl carbon atom to yield the betaine intermediate, which then collapses with elimination of the phosphine oxide and formation of the alkene (**16**). The driving force of the Wittig reaction is the production of the very strong phosphorus–oxygen double bond in the phosphine oxide (Scheme 7).

$$\underset{(\mathbf{1})}{R_3\overset{+}{P}-\overset{-}{C}R'_2} + R''R'''C{=}O \longrightarrow R_3P{=}O + \underset{(\mathbf{16})}{R'_2C{=}CR''R'''}$$

Mechanism

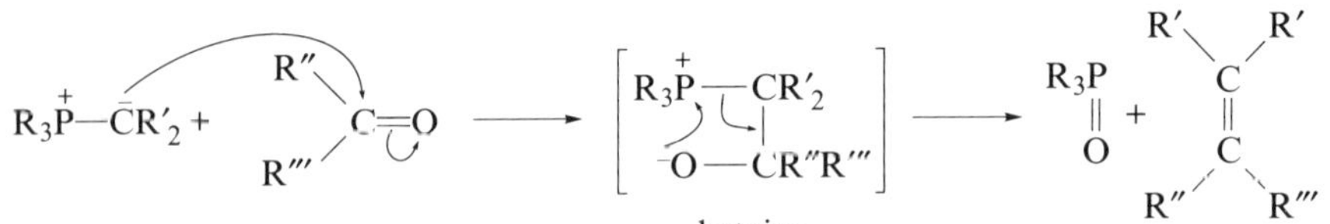

$$R_3\overset{+}{P}-\overset{-}{C}R'_2 + R''R'''C{=}O \longrightarrow \left[\begin{array}{c} R_3\overset{+}{P}-CR'_2 \\ | \qquad | \\ {}^{-}O-CR''R''' \end{array}\right]_{\text{betaine}} \longrightarrow R_3P{=}O + \underset{(\mathbf{16})}{R'_2C{=}CR''R'''}$$

Scheme 7

Sulfur ylides (**2**) generally react differently with carbonyl compounds to give epoxides (**17**) rather than alkenes (**16**) (Scheme 8). The first nucleophilic addition step to form the betaine intermediate (**18**) is identical to the first step of the Wittig reaction. The subsequent decomposition of the betaine, however, is governed by the excellent leaving group capacity of the R_2S moiety resulting from the lower stability of the sulfur–oxygen double bond. So, the betaine intermediate (**18**) yields the dialkyl sulfide; and the epoxide (**17**) in preference to the alkene (**16**) and the sulfoxide (**19**) (Scheme 8).

The reaction of the sulfur ylides with carbonyl compounds provides a valuable synthetic route to a wide range of epoxides. For instance, benzophenone (**20**)

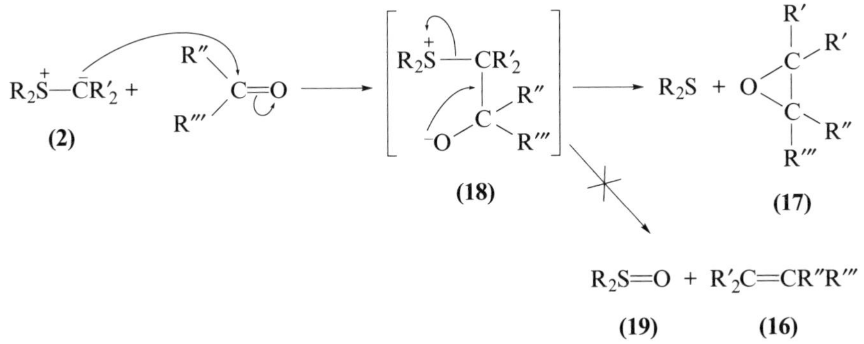

$$R_2\overset{+}{S}-\bar{C}R'_2 + R''R'''C{=}O \longrightarrow \left[R_2\overset{+}{S}-CR'_2-C(R'')(R''')-O^- \right] \longrightarrow R_2S + \text{epoxide}$$

(2) (18) (17)

$$R_2S{=}O + R'_2C{=}CR''R'''$$

(19) (16)

Scheme 8

reacts smoothly with dimethylsulfonium methylide (**21**) to give 1,1-diphenyl epoxide (**22**) (84%) (Scheme 9). The synthesis of epoxides by condensation of sulfur ylides with carbonyl compounds is one of the most widespread applications of sulfur ylides.

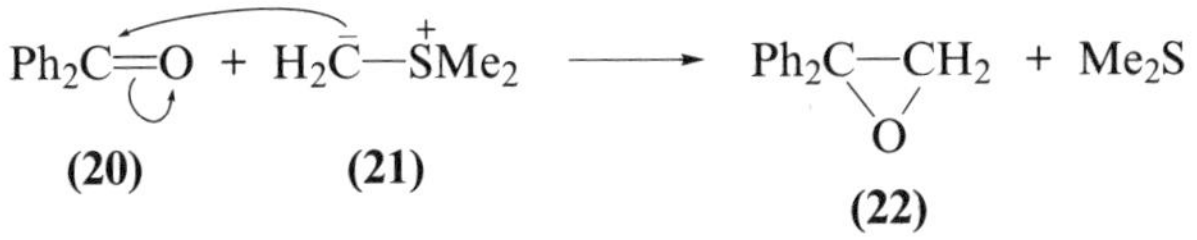

$$Ph_2C{=}O + H_2\bar{C}-\overset{+}{S}Me_2 \longrightarrow Ph_2C\overset{O}{-}CH_2 + Me_2S$$

(20) (21) (22)

Scheme 9

Dimethyl sulfoxide (DMSO) (**23**) can be used in carbon–carbon bond formation; it is a weak acid and with strong bases yields dimsylsodium (**24**) (Scheme 10).

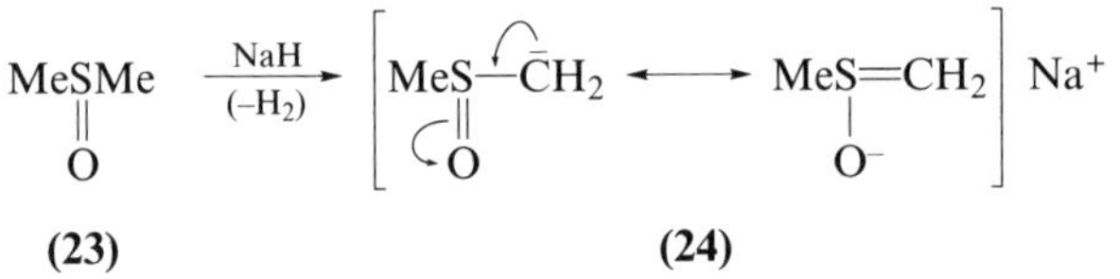

$$MeS(=O)Me \xrightarrow[(-H_2)]{NaH} \left[MeS(=O)-\bar{C}H_2 \longleftrightarrow MeS(-O^-){=}CH_2 \right] Na^+$$

(23) (24)

Scheme 10

Dimsylsodium (**24**) functions as a highly basic sulfur ylide. It can be used to convert phosphonium salts to phosphorus ylides for use in the Wittig reaction. Dimsylsodium also reacts with aldehydes and ketones by nucleophilic addition to form epoxides and with esters by nucleophilic substitution to yield β-ketosulfoxides (**25**) (Scheme 11). The β-ketosulfoxides (**25**) contain acidic α-hydrogens which can be readily removed to allow alkylation, and the products (**26**) suffer reductive desulfuration on treatment with aluminium amalgam to yield ketones (**27**) (Scheme 11) This procedure can, for instance, be applied to the conversion of ethyl benzoate to propiophenone (**28**) (Scheme 12).

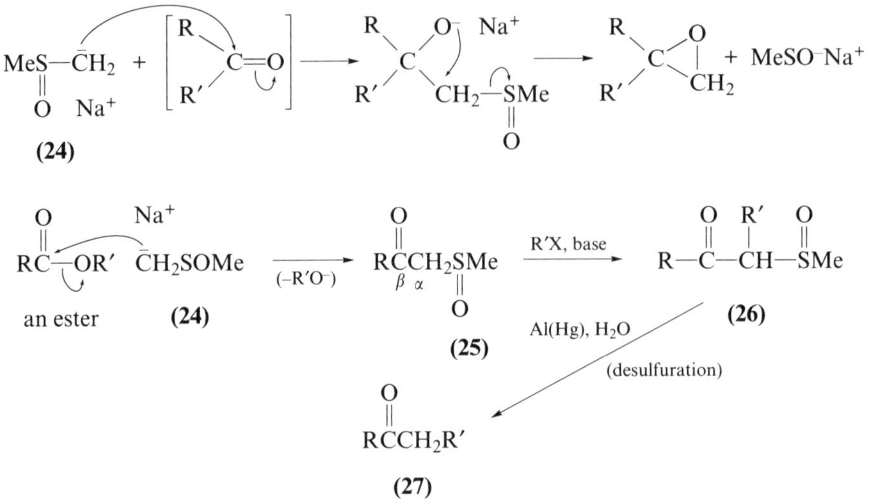

$$MeS(=O)-\bar{C}H_2\ Na^+ + [R(R')C=O] \longrightarrow R(R')C(O^-\ Na^+)-CH_2-S(=O)Me \longrightarrow R(R')C(-O-)CH_2 + MeSO^-Na^+$$

(24)

$$RC(=O)-OR' + \bar{C}H_2SOMe\ Na^+ \xrightarrow[(-R'O^-)]{} RC(=O)CH_2S(=O)Me \xrightarrow{R'X,\ base} R-C(=O)-CH(R')-S(=O)Me$$

an ester **(24)** **(25)** (β, α) **(26)**

$$\xrightarrow[(desulfuration)]{Al(Hg),\ H_2O} RC(=O)CH_2R'$$

(27)

Scheme 11

$$PhC(=O)-OEt + \bar{C}H_2SOMe\ Na^+ \xrightarrow[(-NaOEt)]{} PhC(=O)CH_2SOMe$$

ethyl benzoate **(24)**

$$\xrightarrow{(i)\ MeI,\ base;\ (ii)\ Al(Hg),\ H_2O} PhC(=O)CH_2Me$$

(28)

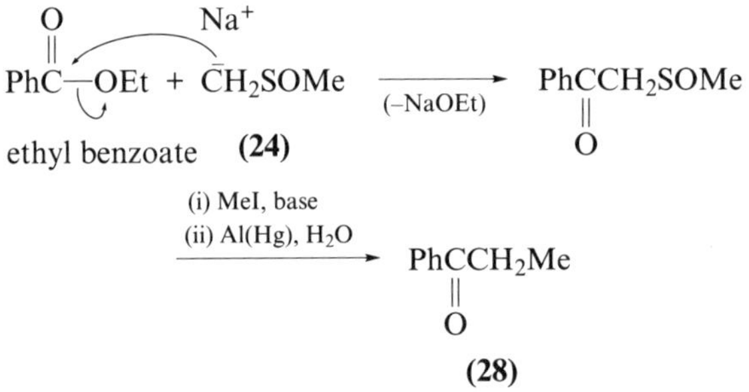

Scheme 12

In certain cases, dimsylsodium (**24**) will undergo selective nucleophilic substitution with bromides; thus, *vic*-dibromides (**29**) and *gem*-dibromides (**30**) are preferentially debrominated to the products (**31**) and (**32**) (Scheme 13).

Dimethyl sulfoxide (**23**) (see Chapter 5, p. 66) is sufficiently nucleophilic to react with methyl iodide to form the corresponding sulfoxonium salt, which on treatment with a powerful base affords the sulfoxonium methylide (**33**) (Scheme 14). The reaction in applicable to other alkyl halides.

Dialkylsulfoxonium ylides like (**33**) are stabilised by the oxygen atom. They are therefore less reactive than the corresponding dialkylsulfonium ylides, e.g. (**21**). The difference is reflected in several important respects. In studies of the stereochemistry of epoxide formation using the rigid 4-*t*-butylcyclohexanone molecule (**34**) (Scheme 15) as substrate, it was found that the more reactive sulfonium ylides like dimethylsulfonium methylide (**21**) reacted very quickly by axial attack to form mainly the kinetically controlled epoxide (**35**).[2a] On the other hand,

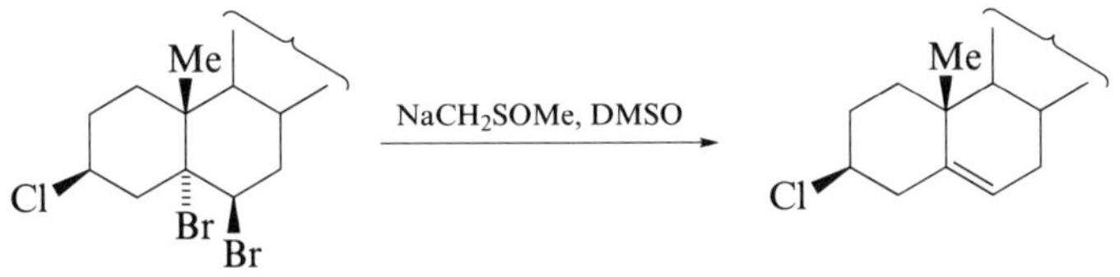

Me, Cl, Br, Br; NaCH$_2$SOMe, DMSO; Me, Cl

(29) **(31)**

Br, Br; NaCH$_2$SOMe, DMSO; H, Br

(30) **(32)**

Scheme 13

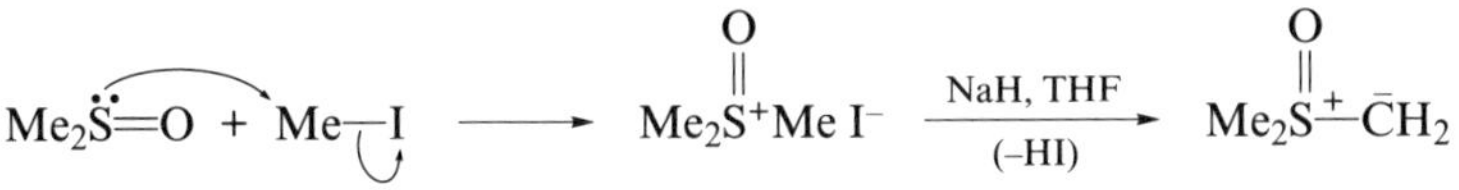

$$Me_2\ddot{S}{=}O + Me{-}I \longrightarrow Me_2S(=O)^+Me\ I^- \xrightarrow[(-HI)]{NaH,\ THF} Me_2S(=O)^+{-}\bar{C}H_2$$

(23) **(33)**

Scheme 14

Me$_3$C, O **(34)**; $Me_2\overset{+}{S}{-}\bar{C}H_2$ **(21)** → Me$_3$C, O **(35)** 83%; $Me_2S(=O)^+{-}\bar{C}H_2$ **(33)** → Me$_3$C, O **(36)** 17%

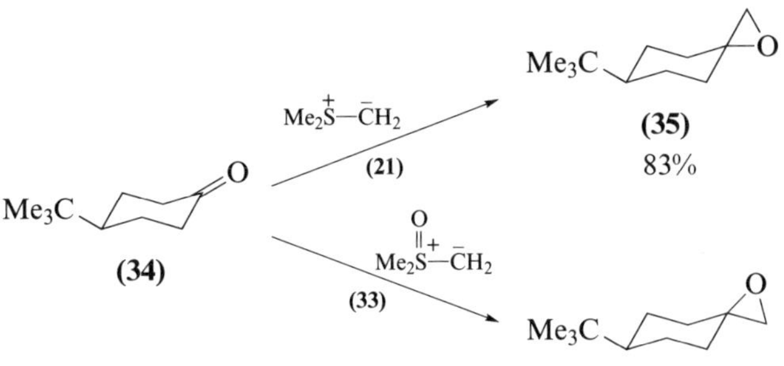

Scheme 15

with the less reactive dimethylsulfoxonium methylide (**33**), the slower reaction afforded the thermodynamically controlled epoxide (**36**). In this case, the more stable sulfoxonium ylide is better leaving group and the slower reaction allows reversible betaine formation, leading to the thermodynamic product (**36**) (Scheme 15).

The two types of sulfur ylides also differ in their reactions with α,β-unsaturated carbonyl compounds. The highly reactive sulfonium ylides react rapidly by 1,2-addition across the carbon–oxygen double bond to yield the epoxides. On the other hand, the less reactive sulfoxonium ylides react by slower conjugate addition (1,4-addition) to give substituted ketocyclopropanes. Thus, dimethylsulfonium methylide (**21**) reacts rapidly with benzylideneacetophenone (chalcone) (**37**) to yield the epoxide (**38**), whereas the sulfoxonium methylide (**33**) gives the keto-cyclopropane (**39**) with the same substrate (Scheme 16).

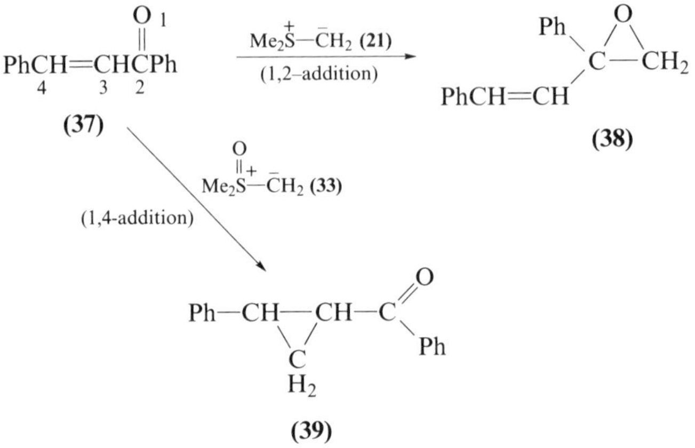

$$\text{PhCH=CHC(=O)Ph}\ (\mathbf{37}) \xrightarrow[\text{(1,2-addition)}]{Me_2\overset{+}{S}-\overset{-}{C}H_2\ (\mathbf{21})} \text{PhCH=CH(Ph)C(O)CH}_2\ (\mathbf{38})$$

$$(\mathbf{37}) \xrightarrow[\text{(1,4-addition)}]{Me_2\overset{+}{S}(=O)-\overset{-}{C}H_2\ (\mathbf{33})} \text{Ph-CH-CH(CH}_2\text{)-C(=O)Ph}\ (\mathbf{39})$$

Scheme 16

The cyclopropanation reaction is greatly facilitated by electron withdrawal from the alkenic double bond. Diethyl isopropylidenemalonate (**40**) consequently reacts with dimethylsulfoxonium methylide (**33**) to give the cyclopropane (**41**) in 91% yield in one hour, whereas ethyl 3-methyl-2-butenoate (**42**) only forms 9% of the cyclopropane (**43**) in the same period (Scheme 17).

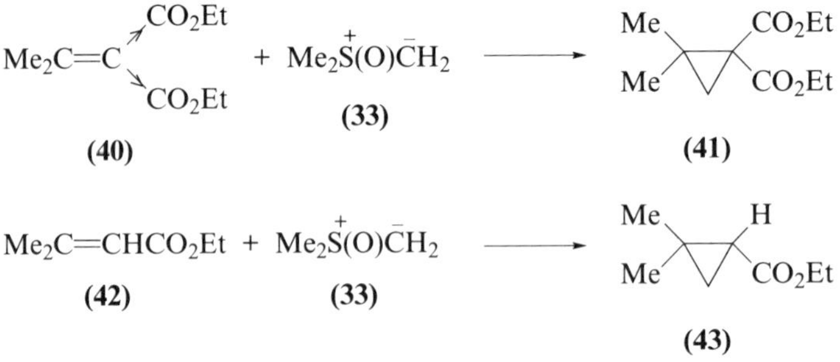

$$Me_2C{=}C(CO_2Et)_2\ (\mathbf{40}) + Me_2\overset{+}{S}(O)\overset{-}{C}H_2\ (\mathbf{33}) \longrightarrow \text{2,2-Me}_2\text{-1,1-(CO}_2\text{Et)}_2\text{-cyclopropane}\ (\mathbf{41})$$

$$Me_2C{=}CHCO_2Et\ (\mathbf{42}) + Me_2\overset{+}{S}(O)\overset{-}{C}H_2\ (\mathbf{33}) \longrightarrow \text{2,2-Me}_2\text{-1-H-1-CO}_2\text{Et-cyclopropane}\ (\mathbf{43})$$

Scheme 17

These observations suggest that the attachment of electron-withdrawing groups to the double bonds of α,β-unsaturated carbonyl compounds facilitates cyclopropanation by stabilisation of an intermediate carbanion (**44**) formed during the nucleophilic addition of the anionic centre to the substrate to form the cyclopropane (**45**) (Scheme 18).

The sulfonium and sulfoxonium ylides also differ in their behaviour with α,β-alkynic ketones (**46**) (Scheme 19). Dimethylsulfonium methylide (**21**) forms the alkynic epoxide (**47**) by 1,2-addition, but the sulfoxonium methylide (**33**) affords

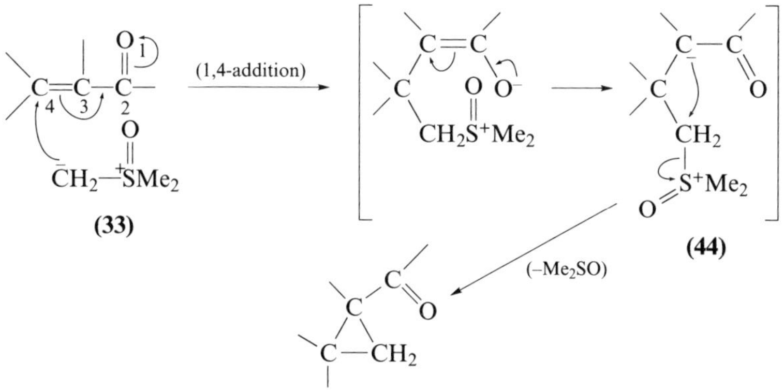

(1,4-addition)

(33)

(44)

($-Me_2SO$)

(45)

Scheme 18

the stabilised ylide (**48**). The latter on thermal dehydration gives the *S*-methylthiabenzene *S*-oxide (**49**) via the sequence of reactions shown in Scheme 19.

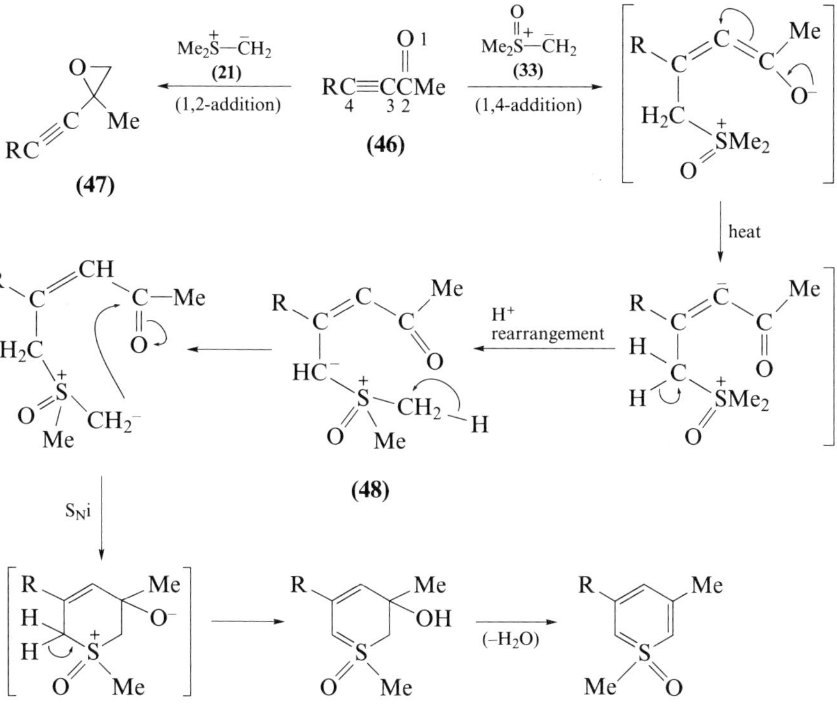

(21) (1,2-addition)

(33) (1,4-addition)

(46)

(47)

heat

H^+ rearrangement

(48)

S_Ni

($-H_2O$)

(49)

Scheme 19

With substrates containing a nucleophilic group sited close to a carbonyl group, on reaction with the sulfonium methylide (**21**) the intermediate epoxides are not isolated and instead heterocycles are generally obtained. For instance, *o*-aminophenyl ketones (**50**) react with the ylide (**21**) to yield the benzopyrroles (**51**) via the epoxides (**52**) (Scheme 20).

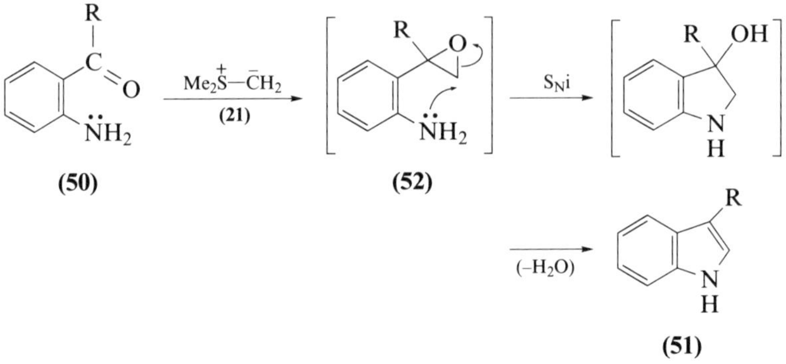

$Me_2\overset{+}{S}-\overset{-}{C}H_2$ (21)

S_Ni

$(-H_2O)$

(50) (52) (51)

Scheme 20

Epoxides may be cleaved by amines. This reaction is used in the synthesis of aminoalcohols (**53**) studied as potential adrenergic drugs (Scheme 21). The amine cleavage of the epoxide and the subsequent nitrous acid deamination of the aminoalcohol are used for the ring enlargement of alicyclic ketones (Tiffeneau–Demjanov reaction). By this sequence of reactions, for example, cycloheptanone (**54**) may be converted to cyclooctanone (**55**) (Scheme 21).

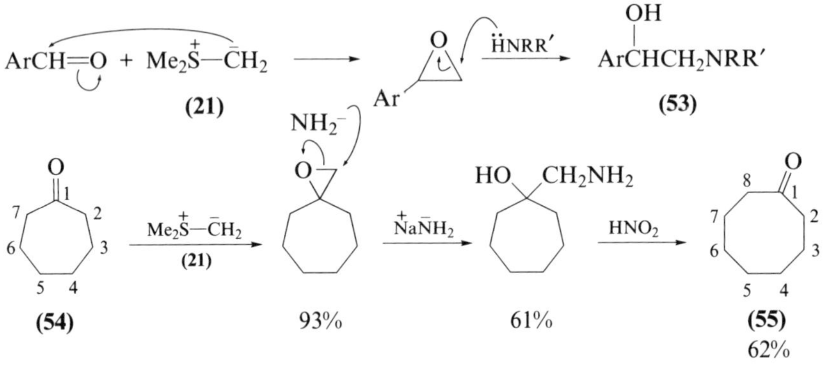

$ArCH{=}O + Me_2\overset{+}{S}-\overset{-}{C}H_2$ (21) ⟶ epoxide (Ar) $\xrightarrow{\ddot{H}NRR'}$ $ArCH(OH)CH_2NRR'$ (53)

(54) $\xrightarrow[(21)]{Me_2\overset{+}{S}-\overset{-}{C}H_2}$ 93% $\xrightarrow{\overset{+}{Na}\overset{-}{N}H_2}$ HO, CH_2NH_2 61% $\xrightarrow{HNO_2}$ (55) 62%

Scheme 21

EPOXIDATIONS WITH STABILISED YLIDES

Ylides stabilised by the presence of electron-withdrawing groups attached to the ylidic carbon atom generally only react with exceptionally reactive carbonyl compounds. For example, dimethylsulfonium carbethoxymethylide (**56**) reacts with 1,2-dicarbonyl compounds like ethyl pyruvate (**57**) to give the epoxide (**58**) (Scheme 22), but does not react with simple aldehydes and ketones.

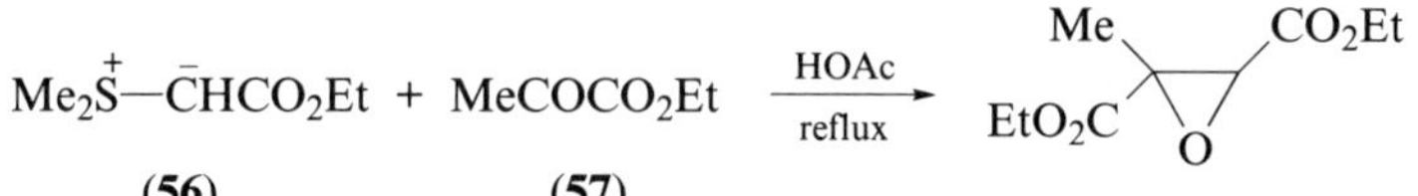

$$\mathrm{Me_2\overset{+}{S}{-}\overset{-}{C}HCO_2Et} + \mathrm{MeCOCO_2Et} \xrightarrow[\text{reflux}]{\text{HOAc}} \text{(58)}$$

(**56**) (**57**) (**58**)

Scheme 22

The reactivity of the ylide increases as the electron-withdrawing power of the substituent is reduced; thus, the amidic ylide (**59**) reacts with benzaldehyde to give the epoxide (**60**) (Scheme 23).

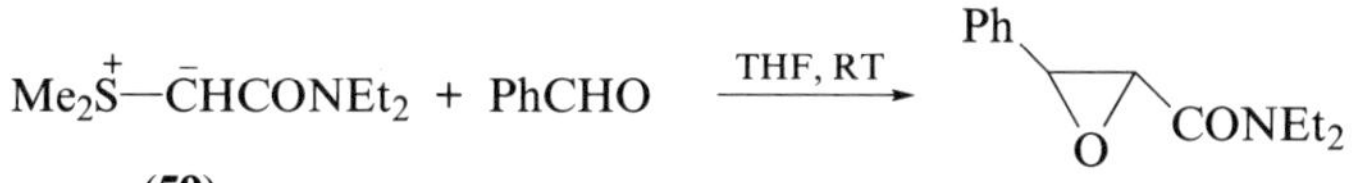

$$\mathrm{Me_2\overset{+}{S}{-}\overset{-}{C}HCONEt_2} + \mathrm{PhCHO} \xrightarrow{\text{THF, RT}} \text{(60)}$$

(**59**) (**60**)

Scheme 23

MICHAEL ADDITION OR CYCLOPROPANATION WITH SULFOXONIUM YLIDES

A wide variety of Michael acceptors have been reacted with sulfoxonium methylides, e.g. α,β-unsaturated ketones, nitriles, sulfones, sulfonamides and nitro compounds, to give good yields of the corresponding cyclopropane derivatives.[1,2a] Steric hindrance may decrease the rate of reaction but need not prevent reaction; for example, the very hindered compound 4,6,6-trimethyl-3-hepten-2-one (**61**) reacts with dimethylsulfoxonium methylide (**33**) to give the cyclopropane (**62**) (Scheme 24).

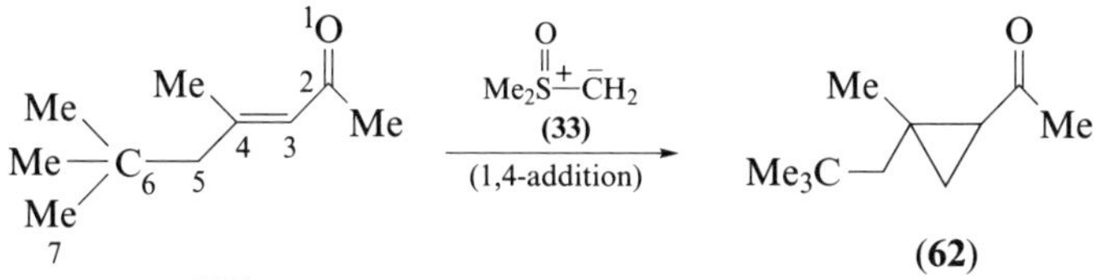

$$\text{(61)} \xrightarrow[\text{(1,4-addition)}]{\mathrm{Me_2\overset{O}{\overset{\|}{S}}{}^{+}{-}\overset{-}{C}H_2}\ \text{(33)}} \text{(62)}$$

(**61**) (**62**)

Scheme 24

With α,β-unsaturated esters, competition between cyclopropanation (1,4-addition) and acylation (1,2-addition) becomes acute. With ethyl acrylate (**63**) and ethyl crotonate (**64**), the small R groups on the β-carbon atom permit facile conjugate addition leading to good yields (60–65%) of the corresponding cyclopropane derivatives (**67**) (Scheme 25). On the other hand, the introduction of a large R group, e.g. phenyl, in ethyl cinnamate (**65**) reduces the yield of the cyclopropane to 30%; the major product is now derived from acylation (1,2-addition to the carbonyl group). However, if this carbonyl addition is hindered by using large R′ groups, like *t*-butyl in *t*-butyl cinnamate (**66**), the yield of the cyclopropane derivative (**67**) is dramatically increased to 65% (Scheme 25).

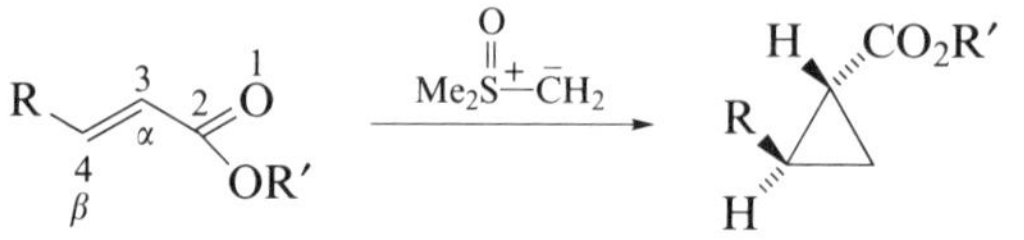

$$\mathrm{RCH{=}CHCO_2R'} \xrightarrow{\mathrm{Me_2\overset{+}{S}(O)\overset{-}{C}H_2}} \text{(67)}$$

	(**67**)
(**63**) R = H, R′ = Et	65%
(**64**) R = Me, R′ = Et	60%
(**65**) R = Ph, R′ = Et	30%
(**66**) R = Ph, R′ = CMe_3	65%

Scheme 25

The biosynthesis of the cyclopropane ring in several natural products probably occurs by the transfer of a methylene group from the methyl group of *S*-adenosylmethionine (**68**) via the sulfonium ylide (**69**) to an alkene, e.g. oleic ester (**70**), as indicated in Scheme 26. Such alkenes are normally unreactive, but the cyclopropanation has been shown to be catalysed by traces of metal chelates, and copper salts to yield the substituted cyclopropane (**71**) (Scheme 26).

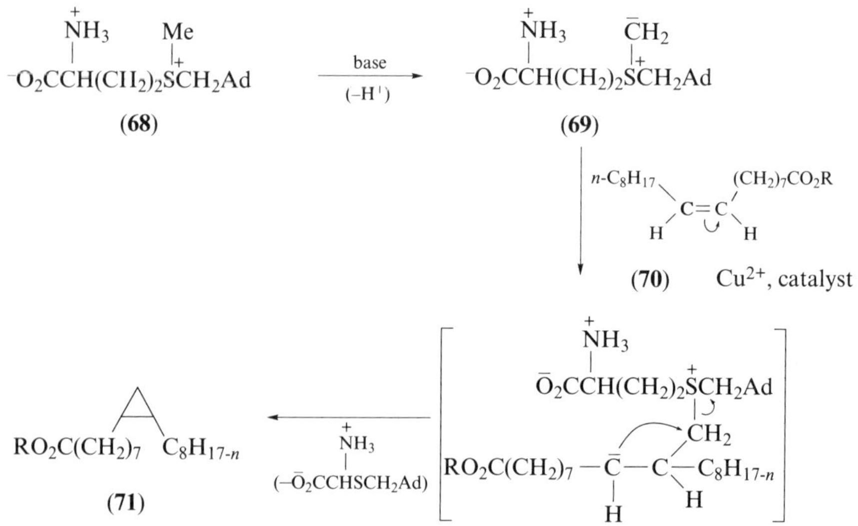

$$\mathrm{^-O_2CCH(\overset{+}{N}H_3)(CH_2)_2\overset{+}{S}(Me)CH_2Ad}\ \text{(68)} \xrightarrow[(-\mathrm{H}^+)]{\text{base}} \mathrm{^-O_2CCH(\overset{+}{N}H_3)(CH_2)_2\overset{+}{S}(\overset{-}{C}H_2)CH_2Ad}\ \text{(69)}$$

$$\text{(69)} + n\text{-}C_8H_{17}CH{=}CH(CH_2)_7CO_2R\ \text{(70)} \xrightarrow{Cu^{2+},\ \text{catalyst}} \left[\mathrm{RO_2C(CH_2)_7{-}\overset{-}{C}H{-}CH(CH_2\overset{+}{S}(CH_2Ad)(CH_2)_2CH(\overset{+}{N}H_3)CO_2^-){-}C_8H_{17}\text{-}n}\right] \xrightarrow{(-\bar{O}_2CCH(\overset{+}{N}H_3)SCH_2Ad)} \mathrm{RO_2C(CH_2)_7\text{–}\triangle\text{–}C_8H_{17}\text{-}n}\ \text{(71)}$$

Scheme 26

Sulfur ylides can also be used in the synthesis of chrysanthemate esters (**72**) from hept-2-enoates (**73**) (Scheme 27). The natural insecticide pyrethrum is a complex chrysanthemate ester, and the formation of *trans*-chrysanthemic acid is consequently important for the synthesis of many synthetic pyrethroid insecticides.

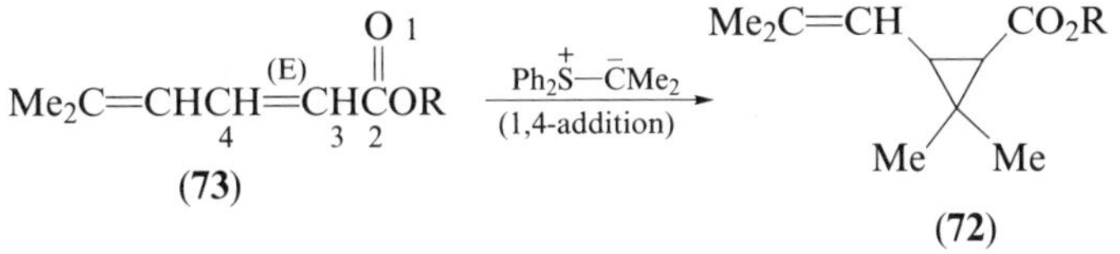

$$\underset{(\mathbf{73})}{Me_2C{=}\overset{4}{C}HCH\overset{(E)}{=}\overset{3}{C}H\overset{2}{C}(\overset{1}{=}O)OR} \xrightarrow[\text{(1,4-addition)}]{Ph_2\overset{+}{S}-\overset{-}{C}Me_2} \mathbf{(72)}$$

(**72**): cyclopropane bearing $Me_2C{=}CH$ and CO_2R, with Me, Me

Scheme 27

Sulfones

Over the last 20 years, the applications of sulfones in synthetic organic chemistry have dramatically increased.[3,4] The carbonyl group is often desired in the target organic molecule, but the sulfonyl group is almost always removed, so that the sulfone functions as a versatile tool in such synthetic transformations.

Sulfones are easily prepared by several high-yielding routes. The most usual procedure is by oxidation of sulfides, generally by reaction with a peroxycarboxylic acid generated *in situ* with hydrogen peroxide and the appropriate carboxylic acid (Scheme 28). The first oxidation step to give the sulfoxide is much faster than the second one to give the sulfone (**74**), and sometimes prolonged reaction or heating may be needed to complete the oxidation to the sulfone. The most common reagent for the oxidation is probably MCPBA, and the presence of metallic catalysts, e.g. tungstic or molybdic acid, facilitates clean sulfone formation with only stoichiometric quantities of hydrogen peroxide.

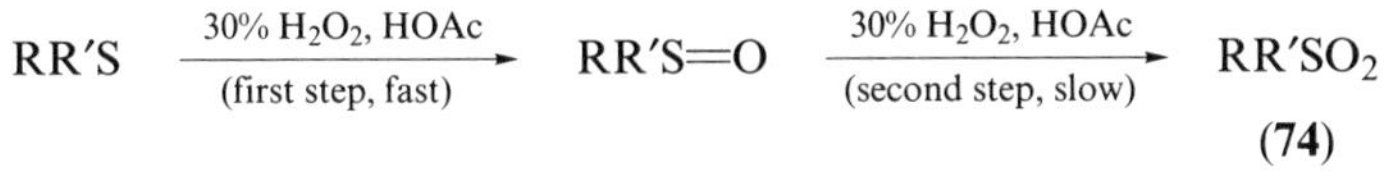

$$RR'S \xrightarrow[\text{(first step, fast)}]{30\%\ H_2O_2,\ HOAc} RR'S{=}O \xrightarrow[\text{(second step, slow)}]{30\%\ H_2O_2,\ HOAc} \underset{(\mathbf{74})}{RR'SO_2}$$

Scheme 28

Oxone® (a mixture of potassium sulfate, potassium hydrogensulfate and potassium hydrogen persulfate) is a commercially available oxidant which is generally used in aqueous ethanol. Oxone® is highly chemoselective and will cleanly oxidise sulfides containing other functional groups; for example, keto, hydroxy and alkenic double bonds are not attacked by this reagent.

Sodium perborate in warm glacial acetic acid will oxidise simple alkyl and aryl sulfides, and amino sulfides, (**75**) are converted to the corresponding nitro sulfones (**76**) (Scheme 29).

Potassium permanganate is another useful oxidant which gives high yields of sulfones under either heterogeneous conditions in dichloromethane, hexane and

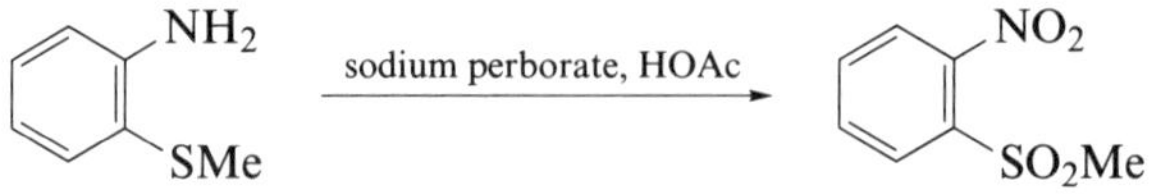

$$\text{(75) } o\text{-}H_2NC_6H_4SMe \xrightarrow{\text{sodium perborate, HOAc}} \text{(76) } o\text{-}O_2NC_6H_4SO_2Me$$

Scheme 29

so on or phase transfer conditions (dichloromethane–water). Potassium permanganate can also effect monooxidation of *gem*-disulfides (**77**) to the thiosulfones (**78**) (Scheme 30).

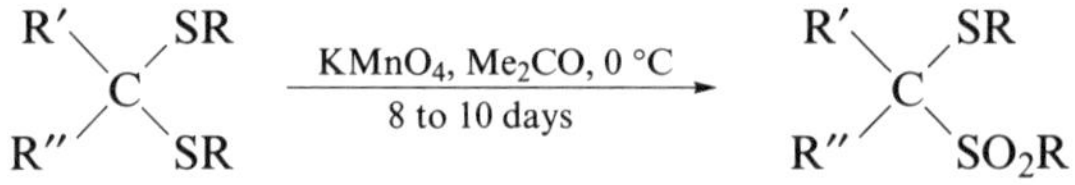

$$\underset{(77)}{R'R''C(SR)_2} \xrightarrow[\text{8 to 10 days}]{KMnO_4,\ Me_2CO,\ 0\ ^\circ C} \underset{\substack{(78)\\ 80\text{–}90\%}}{R'R''C(SR)(SO_2R)}$$

Scheme 30

Sulfones may also be obtained by alkylation of sulfinate salts (**79**). The sulfinate anion is an ambidentate nucleophile; consequently, alkylation can occur either on sulfur to give the sulfone (**74**) or oxygen to yield the sulfinate ester (**80**) (Scheme 31). The reactions of (**79**) with alkyl iodides in polyethylene glycols as solvents generally afford good yields of the sulfones (**74**).

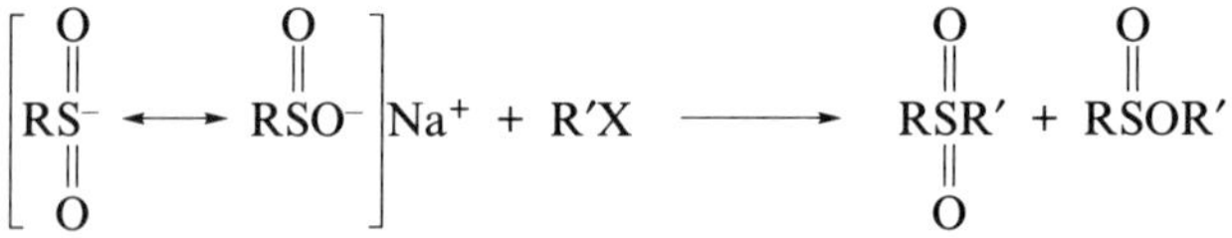

$$\underset{(79)}{\left[RS(=O)_2^- \longleftrightarrow RS(=O)O^-\right]Na^+} + R'X \longrightarrow \underset{(74)}{RS(=O)_2R'} + \underset{(80)}{RS(=O)OR'}$$

Scheme 31

Sulfonic acid derivatives (RSO_2X) are useful starting materials for the preparation of sulfones. Diaryl sulfones (**81**) are easily obtained by a Friedel–Crafts reaction between a nucleophilic arene and an arenesulfonyl chloride (Scheme 32). However, the analogous synthesis of alkyl aryl sulfones from alkanesulfonyl chlorides is often unsatisfactory owing to side reactions. The route can give good results when the chlorine atom is replaced by a more electronegative moiety such as the trichloromethanesulfonyl group. In this procedure, the alkanesulfonyl

$$ArH + Ar'SO_2Cl \xrightarrow{AlCl_3} \underset{(81)}{ArSO_2Ar'} + [HCl]$$

Scheme 32

chloride is reacted with silver trifluoromethanesulfonate to give the alkanesulfonyl trifluoromethanesulfonate which is finally treated with the arene, as shown in Scheme 33.

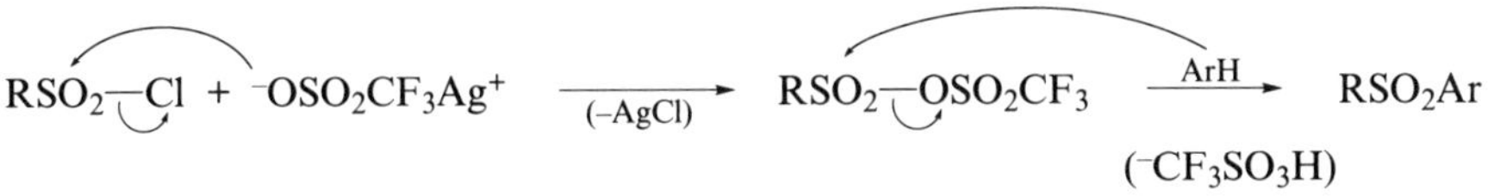

$$RSO_2{-}Cl + {}^{-}OSO_2CF_3Ag^+ \xrightarrow[(-AgCl)]{} RSO_2{-}OSO_2CF_3 \xrightarrow[(^{-}CF_3SO_3H)]{ArH} RSO_2Ar$$

Scheme 33

Certain substituted sulfones may be obtained by special methods; for instance, vinyl sulfones (**82**) may be formed by addition of a sulfonyl carbanion (**83**) to a carbonyl compound followed by elimination (Scheme 34). An example when X = H is the synthesis of methyl styryl sulfones (**84**) by the Knoevenagel condensation of an aromatic aldehyde with a methanesulfonylacetate (**85**) followed by dealkylation–decarboxylation of the intermediate product by treatment with lithium iodide in DMF (Scheme 35).

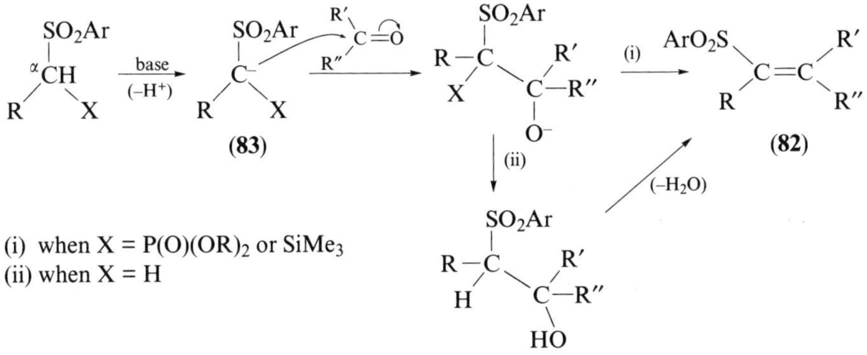

$R(X)C^{\alpha}H(SO_2Ar) \xrightarrow[(-H^+)]{base} R(X)C^{-}(SO_2Ar)$ (**83**) $\xrightarrow{R'R''C{=}O} R(X)C(SO_2Ar){-}C(O^-)R'R'' \xrightarrow{(i)} ArO_2S(R)C{=}CR'R''$ (**82**)

$R(X)C(SO_2Ar){-}C(O^-)R'R'' \xrightarrow{(ii)} R(H)C(SO_2Ar){-}C(HO)R'R'' \xrightarrow{(-H_2O)}$ (**82**)

(i) when X = $P(O)(OR)_2$ or $SiMe_3$
(ii) when X = H

Scheme 34

$ArCH{=}O + H_2C^{\alpha}(SO_2Me)(CO_2Et)$ (**85**) $\xrightarrow[(-H_2O)]{C_5H_5N} ArCH{=}C(SO_2Me)(CO_2Et) \xrightarrow{LiI,\ DMF} [ArCH{=}C(SO_2Me)(CO_2H)] \xrightarrow{(-CO_2)} ArCH{=}CHSO_2Me$ (**84**)

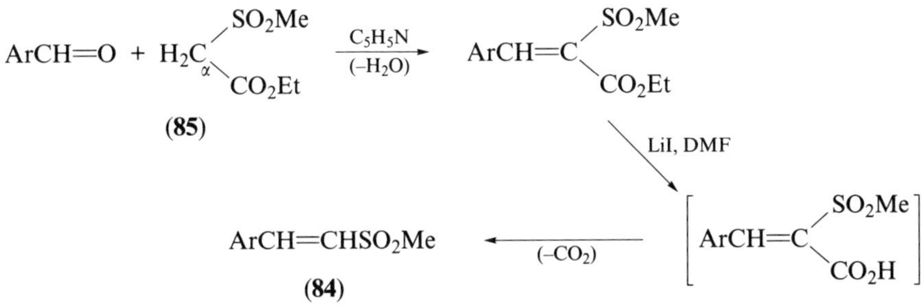

Scheme 35

Hydroxysulfones (**86**) are usually obtained by condensation of a sulfonyl carbanion (**83**) with a carbonyl compound; the resultant hydroxysulfone (**86**) by reductive elimination yields an alkene (**87**). The two-stage process is known as the Julia reaction (1973), namely the formation of the alkene double bond by sulfone elimination (Scheme 36).

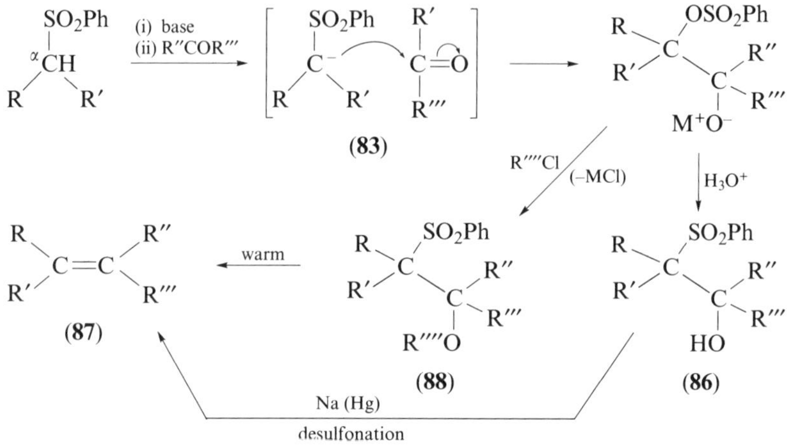

base = $\overset{\delta-}{\text{Bu}}\overset{\delta+}{\text{Li}}$ or LDA (M = Li) or $\overset{\delta-}{\text{R}}\overset{\delta+}{\text{Mg}}$Br (M = Mg)
R'''' = COMe, COPh or SO_2Me

Scheme 36

In certain cases, direct elimination of the hydroxysulfone (**86**) is effective, but generally it is better to go via the acyl derivative (**88**) which is formed *in situ*. The Julia reaction essentially resembles the Wittig and Peterson reactions, in that an α-heteroatom-substituted carbanion and a carbonyl compound react together with the subsequent elimination of two vicinal functional groups to form the

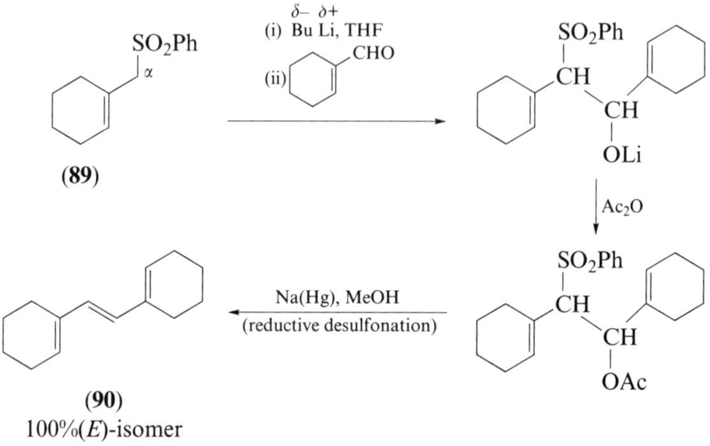

Scheme 37

double bond. The Julia reaction normally leads to predominantly the (*E*) alkene. An example of the Julia reaction is the conversion of the cyclohexenyl phenyl sulfone (**89**) to dicyclohexenylethene (**90**) (Scheme 37).

Halosulfones are important intermediates in the Ramberg–Backlünd reaction (1940) which converts dialkyl sulfones into alkenes (**87**) with extrusion of sulfur dioxide.[5] The reaction requires a leaving group (X) at the α-position and an α′-hydrogen atom; it is generally carried out with α-chloro sulfones (**91**), although other halogens or sulfonates are effective (Scheme 38)

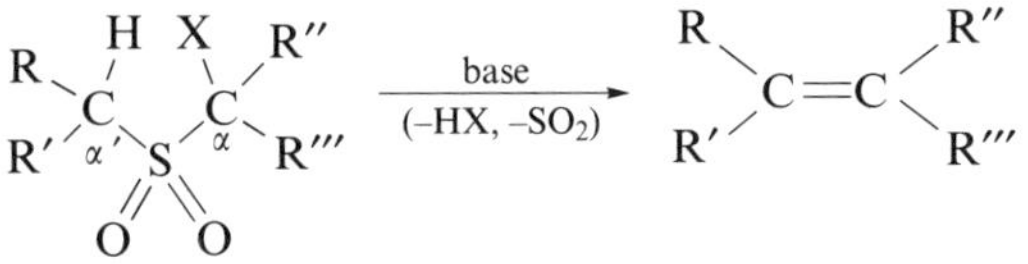

$$RR'C(H)-SO_2-C(X)R''R''' \xrightarrow[(-HX,\ -SO_2)]{\text{base}} RR'C{=}CR''R'''$$

(**91**) X = Cl (**87**)

Scheme 38

The generally accepted mechanism of the Ramberg–Backlünd reaction (Scheme 39) involves the deprotonation of the α-chlorosulfone by the base to give the sulfonyl carbanion (**92**). The latter undergoes an intramolecular nucleophilic attack on the α-carbon atom with elimination of the chloride anion and formation of the episulfone (**93**), which is unstable and extrudes sulfur dioxide to yield the alkene (Scheme 39).[6] The Ramberg–Backlünd reaction, unlike the Julia reaction, yields mainly the (*Z*)-alkene from acyclic starting materials.

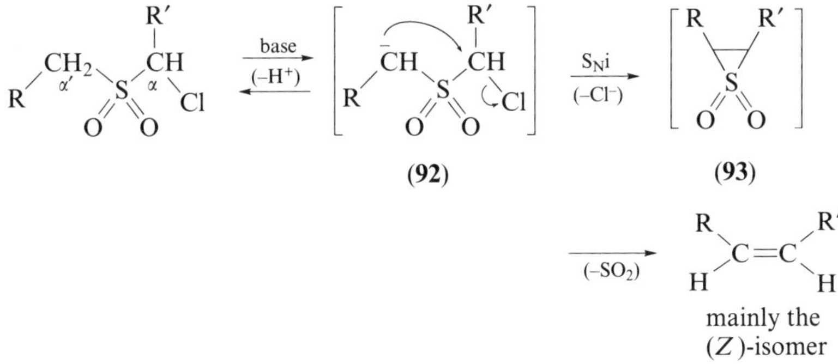

$$RCH_2-SO_2-CH(R')Cl \underset{}{\overset{\text{base}}{\underset{(-H^+)}{\rightleftharpoons}}} \left[R\bar{C}H-SO_2-CH(R')Cl\right] \xrightarrow[(-Cl^-)]{S_Ni} \left[\text{episulfone}\right] \xrightarrow{(-SO_2)} RHC{=}CHR'$$

(**92**) (**93**)

mainly the (*Z*)-isomer

Scheme 39

α-Halosulfones are usually obtained by the α-halogenation of the appropriate sulfones by treatment with carbon tetrachloride and potassium hydroxide in aqueous *t*-butanol; the reaction involves the *in situ* formation and chlorination of the derived sulfonyl carbanion.[5]

In the Ramberg–Backlünd reaction, the stereochemistry of the alkene is rather sensitive to the nature of the base used. With weak bases, e.g. dilute sodium hydroxide, there is preferential formation of the (*Z*)-alkene, but with stronger bases, e.g. potassium *t*-butoxide in DMSO, unexpectedly high yields of the (*E*)-isomer are sometimes obtained. The reaction also goes with cyclic halosulfones; thus, it may be applied to prepare the bicyclic alkenes (**94**) and (**95**) (Scheme 40). The formation of (**95**) is an example of the use of the Ramberg–Backlünd reaction for the synthesis of strained alkenes.

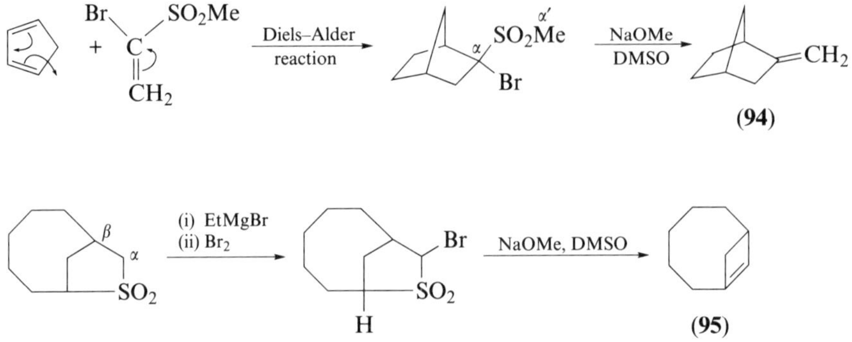

Scheme 40

SULFONYL CARBANIONS

The chemistry of sulfones is dominated by the reactions of sulfonyl carbanions. The sulfone group has a unique ability to facilitate deprotonation of attached alkyl, alkenyl and aryl groups and will permit multiple deprotonation to yield polyanions. These properties, combined with the relative intertness of the sulfone (SO_2) group to nucleophilic attack, have made the SO_2 group the first choice for stabilisation of carbanions and account for the extensive application of sulfones in synthesis. Sulfonyl carbanions can be generated and reacted under a wide variety of conditions extending from aqueous phase transfer reactions using sodium hydroxide as base to the use of alkyllithiums in polar aprotic solvents. The reactivity of sulfonyl carbanions depends on the nature of the metal counterion (Li^+, Na^+, K^+ and Mg^{2+} are the most important ones) and the presence of additives, e.g. TMEDA, HMPA and Lewis acids.

Simple sulfonyl carbanions which do not contain additional carbanion-stabilising groups, e.g. carbonyl groups or heteroatoms, can be readily alkylated in high yield by modern techniques with the use of alkyllithiums and lithium amide bases. A number of allylic halides have been successfully used. In allylic halides, the halogen directly attached to the double-bonded carbon is relatively inert towards nucleophilic attack (Scheme 41). In this way, sulfones (**96**) can be transformed via desulfonation into vinyl halides (**97**) or into ketones (**98**) by hydrolysis (Scheme 41). In contrast to ordinary alkyl sulfones, triflones (**99**) can be alkylated under mildly basic conditions (potassium carbonate in boiling acetonitrile) (Scheme 42). The ease of carbanion formation from triflones (**99**) arises from the additional electron-withdrawing (–I) effect of the trifluoromethyl moiety.

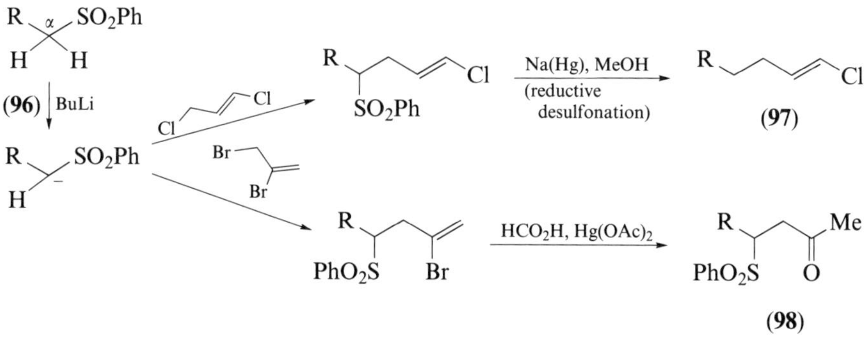

Scheme 41

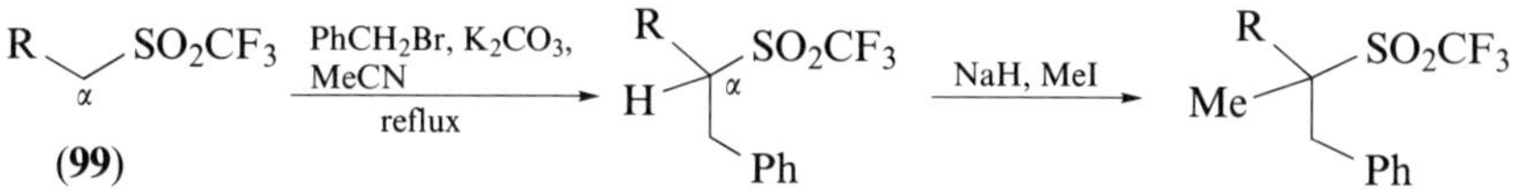

Scheme 42

Suitable ω-haloalkyl sulfones can undergo intramolecular alkylation (cyclisation), which provides a good route to cyclic sulfones; for instance, the alkyl halide (**100**) cyclises to give the cyclopropyl sulfone (**101**) (Scheme 43). Three-, Four- and five-membered rings can be formed in excellent yield by this procedure. Cyclisation to the three-membered ring is extremely rapid as compared with five- and six membered ring formation. The method can also be used in the synthesis of large rings. An illustrative example is the conversion of the phenyl sulfone (**102**) to the 14-membered cyclic sulfone (**103**) (Scheme 44).

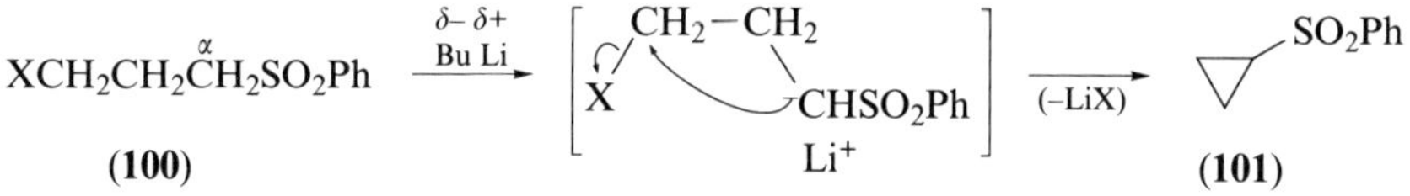

Scheme 43

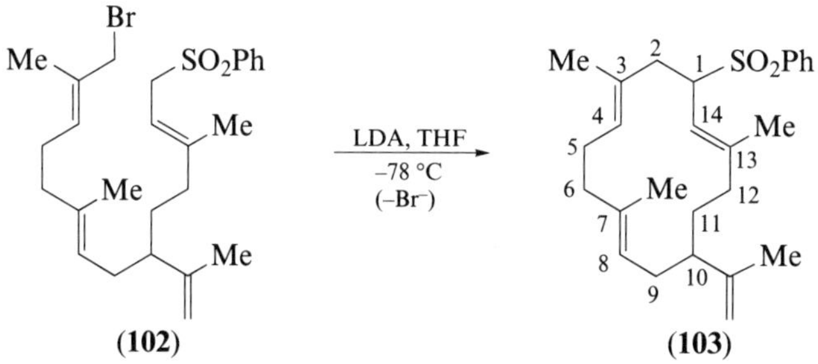

Scheme 44

Sulfonyl carbanions undergo aldol-type reactions with aldehydes and ketones; these are very important owing to the ability of the resultant *β*-hydroxysulfones to participate in elimination reactions to form alkenes (the Julia reaction, see p. 197). Early experiments used Grignard reagents to form the magnesium sulfonyl carbanions. By this procedure, methyl phenyl sulfone (**104**) is transformed into the *β*-hydroxysulfone (**105**) which subsequently can be either oxidised to the ketone (**106**) or dehydrated to the alkene (**107**) (Scheme 45).

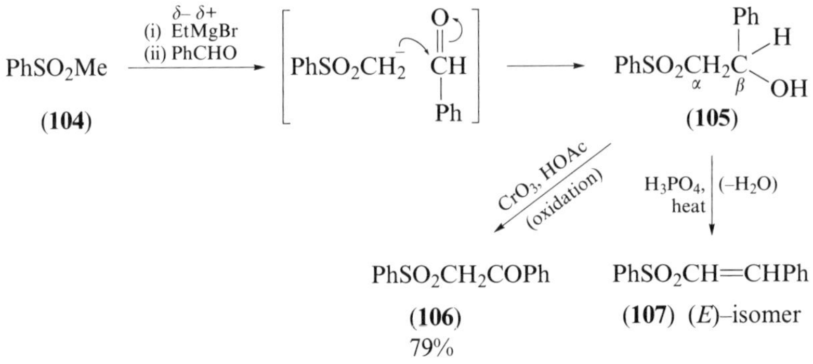

$PhSO_2Me$ (**104**) —(i) EtMgBr ($\delta- \ \delta+$), (ii) PhCHO→ [$PhSO_2CH_2^-$ ... $PhCH{=}O$] → $PhSO_2CH_2C(Ph)(H)OH$ (α, β) (**105**)

(**105**) —CrO_3, HOAc (oxidation)→ $PhSO_2CH_2COPh$ (**106**) 79%

(**105**) —H_3PO_4, heat ($-H_2O$)→ $PhSO_2CH{=}CHPh$ (**107**) (*E*)–isomer

Scheme 45

Later work has shown that Grignard reagents are not as good a base as butyllithium in THF. A recent example of *β*-hydroxysulfone formation involves an intramolecular cyclisation (Scheme 46).

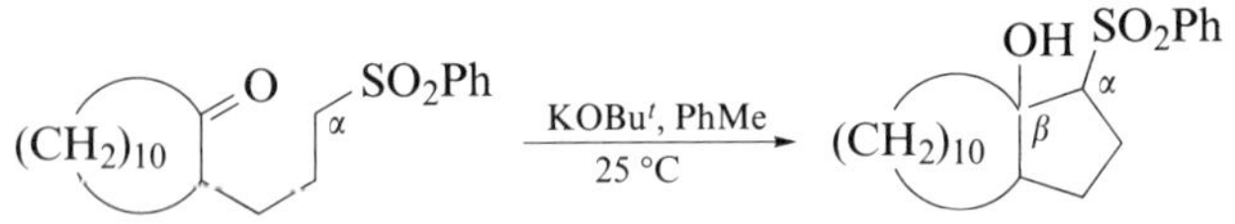

$(CH_2)_{10}$ ring ketone with SO_2Ph side chain (α) —$KOBu^t$, PhMe, 25 °C→ $(CH_2)_{10}$ fused cyclopentane with OH (β) and SO_2Ph (α)

Scheme 46

The Michael reaction of sulfonyl carbanions with *α*,*β*-unsaturated esters provides a useful preparation of cyclopropane esters, as shown in Scheme 47. In this sequence, the sulfonyl carbanion undergoes conjugate 1,4-addition to the *α*,*β*-unsaturated carbonyl compound followed by intramolecular elimination of the benzenesulfinate anion (Scheme 47).

The metallation of aromatic sulfones like diphenyl sulfone (**108**) has been known for a long time and can be used to obtain the corresponding carboxylic acid, e.g. (**109**) (Scheme 48). Interestingly, with the 2, 4, 6-trimethylphenyl sulfone (**110**) metallation occurs preferentially at the *o*-benzylic position leading to the formation of the carboxylic acid (**111**) (Scheme 48). Lithosulfones of type (**112**) on heating undergo the Truce–Smiles rearrangement to give the sulfinate, e.g. (**113**) (see Chapter 5, p. 76) (Scheme 49).

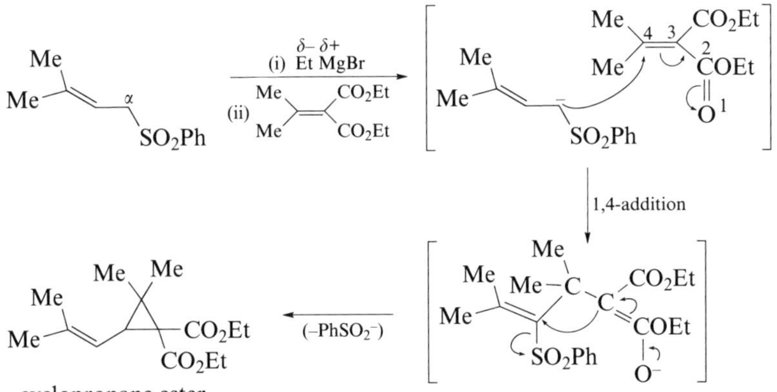

Scheme 47

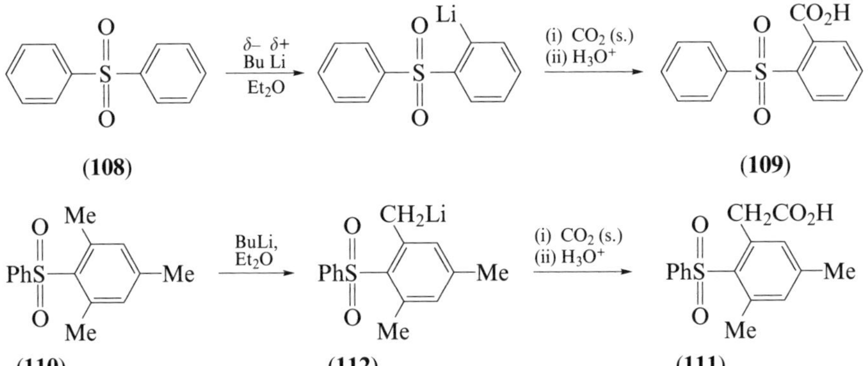

(108)

(109)

(110)

(112)

(111)

Scheme 48

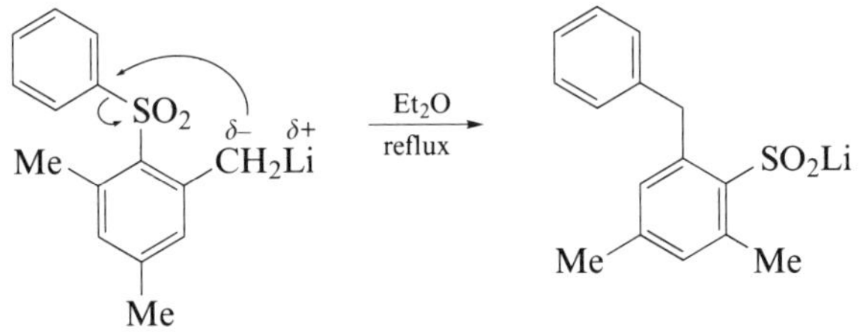

(112)

(113)

Scheme 49

The Truce–Smiles rearrangement is general for diaryl sulfones containing an *o*-methyl group and involves benzylic deprotonation to the anion which subsequently rearranges to the sulfinate. If the sulfinate contains a suitably placed leaving group, further reactions may occur; for example, the chlorophenyl sulfone

(**114**) forms the sulfinate (**115**), but this then suffers an intramolecular nucleophilic substitution with elimination of the chloride anion and formation of the heterocycle (**116**) (Scheme 50).

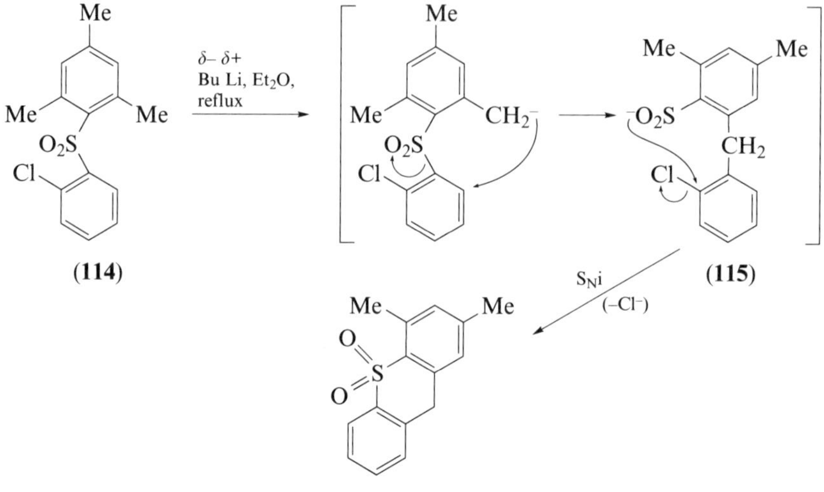

Scheme 50

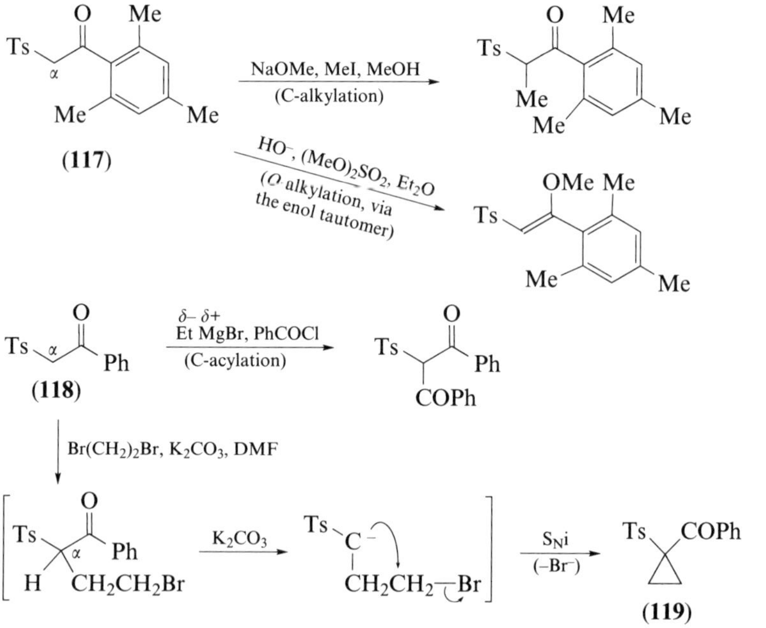

Scheme 51

Ketosulfones like (**117**) and (**118**) are readily alkylated and acylated. The alkylation may occur either at carbon or at oxygen depending on the reaction conditions (Scheme 51). Furthermore, one-pot alkylation using suitable dihaloalkanes occurs to yield cyclic products such as the cyclopropyl sulfone (**119**), as shown in Scheme 51.

Tosylmethyl isocyanide (TsMIC) (**120**) is an easily prepared sulfone with a wide repertoire of useful reactions; thus, TsMIC may be converted to ketones (**121**), alkanes (**122**) and alicyclic ketones (**123**) and (**124**), as shown in Scheme 52. The alicyclic compounds formed may be either small rings, e.g. cyclobutanone (**123**), or medium-sized rings like the 10-membered compound (**124**). In these reactions, the utility of TsMIC (**120**) is dependent on the good leaving group capacities of the tosyl and isonitrile moieties. The reagent (**120**) can also be used to synthesise heterocycles, e.g. oxazoles (**125**) (Scheme 53).

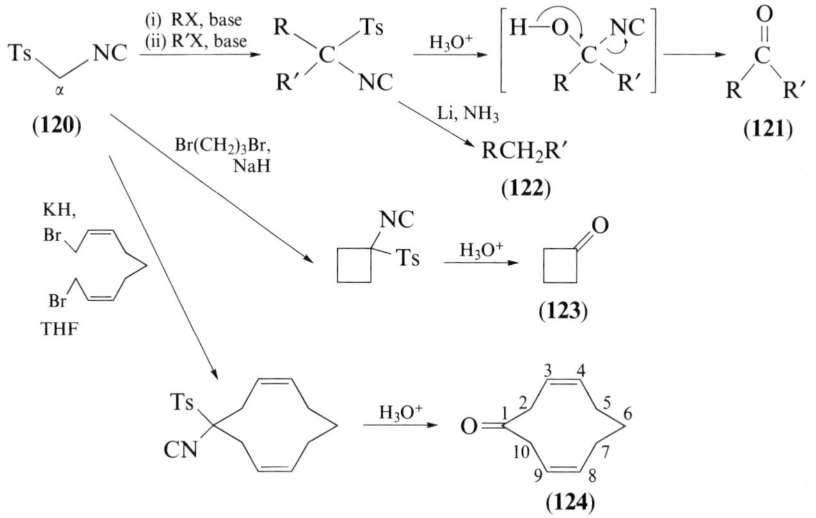

Ts NC α (**120**) (i) RX, base (ii) R′X, base R Ts C R′ NC H_3O^+ [H—O NC C R R′] O C R R′ (**121**)

Li, NH_3 RCH_2R' (**122**)

$Br(CH_2)_3Br$, NaH NC Ts H_3O^+ O (**123**)

KH, Br Br THF Ts CN H_3O^+ O 1 2 3 4 5 6 7 8 9 10 (**124**)

Scheme 52

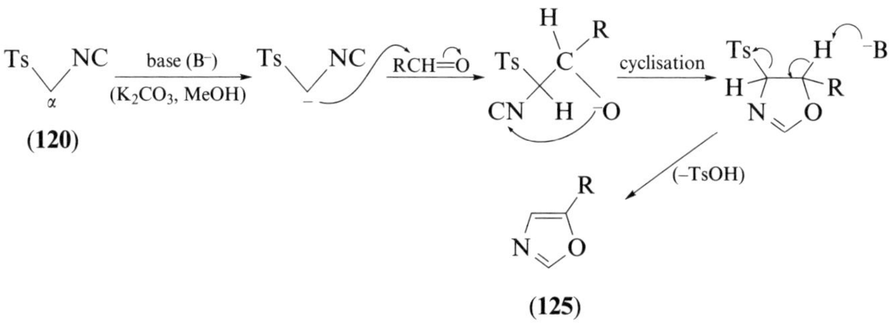

Ts NC α (**120**) base (B^-) (K_2CO_3, MeOH) Ts NC RCH=O Ts H R C CN H O cyclisation Ts H H $^-$B R N O (–TsOH) R N O (**125**)

Scheme 53

α,β-Unsaturated sulfones are valuable Michael acceptors. Generally, nitrogen, oxygen and sulfur nucleophiles will add to alkenyl, allenyl and alkynyl sulfones. For example, with thiols addition occurs with alkenyl (**126**) and alkynyl (**127**) sulfones to give the adducts (**128**), (**129**) and (**130**), respectively (Scheme 54). In view of the –I inductive effect of the sulfonyl group, the normal adduct from the alkenyl sulfone would be the compound (**129**), and, in agreement, this is generally the only significant adduct formed. However, if X is a strongly electron-withdrawing moiety such as NO_2 or SO_2Me, then some of the other adduct (**130**) is also formed as a result of the competing –I effect of the nitro or methanesulfonyl group on the polarisation of the alkynic triple bond.

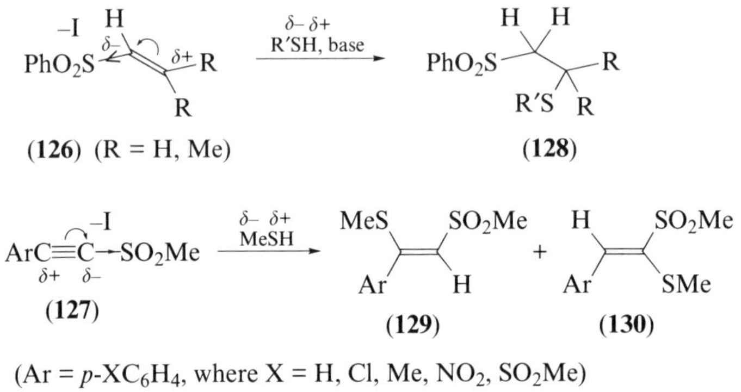

(**126**) (R = H, Me) → (**128**) [$\delta-$ $\delta+$ R′SH, base]

(**127**) → (**129**) + (**130**) [$\delta-$ $\delta+$ MeSH]

(Ar = p-XC_6H_4, where X = H, Cl, Me, NO_2, SO_2Me)

Scheme 54

The Michael addition of enolates to vinyl sulfones allows one-pot tandem reactions to be devised to achieve cyclisations as illustrated in Scheme 55. In the first reaction, the enolate (**131**) is transformed into the cyclic β-hydroxy sulfone (**132**) in good yield. In the second reaction, the enolate (**133**) yields the tricyclic compound (**134**) as a result of a double Michael reaction followed by intramolecular

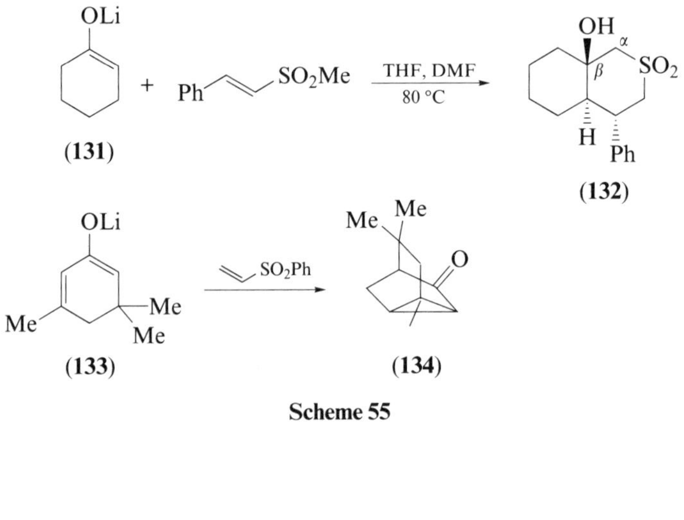

(**131**) + Ph–CH=CH–SO_2Me → (**132**) [THF, DMF, 80 °C]

(**133**) → (**134**) [CH_2=CHSO_2Ph]

Scheme 55

elimination of the benzenesulfonyl group. The Michael addition of enolates to vinyl sulfones is facilitated by the introduction of carbanion-stabilising groups at the α-vinylic position, as illustrated in Scheme 56.

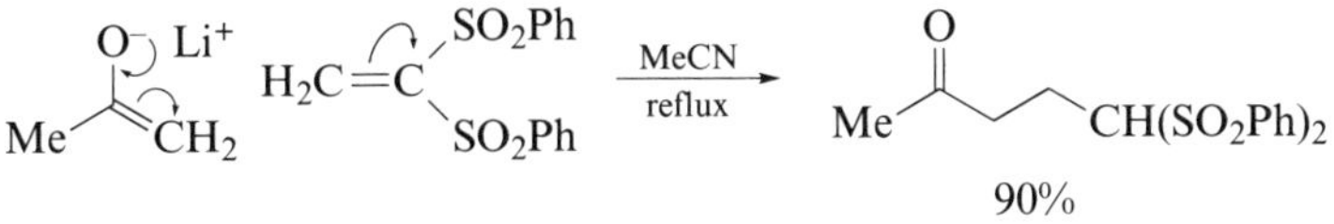

Scheme 56

CYCLOADDITION REACTIONS

Unsaturated sulfones in which the sulfone moiety is directly attached to the alkenyl or alkynyl group undergo a range of cycloaddition reactions, including [2 + 2], [3 + 2] and [4 + 2] processes.

[2 + 2] Cycloadditions

In Scheme 57, the nucleophilic enamine (**135**) adds to the strained cyclic vinyl sulfone (**136**) to give the bicyclic sulfone (**137**); a similar [2+2] cycloaddition occurs between (**138**) and the ynamine (**139**) to form the cyclic phenyl sulfone (**140**). The regiospecific reactions probably involve stepwise attack of the nucleophilic enamine (or ynamine) on the electrophilic sulfone. The four-membered ring products (**137**) and (**140**) are obtained in good yields and can be readily hydrolysed by dilute acid to form the corresponding cyclobutanones.

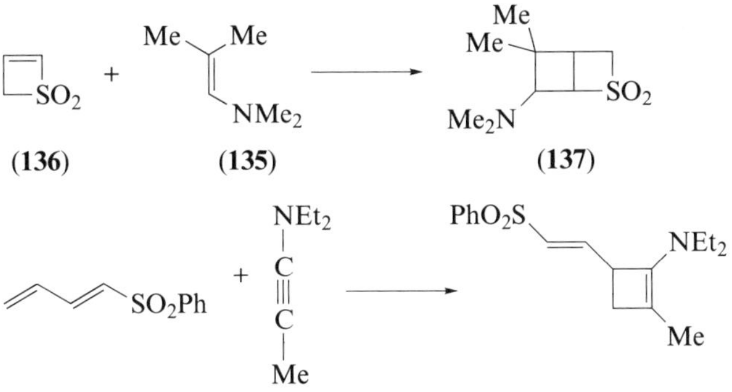

Scheme 57

[3 + 2] Cycloadditions

The regiochemistry of cycloaddition to unsaturated compounds containing incipient 1,3-dipoles (**141**), such as diazoalkanes, nitrenes and nitrile oxides, is quite dependent on the nature of the substrate and the R group in the vinyl phenyl sulfone. This is demonstrated in Scheme 58, showing how the yields of the adducts (**142**) and (**143**) vary with the nature of the reactants. In view of the electron-withdrawing (–I) character of the benzenesulfonyl moiety, the adduct (**142**) would be expected to be the major product, and this is indeed the predominant adduct in all cases except when R = H and X = O, when the adduct (**143**) predominates (Scheme 58).

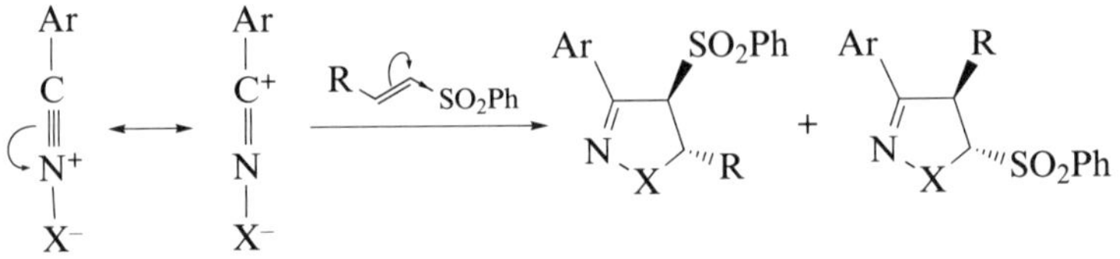

(**141**) (**142**) (**143**)

X	R	(142) Yield (%)	(143) Yield (%)
NPh	H	100	0
	Me	65	35
	Ph	90	10
O	H	9	91
	Me	90	10
	Ph	60	40

Scheme 58

In the analogous [3 + 2] cycloadditions of the more polarisable alkynyl sulfones, the trend in favour of forming the adduct (**144**), as compared with (**145**), is much more pronounced; indeed, the product (**144**) is the sole adduct obtained in the majority of cases (Scheme 59).

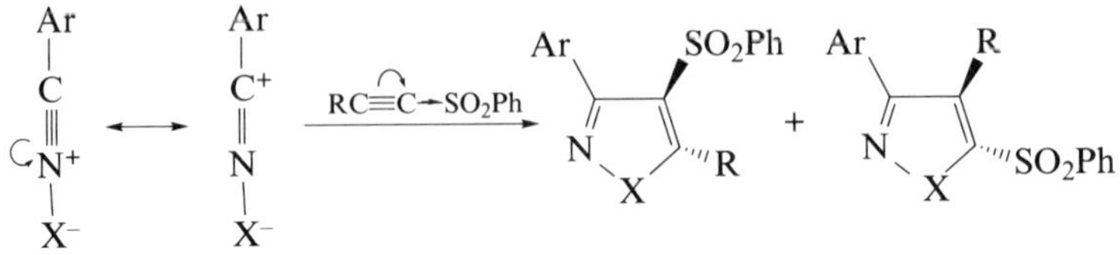

(**141**) (**144**) (**145**)

X	R	(144) Yield (%)	(145) Yield (%)
NPh	H	70	30
	Me	100	0
	Ph	100	0
O	H	80	20
	Me	100	0
	Ph	100	0

Scheme 59

[4+2] Cycloadditions

In these reactions, phenyl vinyl sulfones can be employed as 2π-components; the cycloaddition occurs smoothly with a variety of dienes and the cycloadducts may be subsequently alkylated and desulfonated, providing a useful synthetic route to bicyclic compounds. For example, cyclopentadiene (**146**) reacts with phenyl vinyl sulfone (**147**) under forcing conditions to yield the adduct (**148**), which by alkylation and desulfonation affords bicycloheptane (**149**) (Scheme 60).

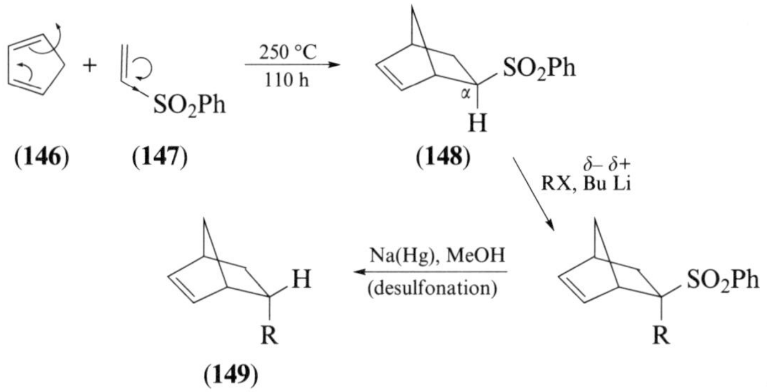

(146) + (147) —250 °C, 110 h→ (148) —RX, Bu Li→ —Na(Hg), MeOH (desulfonation)→ (149)

Scheme 60

The cycloadditions of the majority of phenyl vinyl sulfones with simple activated dienes occur between 80 °C and 150 °C. Monosubstituted sulfones usually demand more drastic conditions for cycloaddition but can give good yields of the adducts

Trifluoromethanesulfonylalkynes (**150**) and activated alkynyl sulfones (**151**) both participate in [4+2] cycloaddition reactions with dienes under relatively mild conditions. For instance, with cyclopentadiene (**146**) and butadiene (**152**) they yield the adducts (**153**) and (**154**), respectively (Scheme 61).

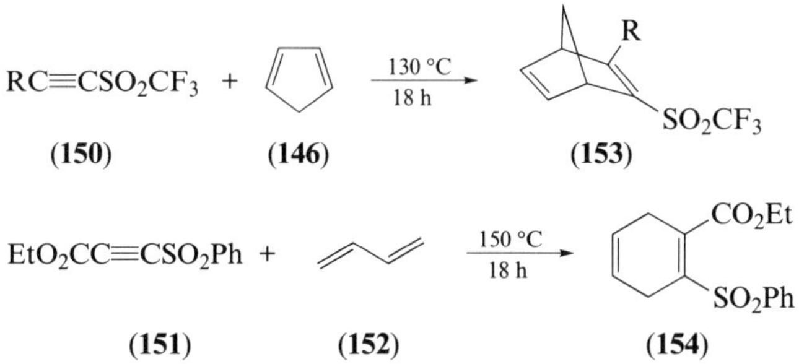

$RC{\equiv}CSO_2CF_3$ (150) + (146) —130 °C, 18 h→ (153)

$EtO_2CC{\equiv}CSO_2Ph$ (151) + (152) —150 °C, 18 h→ (154)

Scheme 61

Dienyl sulfones (**155**) can react effectively as 4π-components in [4+2] cycloadditions, even with electron-poor 2π-compounds. The dienyl sulfone (**155**) is generated *in situ* from 3,4-dibromosulfolan (**156**) via the sulfolene (**157**). The sulfone (**155**) will subsequently react with electron-poor 2π-alkenes like (**158**) (R = Ph, CN, CHO, COMe) to yield the cycloadducts (**159**) (Scheme 62). The [4+2] cycloaddition is highly stereo- and regiospecific, so that a high yield of the 1,4-*endo*-adduct (**159**) is isolated.

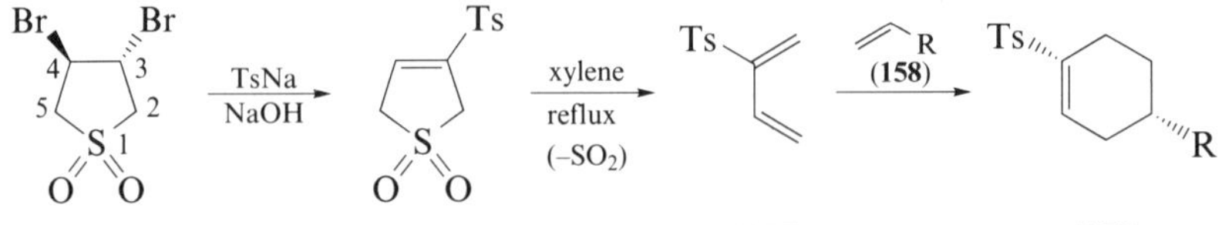

(**156**) —TsNa, NaOH→ (**157**) —xylene, reflux, ($-SO_2$)→ (**155**) —(**158**)→ (**159**)

Scheme 62

CYCLIC SULFONES

The three-, four- and five-membered cyclic sulfones have some interesting reactions. Three-membered ring sulfones, called thiirane-1,1-dioxides or episulfones, can be prepared by reaction of a diazoalkane, e.g. (**160**), with sulfur dioxide (Scheme 63). The reaction affords a mixture of the *cis*-episulfone (**161**) and *trans*-episulfone (**162**), and both isomers can be isolated by fractional crystallisation at low temperature. Another method of preparation of episulfones is by treatment of a diazoalkane with a sulfonyl chloride (containing α-hydrogen atoms) and a tertiary amine (Scheme 64). Both these syntheses involve the formation of the highly reactive sulfene intermediate (**163**) (see Chapter 7, p. 114); episulfones on warming eliminate sulfur dioxide to form the alkene (**164**) as indicated in Scheme 64.[7]

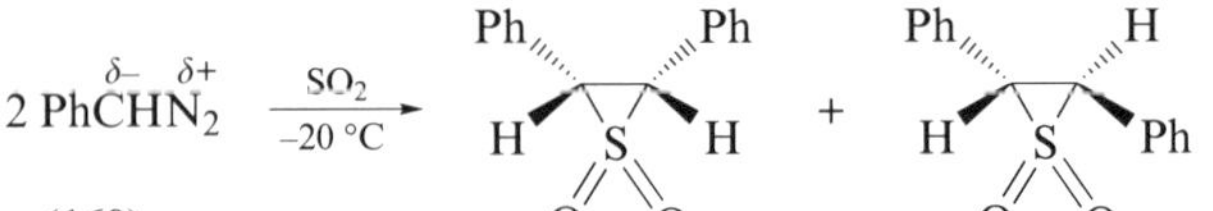

$$2\ \mathrm{Ph\overset{\delta-}{C}H\overset{\delta+}{N_2}}\ \xrightarrow[-20\ ^\circ\mathrm{C}]{SO_2}$$ (**160**) → (**161**) 45 % + (**162**) 55 %

Scheme 63

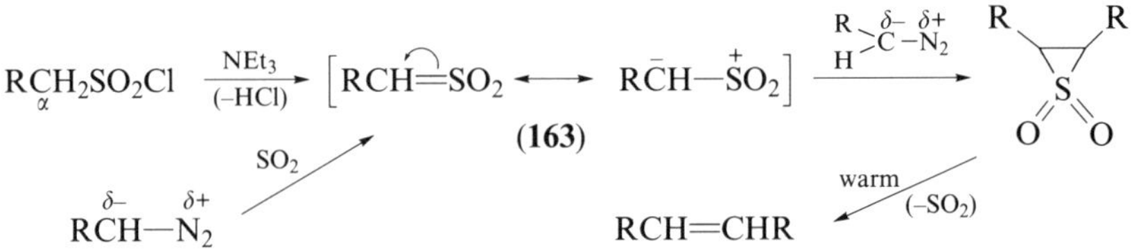

$\mathrm{RCH_2SO_2Cl}$ —NEt_3, (–HCl)→ [$\mathrm{RCH{=}SO_2} \longleftrightarrow \mathrm{R\bar{C}H{-}\overset{+}{S}O_2}$] (**163**) ← SO_2 — $\mathrm{R\overset{\delta-}{C}H{-}\overset{\delta+}{N_2}}$

(**163**) —$\mathrm{RHC{=}N_2}$→ episulfone —warm, ($-SO_2$)→ RCH=CHR (**164**)

Scheme 64

The four-membered ring sulfones, known as thietane-1,1-dioxides (**165**), are usually obtained by the cycloaddition of sulfenes (**163**) with electron-rich alkenes (Scheme 65).[7] The yields of the thietane-1,1-dioxides (**165**) are generally high, although some acyclic products may also be formed. The *trans*-sterochemistry of the alkene is retained in the product and the orientation of the aryl substituent varies with the nature of the group and the reaction conditions. The cycloaddition is probably a non-concerted reaction and occurs with other types of electron-rich alkenes such as vinyl ethers (**166**) and norbornenes (**167**) to give the four-membered sulfones (**168**) and (**169**) (Scheme 66).[8] The cycloaddition to form (**169**) is stereospecific, yielding only the *exo*-isomer owing to preferential attack by the sulfene on the least hindered *exo*-side of the enamine (**167**).

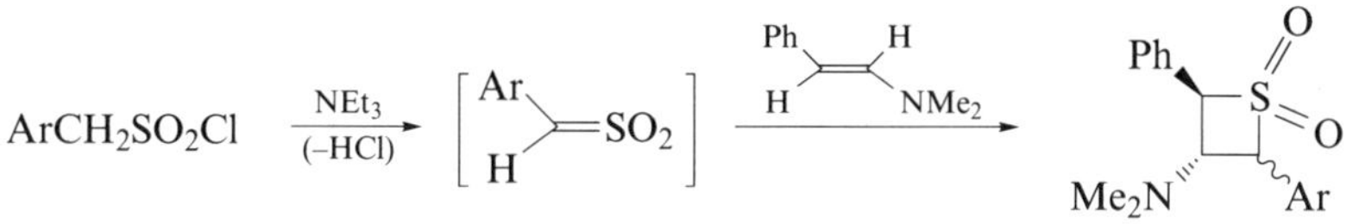

$ArCH_2SO_2Cl$ —NEt_3 (–HCl)→ [Ar(H)C=SO_2] (**163**) —Ph(H)C=C(H)NMe_2→ (**165**)

Scheme 65

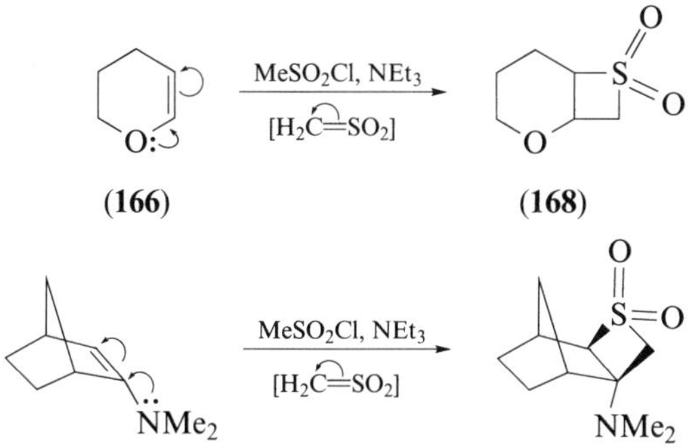

(**166**) —$MeSO_2Cl$, NEt_3 [$H_2C{=}SO_2$]→ (**168**)

(**167**) —$MeSO_2Cl$, NEt_3 [$H_2C{=}SO_2$]→ (**169**)

Scheme 66

The chemistry of thietane dioxides is dominated by extrusion of sulfur dioxide (desulfonation), which occurs on heating to yield cyclopropanes (Scheme 67). The relative amounts of the *cis*-cyclopropane and *trans*-cyclopropane are dependent on the nature of the R groups; with small groups like methyl, ethyl and so on,

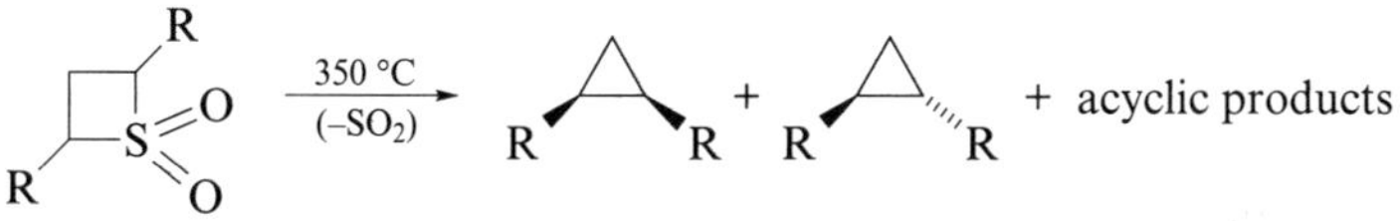

—350 °C (–SO_2)→ + acyclic products

Scheme 67

almost equal amounts of the isomers are formed, but with large groups the *trans*-isomer is favoured.

Alkylation and acylation are possible with thietane dioxides such as (**170**) and (**171**) to afford the derivatives (**172**) and (**173**), respectively (Scheme 68).

(**170**) → [(i) Bu Li ($\delta-$ $\delta+$); (ii) RX] → (**172**)

(**171**) → [(i) Bu Li ($\delta-$ $\delta+$); (ii) RCO_2Et] → (**173**)

Scheme 68

The cycloaddition of sulfenes with enamines provides a good synthetic route to thiete 1,1-dioxides (**174**) (see Chapter 7, p. 115) (Scheme 69).[8] Thiete-1,1-dioxides like (**174**) react with alkylamines by conjugate addition to yield the corresponding amino derivatives (**175**), and by heating in the presence of a reactive alkene, e.g. norbornene (**176**), they form the [2+4] cycloadducts, e.g. (**177**) (Scheme 69).

R_2N–CH=CH$_2$ → [$MeSO_2Cl$, NEt_3; $[H_2C{=}SO_2]$] → → [Bu Li ($\delta-$ $\delta+$)] → [ ] → ($-C_4H_{10}$, $-LiNR_2$) → (**174**)

(**174**) → [$R\ddot{N}H_2$] → (**175**)

(**174**) → [heat, 180–190 °C] → [ ] + (**176**) → (**177**)

Scheme 69

Five membered ring sulfones (thiolene-1,1-dioxides, sulfolenes or dihydrothiophene-1,1-dioxides) can be obtained by peracid oxidation of tetrahydrothiophene. The extrusion of sulfure dioxide from dihydrothiophene-1,1 dioxides or sulfolenes like (**178**) has been much studied as it provides a synthetic route to dienes, e.g. butadiene (**152**) (Scheme 70). The reverse reaction provides a method of synthesis of sulfolene (**178**) from butadiene (**152**) and sulfur dioxide. Thiolene dioxides or sulfolenes can be generally prepared by addition of sulfur dioxide to conjugated dienes; for example, 1,4-dimethylbutadiene (**179**) in the presence of sulfure dioxide, triethylamine and formic acid affords 2,5-dimethylsulfolene (**180**) (Scheme 71).

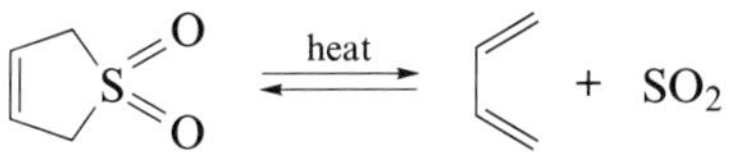

heat
+ SO_2

(178) **(152)**

Scheme 70

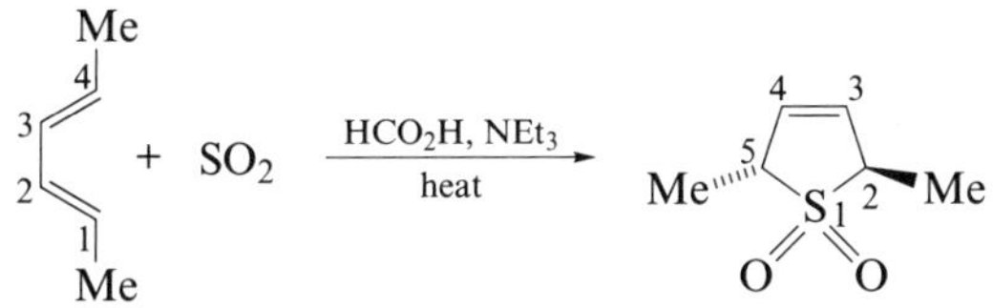

+ SO_2 HCO_2H, NEt_3 heat

(179) **(180)**

Scheme 71

Sulfur dioxide extrusion (desulfonation) from sulfolenes provides a useful procedure for generation of the diene in the presence of a dienophile, thus facilitating *in situ* Diels–Alder cycloadditions as illustrated by Scheme 72. The reaction in Scheme 72 involves the *in situ* formation of 2-cyanobutadiene (181) which then undergoes [4+2] cycloaddition with the dienophile maleimide to yield the adduct (182).

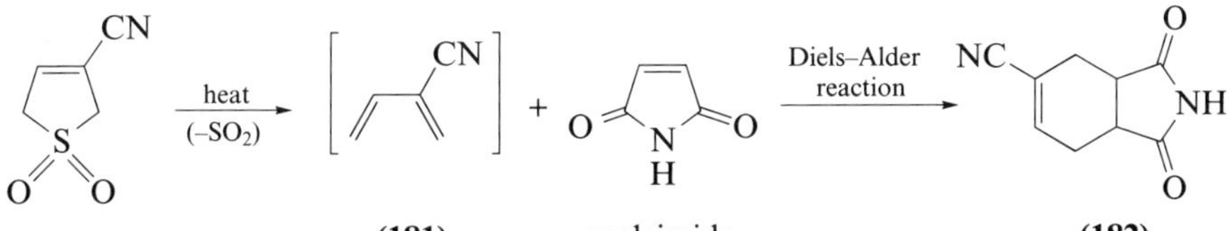

heat ($-SO_2$) + Diels–Alder reaction

(181) maleimide **(182)**

Scheme 72

In synthetic organic chemistry the sulfonyl (SO_2) moiety acts as an activating group for carbon–carbon single-bond and double-bond formation; however, the sulfonyl group is often not required in the final target molecule. Methods for the removal of the sulfonyl group known as desulfonations are consequently important procedures in organic synthesis. The majority of the common methods of desulfonation are reductive, namely the replacement of the sulfonyl group by hydrogen. To achieve this reduction, one can use one of various alkali metals, e.g. lithium or sodium, in liquid ammonia or an amine (Scheme 73). In addition, ultrasonically dispersed potassium in toluene cleaves many cyclic sulfones to give excellent yields of the acyclic sulfones after methylation of the intermediate sulfinate. In this process, cleavage of the carbon–sulfur bond occurs selectively at the more substituted carbon atom (Scheme 73). Overall, the most widely used desulfonation procedure involves treatment of the sulfonyl compound with sodium amalgam under mild conditions, often in the presence of a little disodium hydrogen phosphate. Aluminium amalgam may also sometimes be used for desulfonation.

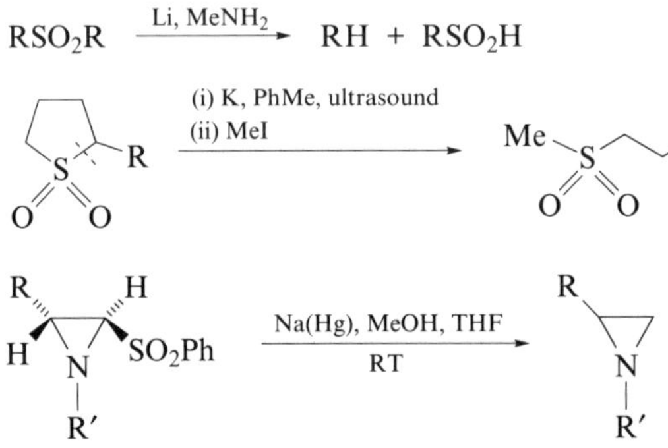

$$RSO_2R \xrightarrow{\text{Li, MeNH}_2} RH + RSO_2H$$

(i) K, PhMe, ultrasound
(ii) MeI

Na(Hg), MeOH, THF
RT

Scheme 73

Oxidative desulfonation is achieved by reaction of a sulfonyl carbanion with a source of electrophilic oxygen. One useful reagent is MoOPH, a well-known oxidant for enolates which will convert a sulfone (**183**), as the sulfonyl carbanion, directly to the ketone (**184**) (Scheme 74).

Sulfonyl carbanions of primary and secondary sulfones may also be converted to the ketones by reaction with either BTSP (**185**) (method 1) or chlorodimethoxyborane (**186**) (method 2). With the latter reagent, subsequent treatment of the intermediate boric ester with sodium perbenzoate affords the ketone (Scheme 74). The use of the sodium salt of the peroxycarboxylic acid avoids side reactions and hence increases the yield of the ketone.

Sulfur can be removed from many organosulfur compounds (desulfuration) by treatment with Raney nickel. The reagent is obtained by the action of sodium hydroxide on a nickel–aluminium alloy and the activity of the reagent varies with the method of preparation. The desulfuration of organosulfur compounds with Raney nickel has been reviewed;[9] see also Chapter 3, p. 34. The mechanism of Raney nickel desulfuration in boiling ethanol or dioxan involves hydrogenolysis of the carbon–sulfur bonds (indicated by the dashed lines), which is probably achieved by the initial formation of free radicals and their subsequent hydro-

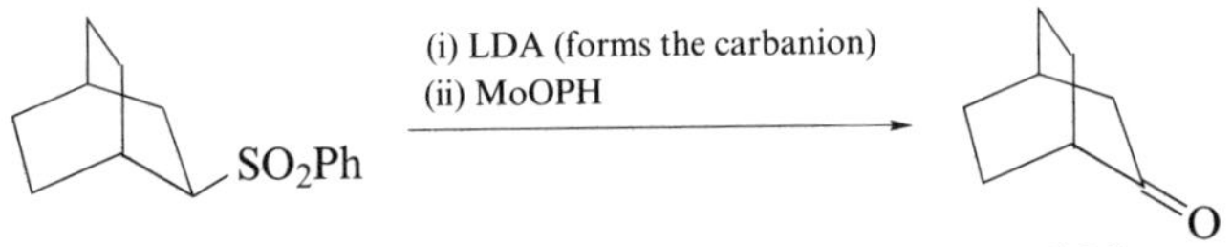

(i) LDA (forms the carbanion)
(ii) MoOPH

(183) SO_2Ph → (184) 95%

$\delta- \ \delta+$ Bu Li; BTSP (185) (method 1); (method 2) (−LiCl) $(MeO)_2B$—Cl (186) THF; $(-Me_3SiOSiMe_3)$; $PhCO_3Na$

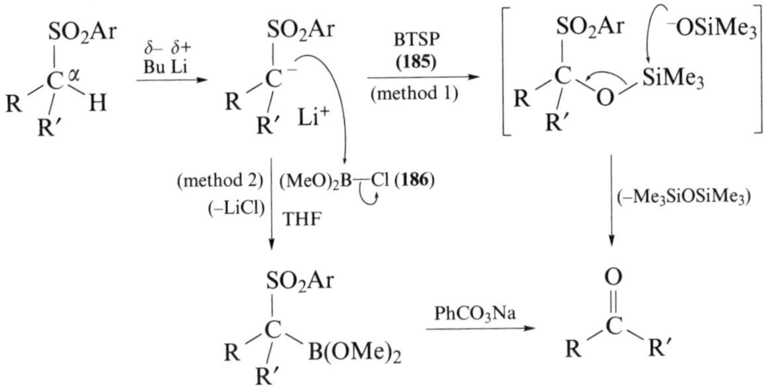

Scheme 74

genation by the activated hydrogen adsorbed on the surface of the nickel catalyst (Scheme 75). An example is provided by the desulfuration of the cyclic sulfide; (**187**) to yield 7-acetamido-8-benzamidononanoic acid (**188**) (Scheme 76).

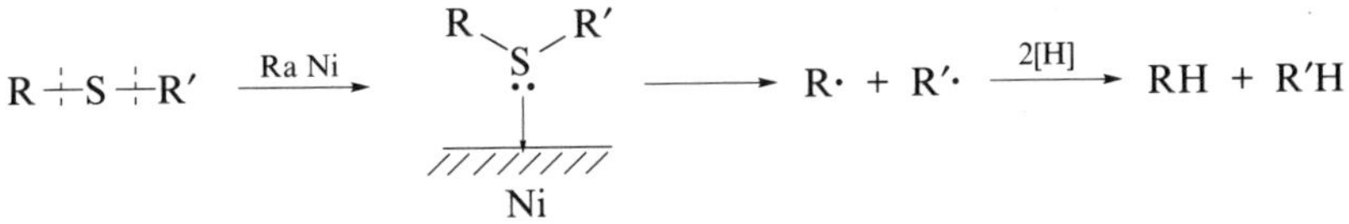

$$R-S-R' \xrightarrow{\text{Ra Ni}} R\text{-}\ddot{S}\text{-}R' \text{ (on Ni)} \longrightarrow R\cdot + R'\cdot \xrightarrow{2[H]} RH + R'H$$

Scheme 75

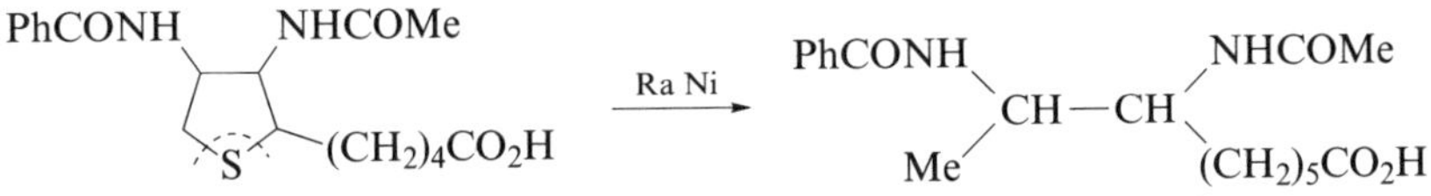

(187) PhCONH, NHCOMe, $(CH_2)_4CO_2H$ $\xrightarrow{\text{Ra Ni}}$ (188) PhCONH—CH(Me)—CH(NHCOMe)—$(CH_2)_5CO_2H$

Scheme 76

TOLUENE-*p*-SULFONYL HYDRAZONES AS SYNTHONS

Ketone toluene-*p*-sulfonylhydrazones of type (**189**) function as a convenient source of vinyllithium reagents (**190**); the latter are obtained by treatment of the hydrazones (**189**) with butyllithium (two equivalents) in a suitable solvent, e.g. ether or hexane (Scheme 77).

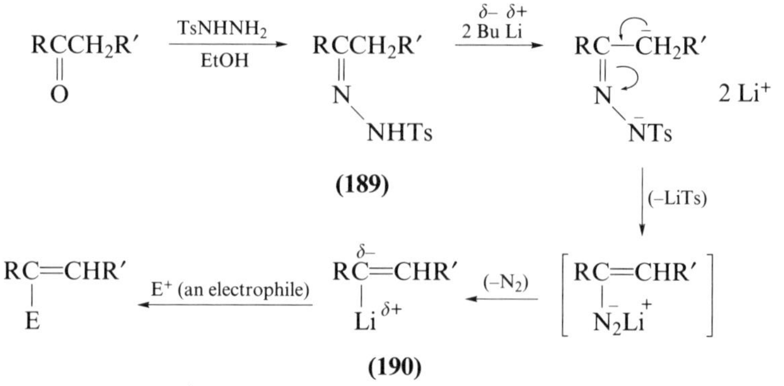

$$RC(=O)CH_2R' \xrightarrow[EtOH]{TsNHNH_2} RC(=N\text{–}NHTs)CH_2R' \ \textbf{(189)} \xrightarrow{\overset{\delta-\ \delta+}{2\ Bu\ Li}} RC(=N\text{–}\bar{N}Ts)\bar{C}HR' \ \ 2\ Li^+$$

$$\xrightarrow{(-LiTs)} \left[RC(N_2^-Li^+)=CHR' \right] \xrightarrow{(-N_2)} \overset{\delta-}{R}C(Li^{\delta+})=CHR' \ \textbf{(190)} \xrightarrow{E^+ \text{ (an electrophile)}} RC(E)=CHR'$$

Scheme 77

Tosylhydrazones (**191**) from unsymmetrical ketones in this case 2-butanone, by reaction with an alkyllithium reagent gave regioselective formation of the tosylhydrazone dianion. The latter, by subsequent treatment with acetone, affords the β-hydroxytosylhydrazone (**192**), which with more alkyllithium gives the homoallylic alcohol (**193**) (Scheme 78).

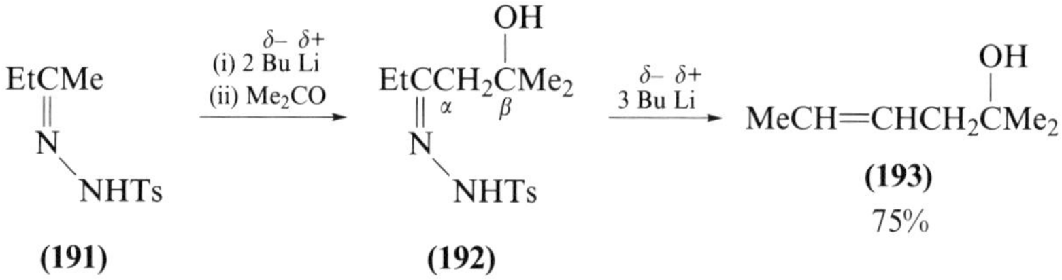

$$EtC(=N\text{–}NHTs)Me \ \textbf{(191)} \xrightarrow{\overset{\delta-\ \delta+}{\text{(i) 2 Bu Li}},\ \text{(ii) } Me_2CO} EtC(=N\text{–}NHTs)CH_2C(OH)Me_2 \ \textbf{(192)} \xrightarrow{\overset{\delta-\ \delta+}{3\ Bu\ Li}} MeCH{=}CHCH_2C(OH)Me_2 \ \textbf{(193)}\ 75\%$$

Scheme 78

When the vinyllithium intermediate (**190**) is treated with water, the procedure provides a useful synthetic method for the conversion of ketones to alkenes (Scheme 79). The method is illustrated by the conversions of the tosylhydrazones of phenyl isopropyl ketone (**194**) and dipropyl ketone (**195**) to the alkenes (**196**) and (**197**), respectively (Scheme 79). In this method, experiments have demonstrated that the hydrogen is derived from the water, as indicated in Scheme 79, and that TMEDA is an excellent solvent. The vinyllithium intermediate (**190**) may be trapped by other electrophiles; thus, with carbon dioxide and DMF, the reaction affords α,β-unsaturated carboxylic acids and aldehydes like (**198**) and (**199**) (Scheme 80).

Other bases may be employed, e.g. lithium hydride, sodium hydride, sodium amide or sodium in ethylene glycol; with sodium in ethylene glycol, the reaction is called the Bamford–Stevens reaction. Aldehyde tosylhydrazones (**200**) do not form dianions with organolithiums, but the reagent adds to the carbon–nitrogen double bond to give the dilithium derivative (**201**) which decomposes to the

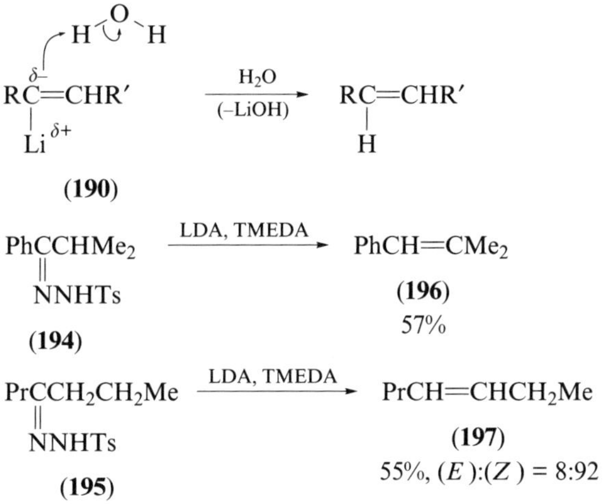

$$\underset{(190)}{R\overset{\delta-}{C}(Li^{\delta+})=CHR'} \xrightarrow[(-LiOH)]{H_2O} RC(H)=CHR'$$

$$\underset{(194)}{PhC(=NNHTs)CHMe_2} \xrightarrow{LDA,\ TMEDA} \underset{(196)\ 57\%}{PhCH=CMe_2}$$

$$\underset{(195)}{PrC(=NNHTs)CH_2CH_2Me} \xrightarrow{LDA,\ TMEDA} \underset{(197)\ 55\%,\ (E):(Z) = 8:92}{PrCH=CHCH_2Me}$$

Scheme 79

$$\underset{(190)}{R\overset{\delta-}{C}(Li^{\delta+})=CHR'} \xrightarrow[TMEDA]{O=C=O} RC(CO_2Li)=CHR' \xrightarrow{H_3O^+} \underset{(198)}{R\overset{\alpha}{C}(CO_2H)=\overset{\beta}{C}HR'}$$

$$(190) \xrightarrow[TMEDA]{DMF} \left[RC(=CHR')C(O^-)(H)NMe_2 \right] \longrightarrow \underset{(199)}{RC(CHO)=CHR'}$$

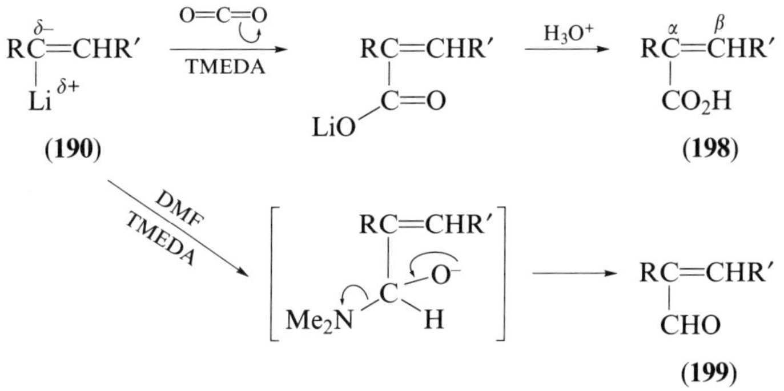

Scheme 80

organolithium compound (**202**). Compound (**202**) on aqueous work-up affords the alkane (**203**) (Scheme 81). This procedure may be used generally for the

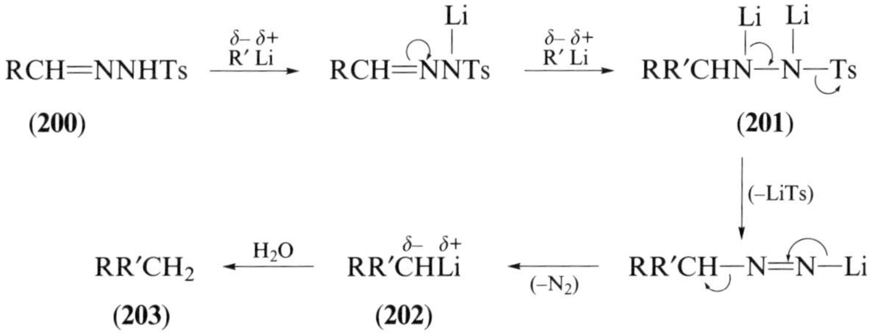

$$\underset{(200)}{RCH=NNHTs} \xrightarrow{\delta-\ \delta+\ R'Li} RCH=N(Li)NTs \xrightarrow{\delta-\ \delta+\ R'Li} \underset{(201)}{RR'CHN(Li)-N(Li)-Ts}$$

$$\xrightarrow{(-LiTs)} RR'CH-N=N-Li \xrightarrow{(-N_2)} \underset{(202)}{RR'\overset{\delta-}{C}H\overset{\delta+}{Li}} \xrightarrow{H_2O} \underset{(203)}{RR'CH_2}$$

Scheme 81 *(continued)*

Scheme 81 *(continued)*

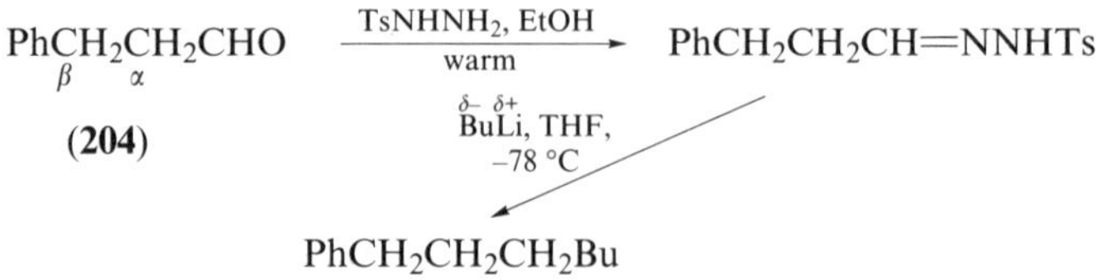

$$PhCH_2CH_2CHO \xrightarrow[\text{warm}]{TsNHNH_2,\ EtOH} PhCH_2CH_2CH{=}NNHTs$$

β α

(204)

$\overset{\delta-\ \delta+}{BuLi}$, THF, −78 °C

$$PhCH_2CH_2CH_2Bu$$

(205)

Scheme 81

synthesis of alkanes.[10] A specific example is provided by the conversion of β-phenylpropionaldehyde (**204**) to 1-phenylheptane (**205**) (Scheme 81).

References

1 B. M. Trost and L. S. Melvin, *Sulfur Ylides*, Academic Press, London, 1975.
2 *Comprehensive Organic Chemistry* (Eds D. H. R. Barton and W. D. Ollis), Vol. 3, Pergamon Press Oxford, 1979: (a) C. R. Johnson, p. 247; (b) T. Durst, p. 171.
3 N. S. Simpkins, *Sulfones in Organic Synthesis*, Pergamon Press, Oxford, 1993.
4 P. D. Magnus, *Tetrahedron*, **33**, 2019 (1977).
5 L. A. Paquette, *Org. React*. **25**, 1 (1977).
6 J. M. Clough, in *Comprehensive Organic Synthesis* (Eds B. M. Trost and I. Fleming), Vol. 3, Pergamon Press, Oxford, 1991, p. 861.
7 J. F. King, *Acc. Chem. Res*. **8**, 10 (1975).
8 *The Chemistry of Sulfones and Sulfoxides* (Eds S. Patai, Z. Rappoport and C. Sterling), Wiley, Chichester, 1988.
9 J. S. Pitzey, *Synthetic Reagents*, Vol. 2, Ellis Horwood, Chichester, 1974, Chap. 4, p. 234.
10 A. R. Chamberlain and S. H. Bloom, *Org. React.* **25**, 1 (1977).

11 USES OF ORGANOSULFUR COMPOUNDS

Many organosulfur compounds have major industrial uses. For instance, carbon disulfide and DMSO are important commercial solvents and dithiocarbamates are used in the rubber industry as vulcanisation accelerators; long chain alkanesulfonic or arenesulfonic acids are important synthetic detergents. Xanthates are used in the manufacture of rayon (see Chapter 8, p. 135) and cellophane, and many commercial dyes contain sulfonic acid groups (see Introduction, p. 5). Sulfamic acid derivatives such as saccharin (see Chapter 9, p. 162) acesulfame potassium (see Introduction, p. 5) and cyclamates (see Chapter 9, p. 162) are valuable artificial sweeteners.

A wide variety of organosulfur compounds show useful biological activity; examples include the sulfonamide antibacterials and diuretics, penicillin and cephalosporin antibiotics, the antiulcer drug ranitidine, and agrochemicals such as the thiophosphoryl (P=S) insecticides like malathion, herbicides like chlorsulfuron, many fungicides like the dithiocarbamates (see Chapter 9, p. 148) and captan (see Chapter 9, p. 151), and sulfone acaricides like tetradifon (see Chapter 7, p. 104).

In this chapter, some of these uses are explored in greater detail. Goodyear and Hancock in 1847 discovered that, when natural rubber was heated with a small amount of sulfur, the physical properties of the resultant rubber were improved; the material became tougher and more resistant to changes in temperature. This process of vulcanisation is also useful for the treatment of synthetic rubbers, and as well as sulfur, many sulfur donors such as symmetrical diphenylthiourea, tetraalkylthiuram disulfides (**1**) and 2-mercaptobenzothiazole (**2**) (Figure 1) can be used.[1] These compounds act as accelerators of the process of polymerisation of the diene monomers in synthetic rubbers; for this purpose, the additional presence of zinc oxide and preferably a carboxylic acid, e.g. stearic acid, is required.

Vulcanisation involves the crosslinking of the polymer chains by mono-, di- or polysulfide bridges rather like the disulfide crosslinks in natural proteins (see Chapter 4, p. 51). The process modifies the properties of the resultant polymer, and makes the product more suitable for the manufacture of footwear, hoses, raincoats, tyres and so on.

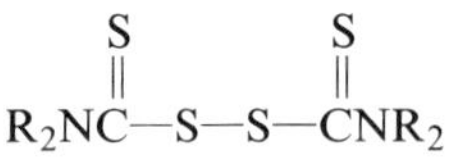

$R_2NC(=S)-S-S-C(=S)NR_2$

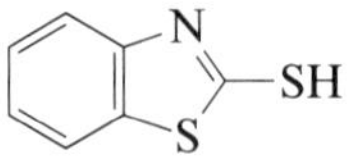

(**1**)

2-mercaptobenzothiazole (SH)

(**2**)

$MeCH(Me)CH_2CH(Me)CH_2CH(Me)CH_2CH(Me)-C_6H_4-SO_3Na$

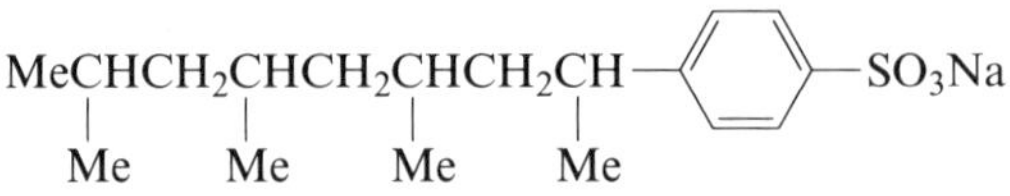

(**3a**)

$Me(CH_2)_nCH_2-C_6H_4-SO_3Na$

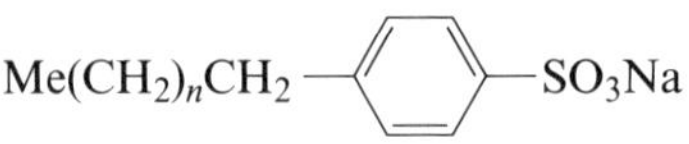

(**3b**) $n = 8–14$

$Me(CH_2)_{10}CH_2(OCH_2CH_2)_8OH$

(**4**)

Figure 1

Accelerators, e.g. zinc oxide and fatty acids, increase the rate of vulcanisation of rubber by sulfur and they reduce the amount of sulfur required from $\simeq 10\%$ to $<3\%$. Certain sulfur-donating accelerators, like thiuram disulfides (**1**) and mercaptobenzothiazole (**2**), will effect vulcanisation without added sulfur to yield products with greatly enhanced ageing properties.[1]

Some bis (sulfonyl) hydrazides and azides are used as 'blowing agents' in the plastics industry to obtain plastic foams.

Detergents

Sodium alkylbenzenesulfonates (ABSs) are important anionic surfactants (see Introduction, p. 5). The alkyl group must be highly lipophilic to dissolve oils and greases. Formerly, branched chain alkyl groups as shown in (**3a**) and based on tetrapropene were used, but these have now been replaced by linear alkyl groups to give linear alkylbenzenesulfonates (LABSs). (**3b**) (Figure 1) because linear alkyl groups are more easily degraded by microorganisms; hence, LABSs, unlike the branched chain analogues (**3a**), do not damage the environment. These compounds are the most important surfactants currently used.

Long chain alkyl sulfates (**5**) are other examples of anionic detergents; these are prepared by sulfation of long chain alkanols containing 12–18 carbon atoms, e.g. lauryl alcohol (**6**), followed by treatment with sodium hydroxide to form the sodium salt (Dreft) (**5**) (Scheme 1).

$$Me(CH_2)_{10}CH_2OH \xrightarrow[\text{(ii) NaOH}]{\text{(i) } H_2SO_4} Me(CH_2)_{10}CH_2OSO_3Na$$

(**6**) (**5**)

Scheme 1

These synthetic detergents do not form scums in hard water, unlike ordinary soaps, and are widely used in laundering, hair and rug shampoos, and dishwasher powders. Non-ionic surfactants are employed to a lesser extent; these compounds, like the polyether (**4**), do not ionise in solution, but are soluble in water owing to hydrogen bonding. They are used as wetting and emulsifying agents and by sulfation can be transformed into the corresponding anionic detergents.

Dyes[2]

Many dyes used in the clothing industry contain sulfonic acid groups which impart water solubility and help to make the dye become fast to the fabric. The dye becomes fast by attaching itself to polar sites in the fibres of, for instance, cotton, wool or silk. Many azo dyes, like Congo Red (see Introduction, p. 5), contain one or more sulfonic acid groups. Another example is Orange II (**7**), synthesised by coupling β-naphthol (**8**) with diazotised sulfanilic acid (**9**) (Scheme 2).

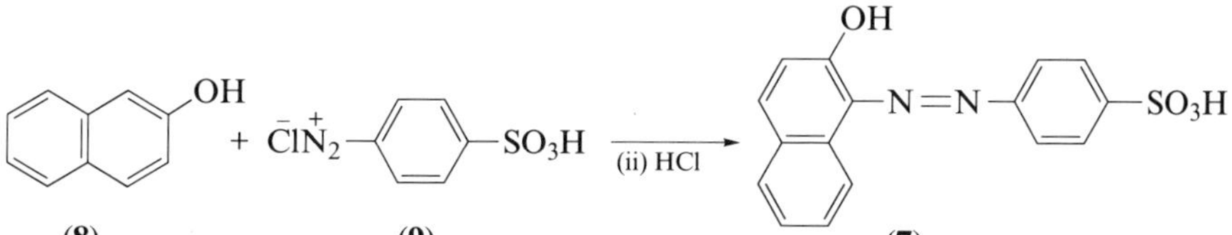

Scheme 2

The azo compounds are important chromophores because of extended electronic delocalisation between the two aromatic rings via the azo bond. The darkness of the dye is enhanced by extensive delocalisation combined with several sulfonic acid groups which function as auxochromes. An example is provided by Naphthol Blue Black B (10), prepared from 8-amino-1-hydroxynaphthalene-3,6-disulfonic acid (H-acid) (**11**) by coupling it in the 7-position with diazotised *p*-nitroaniline in acidic solution and subsequently coupling in the 2-position with diazotised aniline in alkaline solution (Scheme 3). The H-acid (**11**) is a very versatile component in dye manufacture because it can couple with diazonium salts in either the 2-position or 7-position depending on the pH of the reaction medium, as indicated in Scheme 3.

Sulfur compounds are also important examples of vat dyes—water insoluble, coloured solids which can be reduced to colourless, water soluble leuco compounds. The fabric to be dyed is dipped in the colourless leuco solution, and when subsequently exposed to the air, the leuco compound is oxidised to the coloured dye within the fabric. An example is the dibenzothioindigo (**12**) (Figure 2), a brown dye, which is introduced as the water soluble reduced form and subsequently oxidised in air to give the dye (**12**).

Sulfur dyes, another class of vat dyes, are obtained by heating various organic compounds, e.g. amines, aminophenols and nitrophenols, with sodium polysul-

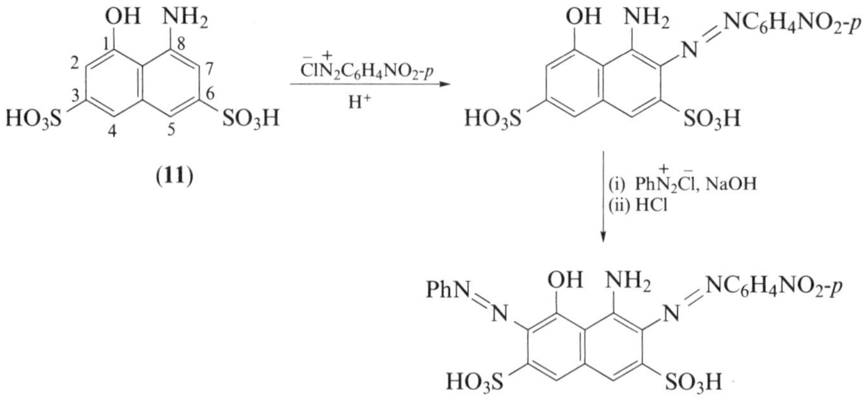

OH NH2
1 8
2 7
3 6
HO3S 4 5 SO3H

(11)

$\overset{-}{Cl}\overset{+}{N_2}C_6H_4NO_2$-*p*
H^+

OH NH2 N=NC6H4NO2-*p*
HO3S SO3H

(i) $Ph\overset{+}{N_2}\overset{-}{Cl}$, NaOH
(ii) HCl

PhN=N OH NH2 N=NC6H4NO2-*p*
HO3S SO3H

(10)

Scheme 3

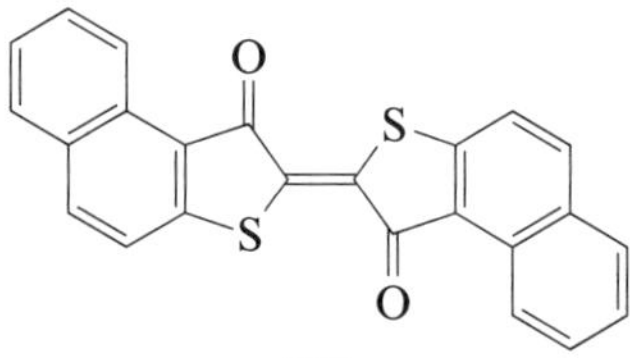

O
S
S
O

(12)

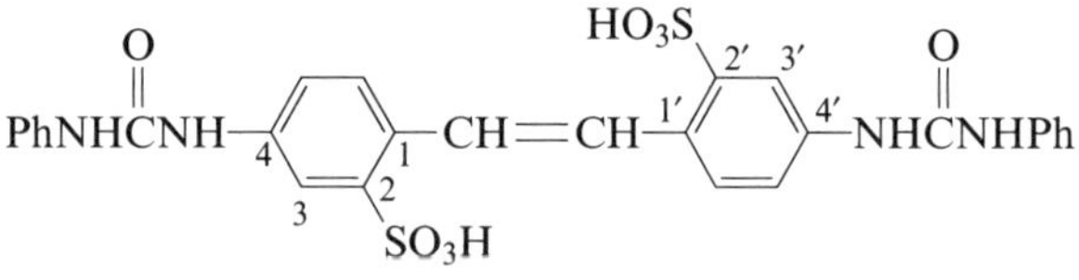

HO3S
O 2′ 3′ O
PhNHCNH— 4 1 —CH=CH— 1′ 4′ —NHCNHPh
3 2
SO3H

(13)

Figure 2

fides.[2] The coloured solids (sulfur dyes) are soluble in an alkaline solution of sodium sulfide; the fabric is dipped in this solution, and on exposure to air the solution is converted to the insoluble dye. Derivatives of 4,4′-diamino-2,2′-stilbenedisulfonic acid, such as the phenylamido derivative (**13**) (Figure 2), are fluorescent whitening agents used to enhance the appearance of white fabrics, e.g. cotton, which otherwise tend to turn yellowish with repeated washing.

Drugs

As mentioned in the Introduction (see p. 4), the sulfa antibacterial drugs were derived from the sulfonic acid dyes, and their discovery by the German doctor Domagk (1932) marks an important milestone in the development of medicinal chemistry. Modern chemotherapy started with the work of Ehrich, who in 1909 introduced the organoarsenic compound salvarsan which was successfully used for the treatment of syphilis. Ehrich observed that certain dyes selectively stained tissues and reasoned that this indicated a specific chemical interaction, and he therefore screened a large number of synthetic dyes against microorganisms. Few had any effect, but the orange-red azo dye prontosil (**14**) was showed by Domagk (1932) to inhibit the growth of *Streptococcus*. Later, Fourneau (1936) of the Pasteur Institute in Paris demonstrated that (**14**) breaks down *in vivo* to give sulfanilamide (**15**), and this was shown to be the active entity in prontosil (Scheme 4).

H_2N — N=N — S(=O)(=O) — NH_2 ; NH_2

(**14**)

metabolism *in vivo* → H_2N — S(=O)(=O) — NH_2 + H_2N — NH_2 ; NH_2

(**15**)

Scheme 4

This discovery led to the synthesis of some 15 000 sulfonamides. Antibacterial activity was demonstrated to be confined to compounds of general structure (**16**) in which both R and R′ groups may be varied, although R is usually hydrogen as shown in compound (**17**).[3,4a] Such sulfonamides are synthesised by condensation of the arenesulfonyl chloride with the appropriate amino compound (see Chapter 7, p. 104) The preparative route is illustrated by the synthesis of the general structures (**16**) and (**17**) (Scheme 5). Specifically, the synthesis is shown by the preparation of sulfapyridine (**18**) and sulfadiazine (**19**) from acetanilide, involving condensation of the N^4-acetamidobenzenesulfonyl chloride with 2 aminopyridine or 2-aminopyrimidine, respectively (Scheme 5). The final stage depends on the greater susceptibility of the acetamido group to hydrolysis as compared with the sulfonamido moiety, so the former can be selectively hydrolysed by rapid treatment with hydrochloric acid.

Most of the sulfa drugs possess the general structure (**17**), and many of the most useful compounds contain a heterocyclic nucleus. Important examples include sulfapyridine (**18**), sulfadiazine (**19**), sulfathiazole (**20**) and sulfamethoxazole (**21**) (Figure 3). Sulfapyridine (**18**) is effective against pneumonia, (**19**) against malaria, (**20**) against bacterial infections (colitis), and may be used instead of corticosteroids, and (**21**) against urinary infections. However, since the introduction of antibiotics, the use of the antibacterial sulfa drugs has markedly declined.

In 1940, Woods observed that the antibacterial action of sulfanilamide (15) was reversed by the addition of *p*-aminobenzoic acid (PABA). Sulfanilamide and

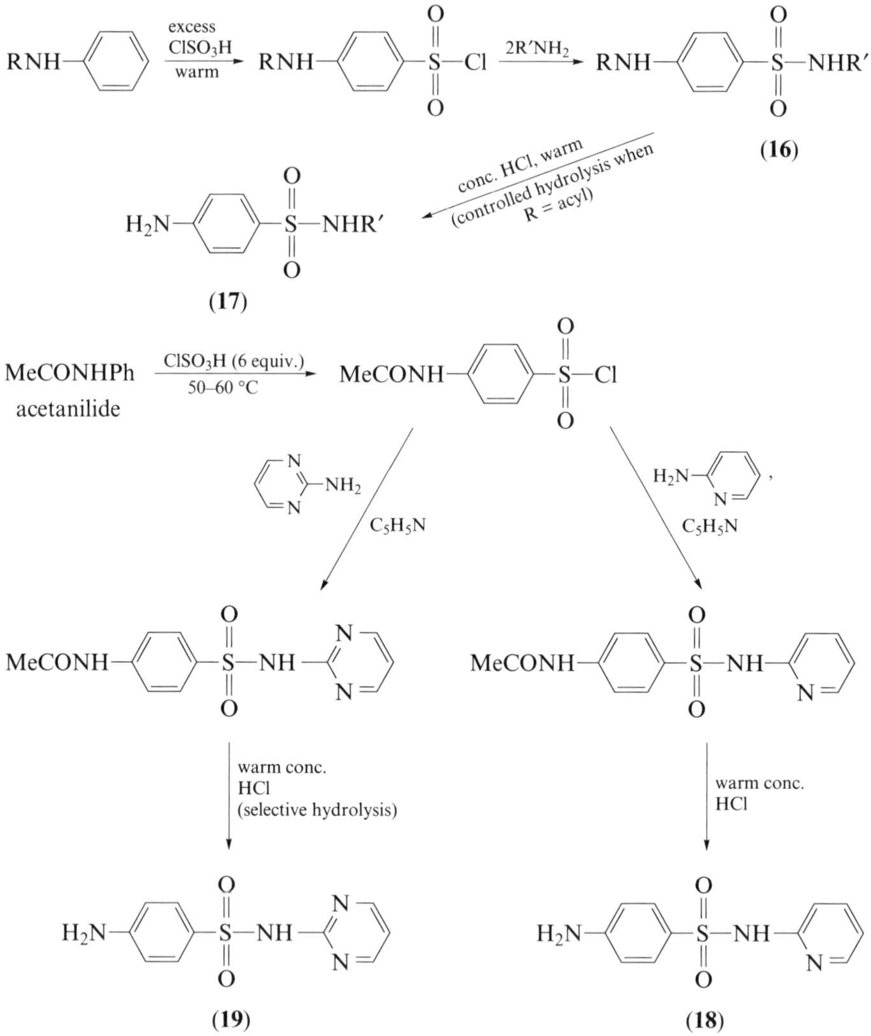

RNH–C_6H_4–H $\xrightarrow[\text{warm}]{\text{excess } ClSO_3H}$ RNH–C_6H_4–SO_2–Cl $\xrightarrow{2R'NH_2}$ RNH–C_6H_4–SO_2–NHR′ (**16**)

(**16**) $\xrightarrow{\text{conc. HCl, warm (controlled hydrolysis when R = acyl)}}$ H_2N–C_6H_4–SO_2–NHR′ (**17**)

MeCONHPh (acetanilide) $\xrightarrow[\text{50–60 °C}]{ClSO_3H \text{ (6 equiv.)}}$ MeCONH–C_6H_4–SO_2–Cl

MeCONH–C_6H_4–SO_2–Cl + 2-aminopyrimidine (NH_2), C_5H_5N → MeCONH–C_6H_4–SO_2–NH–(2-pyrimidinyl) $\xrightarrow{\text{warm conc. HCl (selective hydrolysis)}}$ H_2N–C_6H_4–SO_2–NH–(2-pyrimidinyl) (**19**)

MeCONH–C_6H_4–SO_2–Cl + 2-aminopyridine (H_2N–), C_5H_5N → MeCONH–C_6H_4–SO_2–NH–(2-pyridyl) $\xrightarrow{\text{warm conc. HCl}}$ H_2N–C_6H_4–SO_2–NH–(2-pyridyl) (**18**)

Scheme 5

other sulfa drugs are structurally similar to PABA, in particular with respect to the distance between the amino moiety and the sulfonyl group (6.9 Å as compared with 6.7Å in PABA) (Figure 3). The sulfa drugs are therefore considered to act as PABA mimics. PABA is an essential nutrient for many bacteria which utilise it for the biosynthesis of the vitamin folic acid; consequently, sulfa drugs are antimetabolites for those bacteria that require PABA.

The bacterial enzymes are not able to distinguish between sulfanilamide (**15**) and PABA, and thus the bacterial enzymes are poisoned in the presence of the drug. The microorganism is consequently deprived of the essential folic acid and

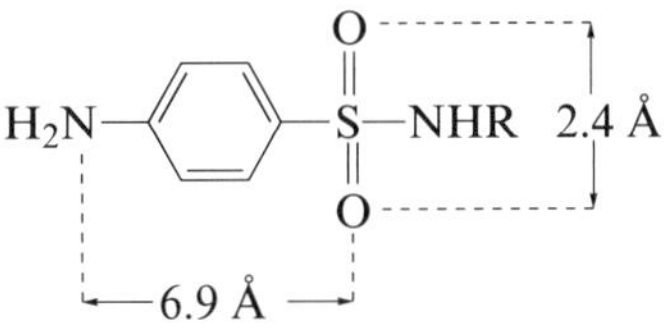

(17)

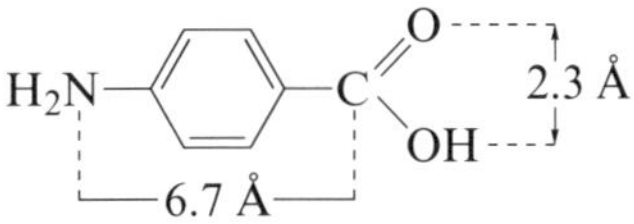

PABA

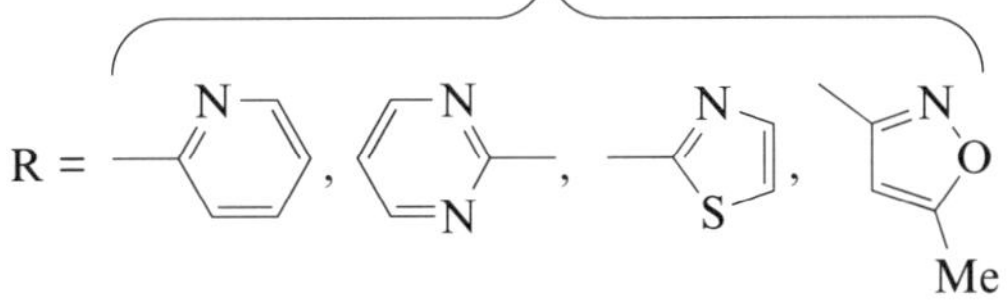

(18) (19) (20) (21)

Figure 3

eventually dies. Mammals are not affected by sulfanilamide, because they derive their folic acid from the diet and not from PABA. Sulfonamides are not only antibacterials; different members of the class differ widely in their pharmokinetic characteristics. Later studies showed that various compounds possessed othertypes of therapeutic properties.[4a] The 'chemical tree' of sulfonamide drugs (Figure 4) thus includes several valuable diuretics, e.g. acetazolamide (**22**) and benzothiadiazines (**23a**), which act by inhibition of carbonic anhydrase and are used in the treatment of hypertension; (**22**) is also used in the treatment of the eye disease glaucoma; the *N*-methylsulfonamide (**23b**) is an antimigraine drug. In addition, the sulfonylurea tolbutamide (**24**) is an oral hypoglycaemic agent useful in the treatment of non-insulin dependent diabetes; probencid (**25**) is a uricosuric agent; sulfonylbisbenzamines, e.g. dapsone (**26**), are effective against

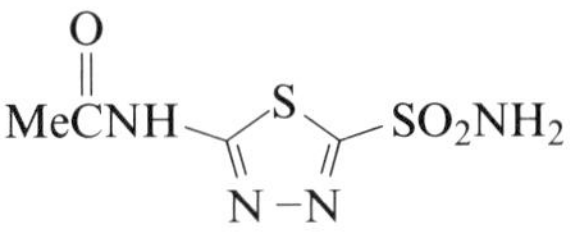

(22)

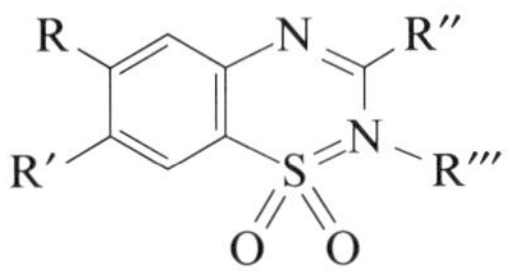

(23a)

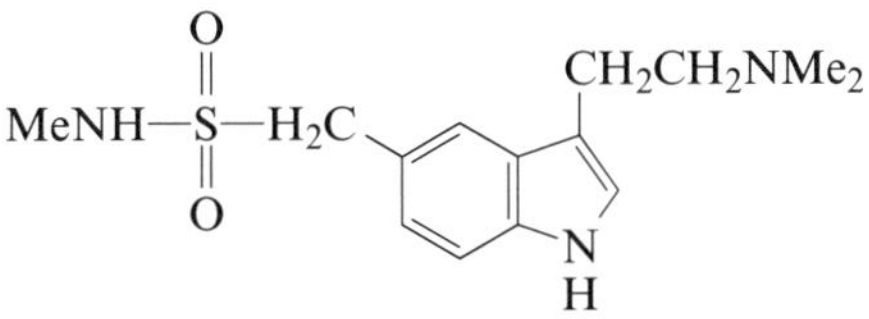

(23b)

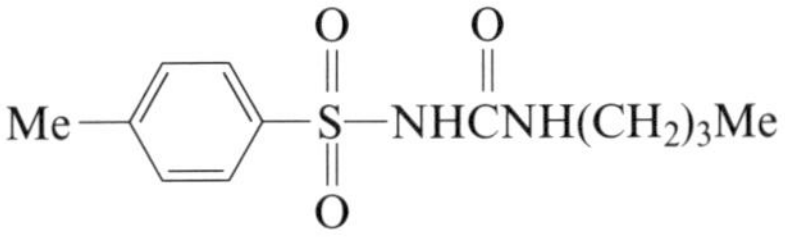

(24)

Figure 4 *(continued)*

Figure 4 *(continued)*

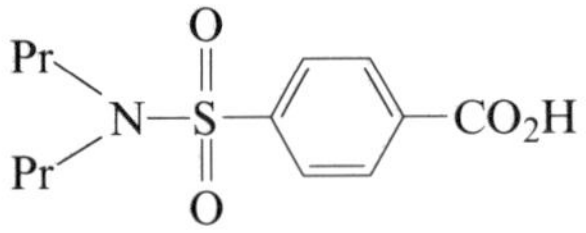

(25) 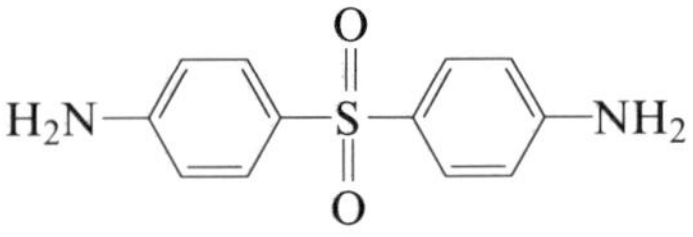 **(26)**

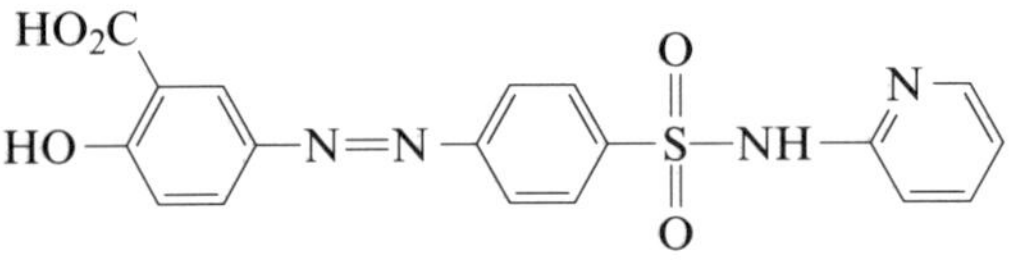

(27) 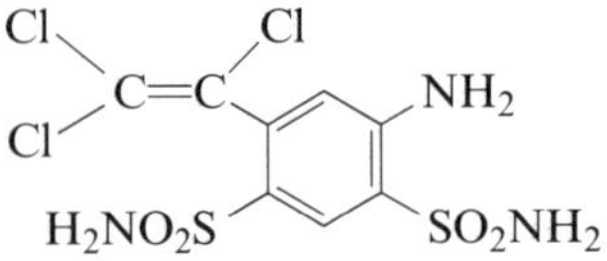 **(28)**

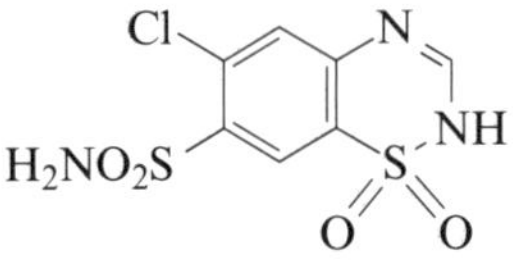

(29)

Figure 4

malaria and leprosy; salazopyrin (**27**) is used for the treatment of ulcerative colitis; and the bissulfonamide (**28**) is active against flukes, lungworms, roundworms and lice. Chlorothiazide (**29**) is a more powerful diuretic than acetazolamide (**22**) and it acts by a different biochemical mechanism, namely the inhibition of the transport of sodium and chloride ions.

N-Chloro- and *N*,*N*-dichlorosulfonamides are widely employed as antiseptics (see Chapter 7, p. 110). Sulfonamides like sulfanilamide (**15**) also exhibit systemic antifungal properties against rust diseases on cereals, but have not been commercially exploited owing to poor mobility within the plant and the danger of phytotoxicity to the crop.[5]

β-Lactam antibiotics[6]

This group of compounds comprises penicillins, cephalosporins and monobactams, all of which contain the *β*-lactam ring (**30**) which is derived from *β*-amino acids (**31**) and is highly reactive owing to steric strain (Scheme 6).

Penicillin (or benzylpenicillin) (**32a**) (R = CH_2Ph) was first dicovered by Sir Alexander Fleming in 1929, who isolated the antibiotic from the fungal strain *Penicillium notatum* (see Introduction, p. 4). Penicillin is valuable to combat bacterial infections in man and animals. It was first manufactured commercially in 1945 and the first semisynthetic penicillins were introduced in 1954. The penicillins (**32**) have a common structural feature, namely a *β*-lactam ring fused to the

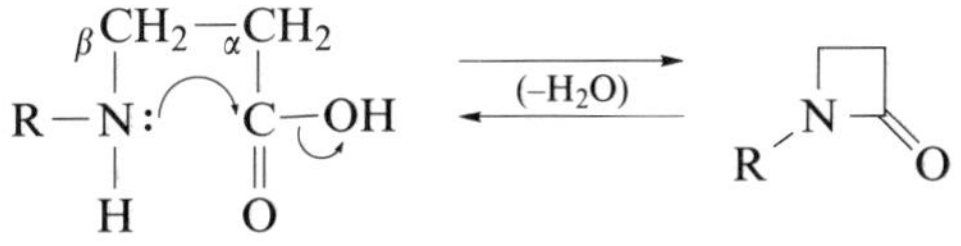

$R{-}NH{-}{}_{\beta}CH_2{-}{}_{\alpha}CH_2{-}C(=O){-}OH \underset{}{\overset{(-H_2O)}{\rightleftharpoons}}$ R–N(β-lactam)=O

(**31**) (**30**)

Scheme 6

five-membered thiazolidine ring; they are thus derivatives of 6-aminopenicillanic acid (6-APA) in which a hydrogen atom is replaced by RCO. In penicillins (**32**) the R group is variable, and some important examples are shown in Figure 5.

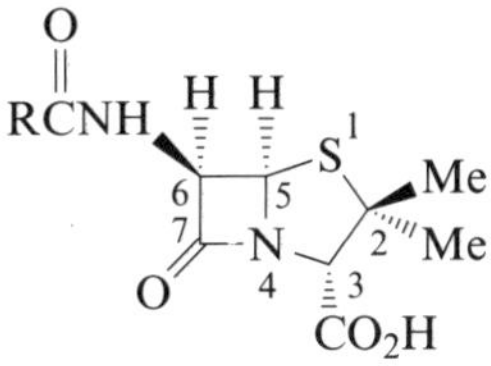

RC(=O)NH, H H, S 1, Me, Me, 2, 3, 4, 5, 6, 7, N, O, CO_2H

(**32**)

	R	Name
(**32a**)	$PhCH_2$	penicillin G
(**32b**)	D-PhCH(NH_2)	ampicillin
(**32c**)	$PhOCH_2$	penicillin V
(**32d**)	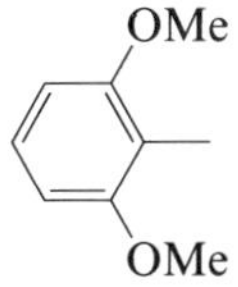 2,6-(OMe)$_2C_6H_3$ (OMe, OMe)	methicillin

Figure 5

The biosynthesis of penicillins has been extensively studied, and the general biosynthetic pathway from α-aminoadipic acid (**33**), cysteine (**34**) and valine (**35**) is shown in Scheme 7. Enzymic removal of the side chain of the product (**36**) affords 6-APA which may then be chemically modified to form the other semisynthetic penicillins in Figure 5.

In cephalosporins,[6] the β-lactam ring is fused to a six-membered dihydrothiazine ring; the first members were isolated from the marine fungus *Cephalosporium acremonium*. To date, however, all the clinically useful cephalosporins are semisynthetic compounds; the formulae (**37a**)–(**37c**) (Figure 6) are some representative examples.

The cephalosporins[7] have a wider spectrum of antibacterial activity than the penicillins and some are active against organisms that are resistant to penicillin.

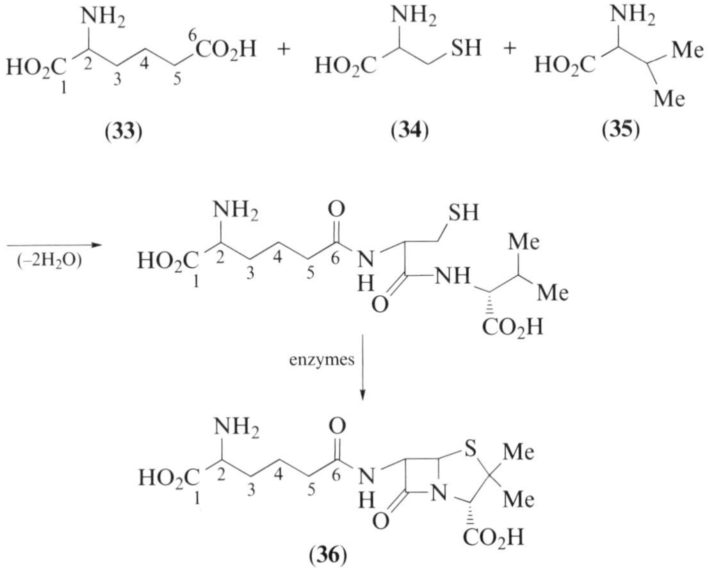

(33) (34) (35)

(36)

Scheme 7

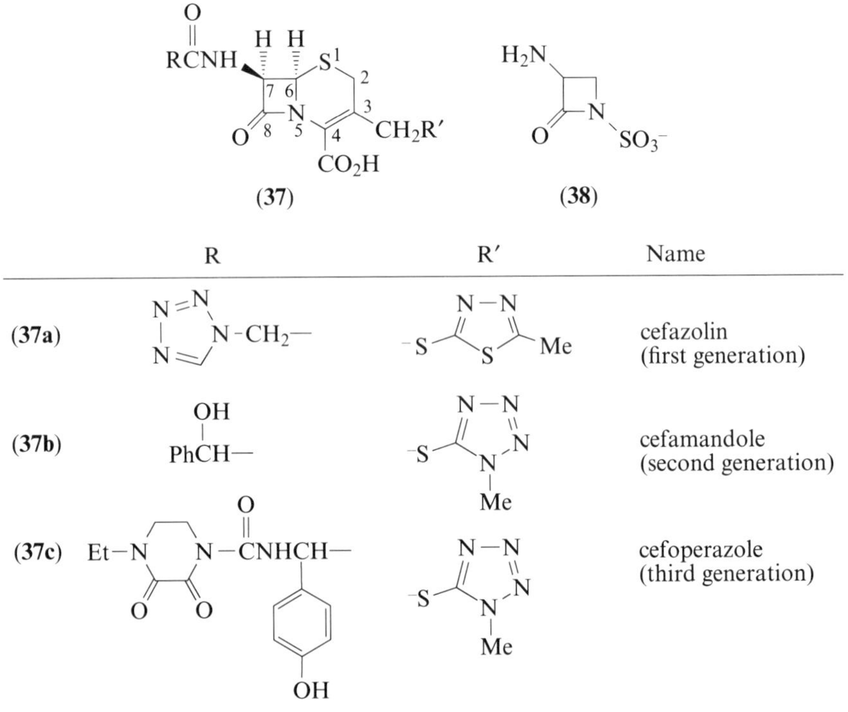

(37) (38)

	R	R′	Name
(37a)	(tetrazol-1-yl)$-CH_2-$	$^{-}S-$(5-methyl-1,3,4-thiadiazol-2-yl)	cefazolin (first generation)
(37b)	$PhCH(OH)-$	$^{-}S-$(1-methyltetrazol-5-yl)	cefamandole (second generation)
(37c)	Et−N(2,3-dioxopiperazin-1-yl)−C(O)NHCH(4-hydroxyphenyl)−	$^{-}S-$(1-methyltetrazol-5-yl)	cefoperazole (third generation)

Figure 6

They are derivatives of 7-aminocephalosporanic acid (**37**) in which the RCO group is replaced by hydrogen and R′ by the acetoxy group. Penicillins and cephalosporins are interconvertible as indicated in Scheme 8, in which the sulfoxide–sulfenic acid rearrangement (see Chapter 4, p. 52) has been applied in the preparation of the pharmaceutically important cephalosporin (**39**) from the semisynthetic penicillin (32).

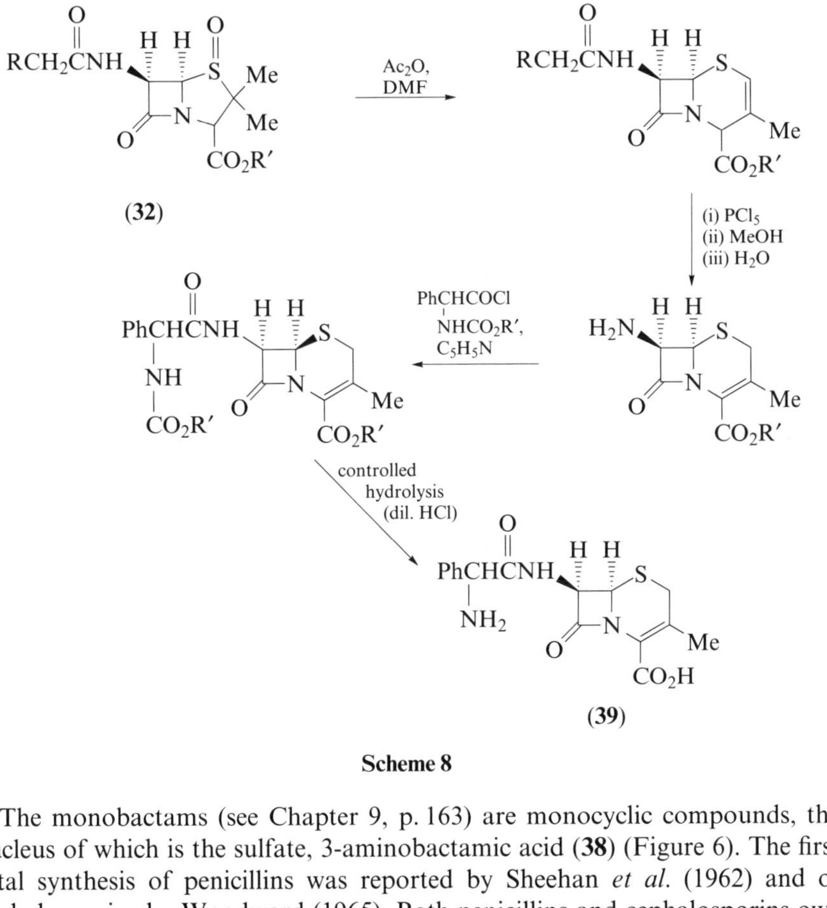

Scheme 8

The monobactams (see Chapter 9, p. 163) are monocyclic compounds, the nucleus of which is the sulfate, 3-aminobactamic acid (**38**) (Figure 6). The first total synthesis of penicillins was reported by Sheehan *et al.* (1962) and of cephalosporins by Woodward (1965). Both penicillins and cephalosporins owe their antibacterial properties to their ability to interfere with the construction of the bacterial cell wall; for example, the biosynthesis of the cell wall is inhibited by the interaction of the antibiotic penicillin (**32**) with an amino group on an essential enzyme (Ez), giving the poisoned penicillinoyl enzyme (**40**) (Scheme 9). The enzyme is thus inactivated by nucleophilic ring opening of the β-lactam, so the biological activity of the β-lactam antibiotics ultimately depends on the high degree of steric strain in the β-lactam ring which makes it susceptible to nucleophilic ring opening by the enzyme. The poisoning of the essential enzyme

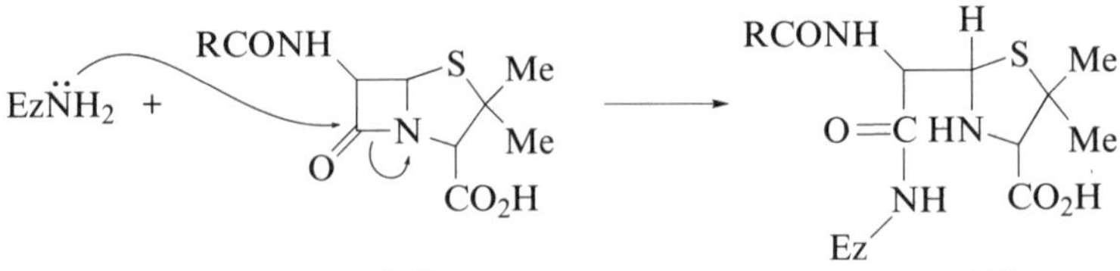

(32) (40)

Scheme 9

destroys the integrity of the bacterial cell wall so that vital cellular contents leak out and the bacteria are eventually killed.

In addition to the sulfonamides and the β-lactam antibiotics, many other types of organosulfur compounds exhibit valuable pharmacological properties, and some important examples are given in Figure 7. These include the thiols captopril (**41**), an antihypertensive, and the antirheumatic drug D-penicillamine (**42**); the thioethers cimetidine (**43**) and ranitidine (**44**) which are very important antiulcer drugs; bithional (**45**), a bactericide, and the corresponding sulfoxide used against liver flukes; and the disulfide disulfiram (**46**) used in the treatment of chronic alcoholism. Thiones are represented by thiopentone (**47**) used in the form of the sodium salt as an intravenous anaesthetic; ethaniamide (**48**) used for the treatment of leprosy and tuberculosis; and quazepam (**49**) used as a sedative and hypnotic. Malathion (**50**) controls head and pubic lice and is also an important agricultural insecticide; methimidazole (**51**) is an antithyroid agent; and thiocarbandin (**52**) is an antitubercular drug. In addition, various heterocyclic organosulfur compounds are useful medicinal agents; for example, certain phenothiazines like chlorpromazine (**53**) are antipsychotic drugs and are also useful anthelmintics used in veterinary medicine. Thiabendazole (**54**) is another anthelmintic which is also an effective agricultural fungicide; and chlormethiazole (**55**) is a valuable hypnotic drug.

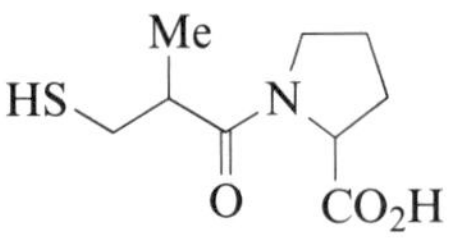

(41) 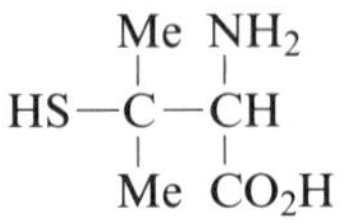 (42) 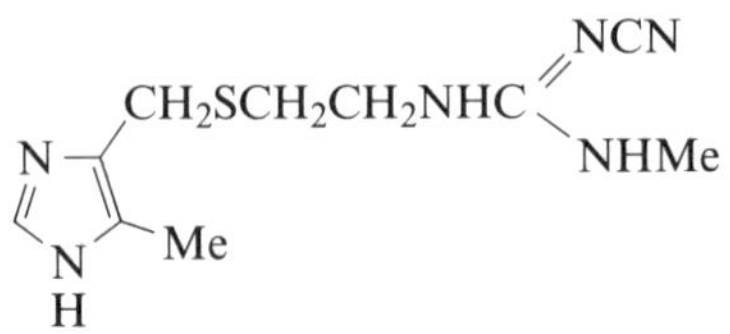 (43)

(44) 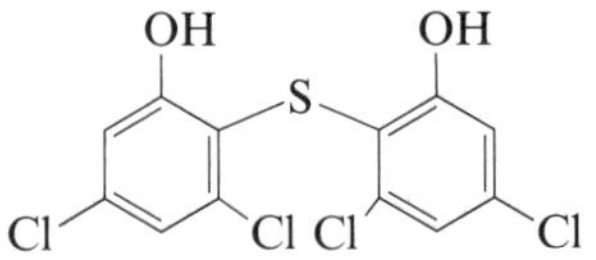 (45)

Figure 7 *(continued)*

Figure 7 *(continued)*

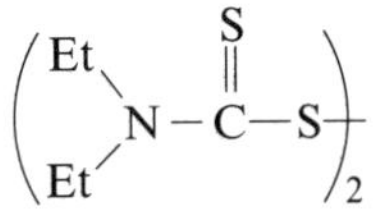

(46) 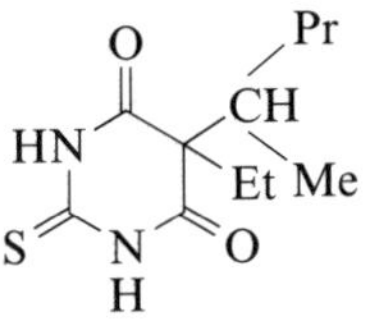 (47) 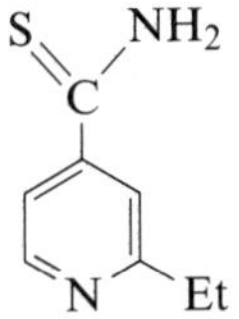 (48) 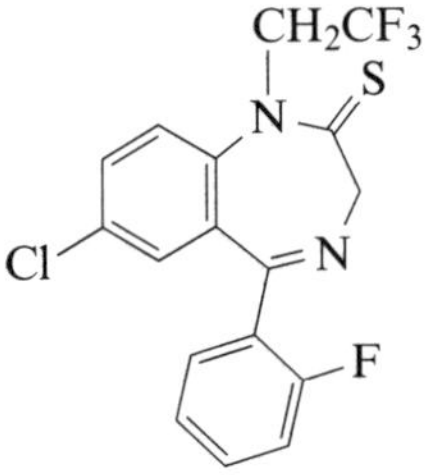 (49)

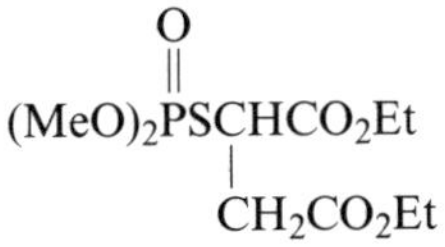

(50) 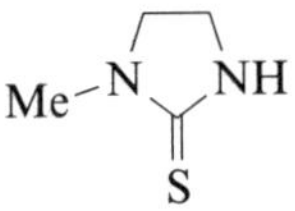 (51) 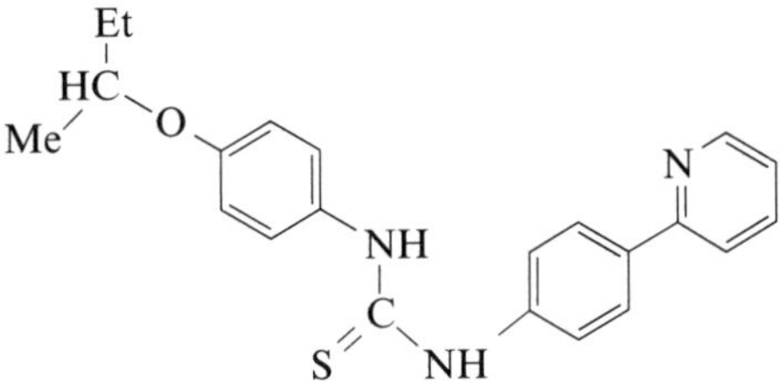 (52)

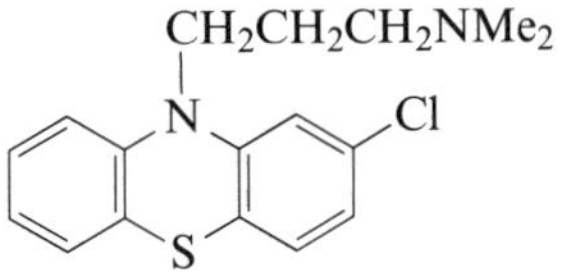

(53) 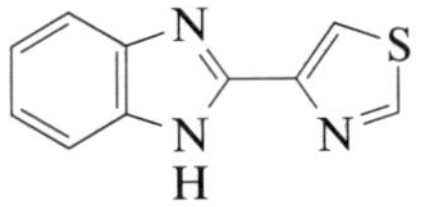 (54) 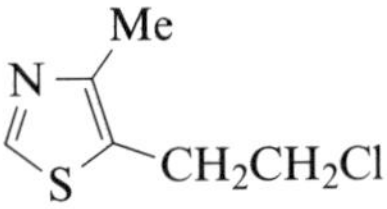 (55)

Figure 7

Some of the more interesting of these drugs include captopril (41), an effective oral antihypertensive[4b] which was the first major drug to be developed by the strategy of inhibiting a specific peptidase, namely the angiotensin-converting enzyme (ACE) which is involved in the conversion of the decapeptide angiotensin I (**56**) to the octapeptide angiotensin II (**57**), a natural pressor substance (Scheme 10).[8]

Asp—Arg—Val—Tyr—Ile—His—Pro—Phe—His—Leu
1 2 3 4 5 6 7 8 9 10

(**56**)

↓ ACE

Asp—Arg—Val—Tyr—Ile—His—Pro—Phe + H—His—Leu—OH
1 2 3 4 5 6 7 8

(**57**)

Scheme 10

The development of ACE inhibitors as antihypertensive drugs derived from the discovery (1970) of the effect of a nonapeptide isolated from the venom of the Brazilian snake *Bothrops javava* in lowering blood pressure. Later work showed that smaller peptides retained activity and led to the discovery of the drug captopril (**41**) which binds tightly to the ACE enzyme; the use of molecular modelling has helped to discover other ACE inhibitors. In (**41**), the thiol (SH) group is essential for the pharmacological activity and plays a vital role in the metabolism of the drug. Another thiol, namely D-penicillamine (**42**), is extensively employed in the treatment of severe rheumatoid arthritis, Wilson's disease (caused by copper accumulation), lead poisoning and chronic hepatitis. The D-isomer is used because the L-isomer showed toxicity to rats. D-Penicillamine[4b] (**42**) is obtained by hydrolytic degradation of the antibiotic penicillin and functions as a metal-chelating agent, hence the efficacy of the drug for the treatment of patients suffering from heavy metal poisoning. The metabolism of (**42**) is largely dependent on the presence of the thiol moiety, as is demonstrated by the structures of the two major metabolites, the disulfides (**58**) and (**59**) (Scheme 11).

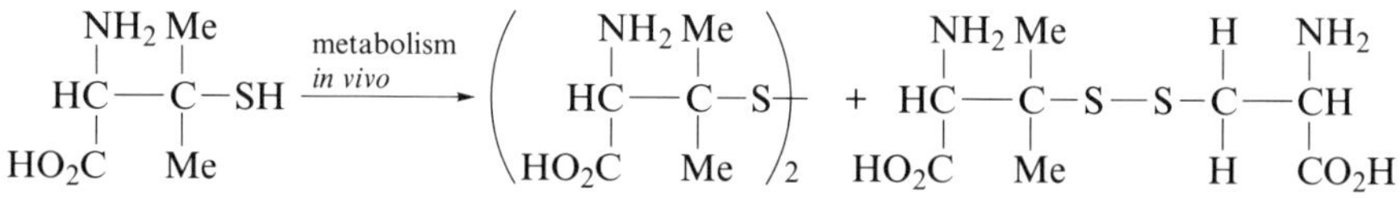

$$HO_2C\text{-}CH(NH_2)\text{-}C(Me)_2\text{-}SH \xrightarrow[\textit{in vivo}]{\text{metabolism}} \left(HO_2C\text{-}CH(NH_2)\text{-}C(Me)_2\text{-}S\right)_2 + HO_2C\text{-}CH(NH_2)\text{-}C(Me)_2\text{-}S\text{-}S\text{-}CH_2\text{-}CH(NH_2)\text{-}CO_2H$$

(**42**) (**58**) (**59**)

Scheme 11

The highly significant antiulcer drug cimetidine (**43**) was derived from studies carried out by Sir James Black and C. R. Ganellin of Smith, Kline and French Ltd from the mid-1960s.[9] The work was based on the known ability of histamine (**60**) (Figure 8) to stimulate gastric acid (GA) secretion which is mediated by H_2 receptors. In the search for effective antihistamines (H_2 blockers), over 200 analogues of (**60**) were synthesised and screened. The research led to the introduction of cimetidine for the clinical treatment of duodenal ulcers and other problems associated with gastric hyperactivity. Cimetidine (**43**) became one of the most widely prescribed drugs in the world, and scientists at Glaxo Research Ltd later demonstrated that the imidazole ring of cimetidine may be replaced by a furan, thiophene or benzene ring without loss of antihistamine activity, provided that the ring always contains an aminoalkyl substituent. The Glaxo research culmi-

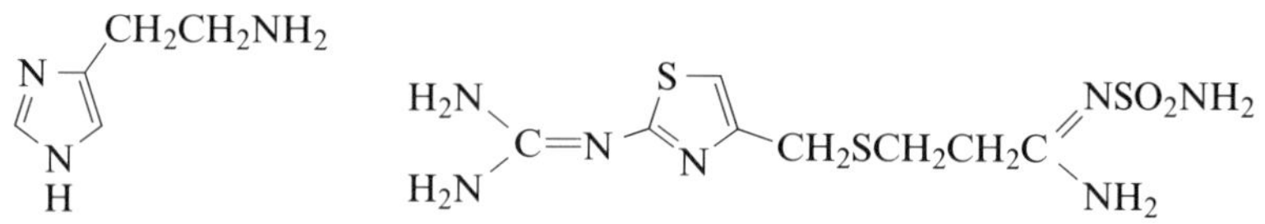

$CH_2CH_2NH_2$ (imidazole ring, N, N–H)

(**60**)

$(H_2N)_2C{=}N$–(thiazole ring, S, N)–$CH_2SCH_2CH_2C(=NSO_2NH_2)NH_2$

(**61**)

Figure 8

nated in the introduction of ranitidine (**44**), which is now the favoured drug since it is associated with fewer side reactions and appears more potent than cimetidine (**43**).

Both cimetidine (**43**) and ranitidine (**44**) are extremely effective drugs, and in many cases their use can avoid the need for surgery in the treatment of ulcer patients. These drugs rank with antibiotics in their worldwide effect in improving the quality of life of many individuals. The majority of active antihistamine drugs contain sulfur, but the element is not essential. In many cases, the presence of sulfur appears to modify the electronic properties of regions of the molecule, so facilitating optimum binding of the compound to the receptor. Cimetidine and ranitidine are both metabolised *in vivo* to the corresponding sulfoxides. Other H_2 receptor antagonists contain thioether, thiourea, thiazole and sulfonamide groups; for example, famotidine (**61**) (Figure 8) contains three sulfur atoms and is the only effective antihistamine sulfonamide drug.

Heparin, a natural anticoagulant, is a sulfated polymeric glycosaminoglycan; the structure of the two repeating units (**62**) and (**63**) are shown in Figure 9. Heparin is a macromolecule with a molecular weight of more than 20 000; the *N*- and *O*-sulfate groups contribute to the anticoagulant activity since the desulfated molecule is inactive.[10] Attempted sulfation and oxidation reactions on chitin do not yield a product of the same biological activity. Heparin in unique among the glycosaminoglycans in possessing anticoagulant activity and is used therapeutically as the sodium salt. Experiments have demonstrated that when heparin is bonded to silicon rubber and other polymers, the resultant surfaces do not cause clotting of contacted blood. Sulfation is also of biochemical significance as it is probably the major route for the metabolism of neuroactive steroids in the brain.

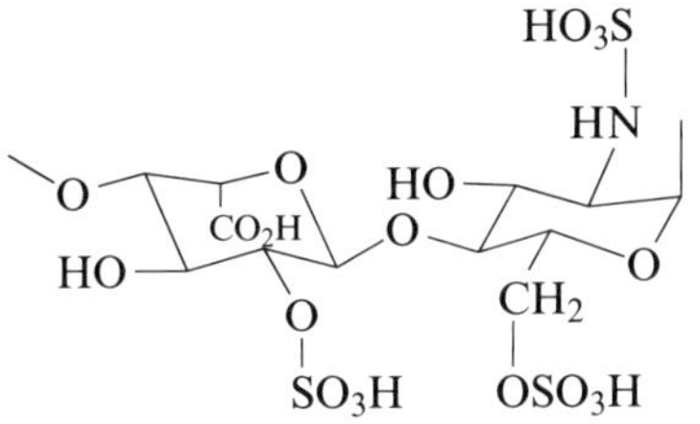

(**62**) 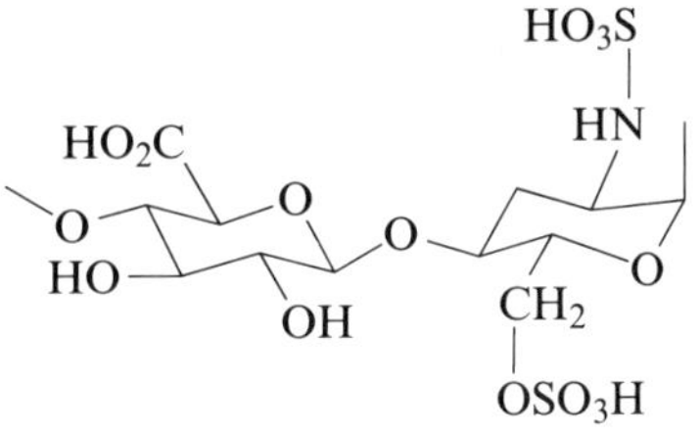 (**63**)

Figure 9

The phosphorothiolothionate malathion (**50**) (Figure 7) is used for the treatment of head and pubic lice and their eggs; it is also an important agricultural insecticide for the control of aphids, red spider mites and other insect pests on a wide range of crops. Like other organophosphorus insecticides containing the thiophosphoryl (P=S) group, malathion is activated *in vivo* by oxidation to the phosphoryl (P=O) analogue, malaoxon (**64**). The selective toxicity of malathion towards insects depends on this oxidation, as well as the fact that in mammals the favoured metabolic pathway of the chemical is detoxification by the action of a carboxyesterase enzyme to give the free carboxylic acid metabolite (Scheme 12).[5,11]

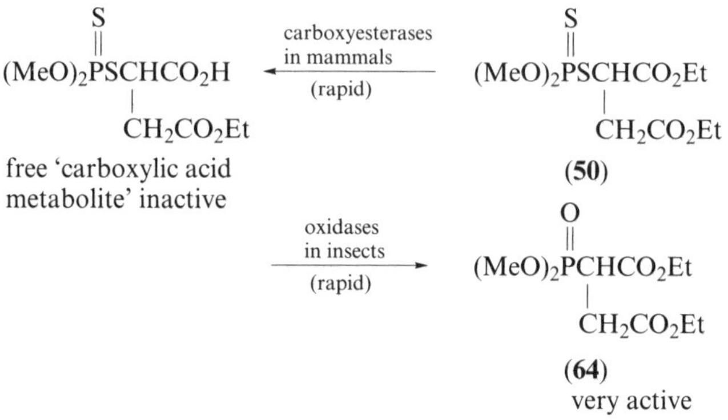

$(MeO)_2P(=S)SCH(CO_2H)CH_2CO_2Et$ ← carboxyesterases in mammals (rapid) ← $(MeO)_2P(=S)SCH(CO_2Et)CH_2CO_2Et$ (**50**)

free 'carboxylic acid metabolite' inactive

(**50**) → oxidases in insects (rapid) → $(MeO)_2P(=O)CH(CO_2Et)CH_2CO_2Et$ (**64**) very active

Scheme 12

Thiabendazole (**54**) (Figure 7) is used as an anthelmintic to kill parasitic worms, as a systemic fungicide in agriculture to control post-harvest diseases of potatoes, apples, pears and oranges, and also as a seed dressing against wheat bunt. Thiabendazole is a member of the benzimidazole group of systemic fungicides which owe their activity to the inhibition of nuclear cell division in sensitive fungi.

Agrochemicals

Several organosulfur compounds in addition to malathion (**50**) and thiabendazole (**54**) are important agricultural pesticides.[5,11] Many organophosphorus compounds containing the thiophosphoryl (P=S) group are extensively employed as insecticides; for example, methyl parathion (**65**) (Figure 10) is a wide spectrum contact insecticide, especially effective against scale and other insects on citrus and coffee; carbosulfan (**66**) an *N*-sulfenylcarbamate, is a valuable soil insecticide extensively used to control pests of sugar cane, maize, rice and coffee. In weed control, herbicides include the sulfonylcarbamate asulam (**67**) (see Chapter 2, p. 26) used to kill docks and bracken in grassland; the sulfonamide oryzalin (**68**) provides pre-emergence weed control in a wide range of crops. Du Pont in 1982 introduced the sulfonylurea herbicides, e.g. chlorsulfuron (**69**), for the selective control of broad-leaved weeds in cereals. The thiolocarbamate butylate (**70**) is valuable for pre-emergence weed control in maize. Sulfur and inorganic sulfur compounds have long been used to combat fungal diseases in agriculture (see Introduction, p. 3), and the dithiocarbamates introduced by Tisdale and Williams (1934) were the first really effective organic fungicides (see Chapter 9, p. 148), e.g. maneb (**71**) used as a foliar spray against the late blight of potato and tomato. *N*-Trichloromethanesulfenyl compounds, e.g. captan (**72**), containing the $NSCCl_3$ group are valuable surface fungicides for the control of a wide spectrum of phytopathogenic fungi (see Chapter 9, p. 151). Compounds (**73**)–(**75**) (Figure 10) are systemic fungicides. The oxathiin carboxin (**73**) is formulated as a seed dressing to control fungal diseases (rusts, smuts and bunts) in cereals; thiophanate methyl (**74**) controls powdery mildews and apple and pear scab; and the *N*,*N*-dimethylsulfamate bupirimate (**75**) is especially effective against apple powdery mildew. Several organosulfur compounds are also useful for the control of mites (acaricides) on agricultural crops; examples are the sulfone tetradifon, the sulfonate chlorfenson and the sulfide; chlorbenside (**76**) (see Chapter 7, p. 105). Such bridged diphenyl sulfones and related compounds control phytophagous mites on a wide range of crops and ornamental plants. Organophosphorus compounds—one of the major groups of insecticides—like malathion (**50**) and parathion (**65**) contain the thiophosphoryl group (P=S) and act as proinsecticides since they are activated by *in vivo* conversion to the phosphoryl (P=O) analogues. The phosphoryl analogues owe their insecticidal action to the interference with the transmission of nerve impulses across the synapse, so that the insect passes into spasms and eventually dies. The phosphoryl compounds poison the enzyme acetylcholinesterase by phosphorylation; the enzyme plays a vital role in nerve impulse transmission. On the other hand, the thiophosphoryl compounds such as (**50**) and (**65**) are not effective phosphorylating agents, hence the key importance of their *in vivo* oxidation to the phosphoryl compounds which are good phosphorylating agents and are the active entities in these insecticides.

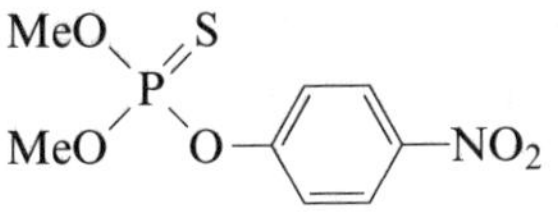

(**65**) 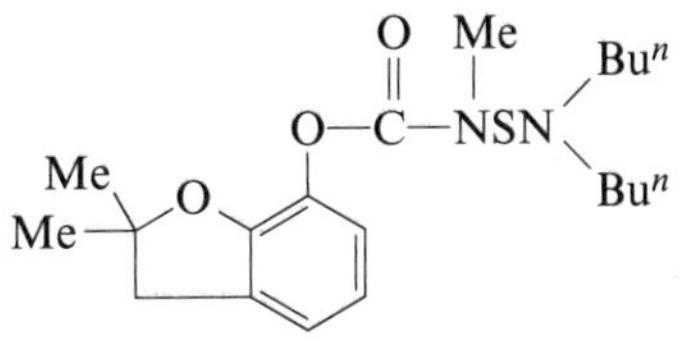 (**66**)

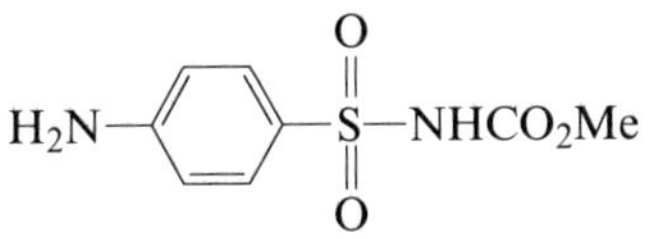

(**67**) 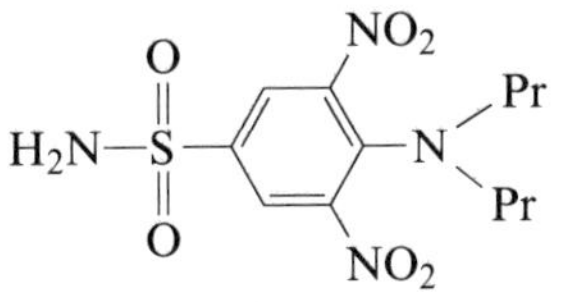 (**68**)

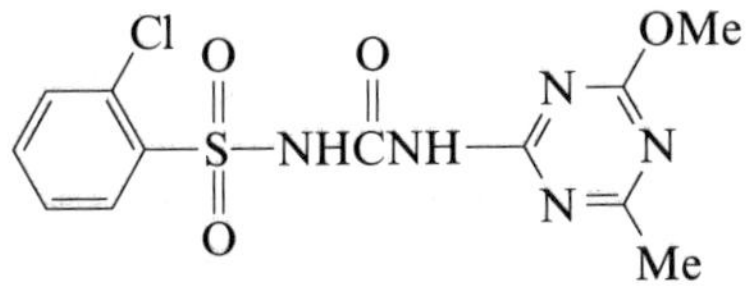

(**69**) 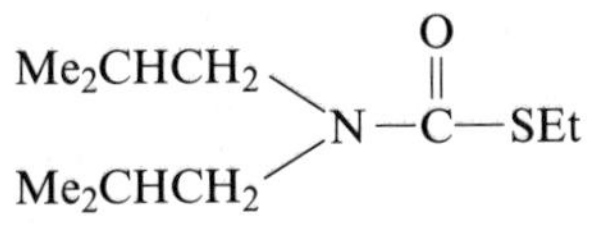 (**70**)

Figure 10 *(continued)*

Figure 10 *(continued)*

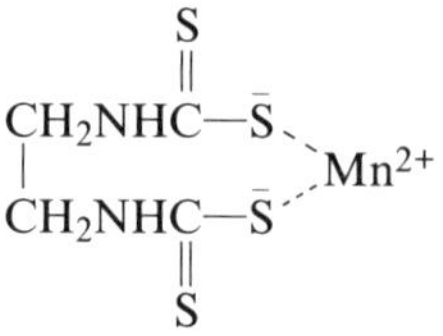

(71) 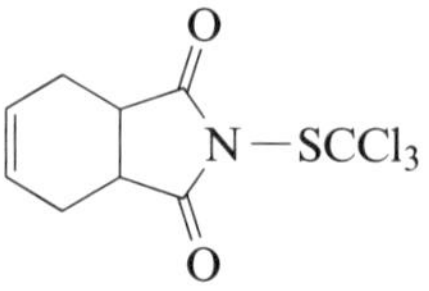 (72) 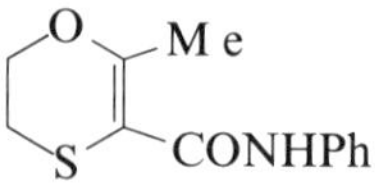 (73)

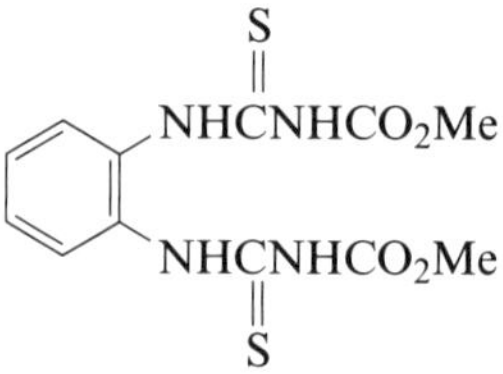

(74) 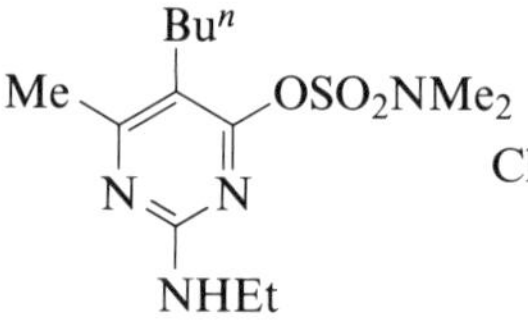 (75) 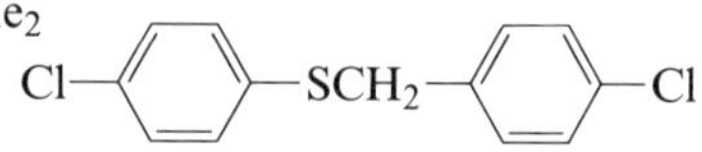 (76)

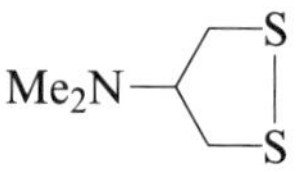

(77) 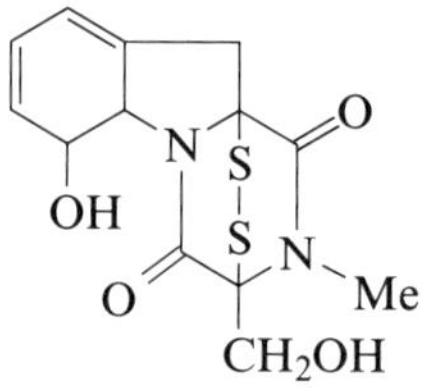 (78)

Figure 10

Carbamates, like carbosulfan (**66**), have a similar mode of action, namely poisoning of the enzyme acetylcholinesterase, except that this is now achieved by carbamoylation rather than phosphorylation. Thiolocarbamates like (**70**) are prepared from the appropriate amines by reaction with phosgene (Scheme 13).

$$RR'NH + COCl_2 \xrightarrow[(-HCl)]{base} RR'NCOCl \xrightarrow[(-NaCl)]{R''SNa} RR'N{-}C(=O)SR''$$

phosgene (**70**)

Scheme 13

Thiolocarbamates, e.g. butylate (**70**) ($R = R' = Me_2CHCH_2$, $R'' = Et$), are valuable soil-applied herbicides for pre-emergence weed control in a wide range of crops. They probably kill weeds by interference with lipid biosynthesis.

Sulfonylureas, e.g. chlorsulfuron (**69**), are an important group of herbicides which will selectively control many broad-leaved weeds in cereals at remarkably low doses ($\approx 20 g/ha^{-1}$; $1 ha = 10^4 m^2$). Chlorsulfuron (**69**), can be synthesised from chlorobenzene by treatment with chlorosulfonic acid, which yields a mixture of the *o*- and *p*-sulfonyl chlorides. The desired *o*-sulfonyl chloride (**79**) is separated

chlorobenzene → (excess $ClSO_3H$) → (79) $-SO_2Cl$ → ($2NH_3$) → $-SO_2NH_2$ → ($COCl_2$, (−2HCl)) → $-SO_2N{=}C{=}O$ → (2-amino-4-methoxy-6-methyl-1,3,5-triazine) → (69)

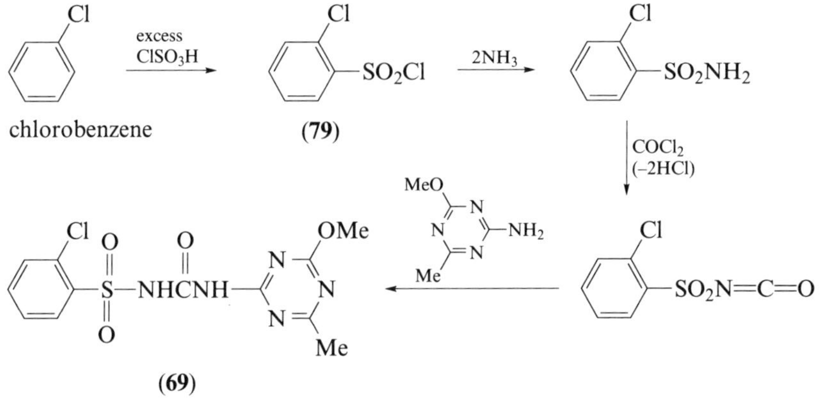

Scheme 14

from the mixture by fractional distillation since it has a lower boiling point, and is subsequently converted to (**69**) as indicated in Scheme 14.

The selective toxicity of sulfonylureas to certain weeds without damage to the cereal crop arises from their rapid metabolism in the crop plant to inactive compounds, whereas in sensitive weeds the metabolism is much slower. The very high herbicidal activity suggests a specific biochemical mode of action, which is concluded to be the inhibition of plant cell division. Sulfonylureas block the enzyme acetolacetate synthase (ALS), which catalyses the biosynthesis of the essential branched chain amino acids valine, leucine and isoleucine.

In terms of tonnage, the dithiocarbamates, e.g. maneb (**71**), are the most used organic fungicides. They are prepared by the reaction of amines with carbon disulfide (see Chapter 9, p. 148). Other examples include thiram, used to control damping-off diseases; thiram forms metal chelates,, and it is considered that the 1:1 copper chelate is the essential toxic entity since this compound has the ability to penetrate the fungal cell. Chelation also deprives the fungus of essential trace metals, which may be a contributory factor in the fungitoxicity of dithiocarbamates. The ethylenebisdithiocarbamates (see Chapter 9, p. 153) such as maneb (**71**) and the mixed manganese–zinc complex mancozeb are the most widely used of this group of fungicides: they are valuable protective fungicides against potato and tomato blight. The bisdithiocarbamates probably have a different mode of fungicidal action from the dithiocarbamates like thiram. It is believed that they owe their fungitoxicity to oxidation on the leaf surface to other compounds, notably ethylenebisisothiocyanate, which kills the fungus by interaction with vital thiol enzymes.

Captan (**72**), an example of the *N*-trichloromethanesulfenyl fungicides, is a useful protective foliar fungicide and seed dressing. It is prepared from butadiene and maleic anhydride (see Chapter 9, p. 151). The fungicidal activity arises from interaction with cellular thiols to give thiophosgene, which is probably the ultimate toxicant (Scheme 15).

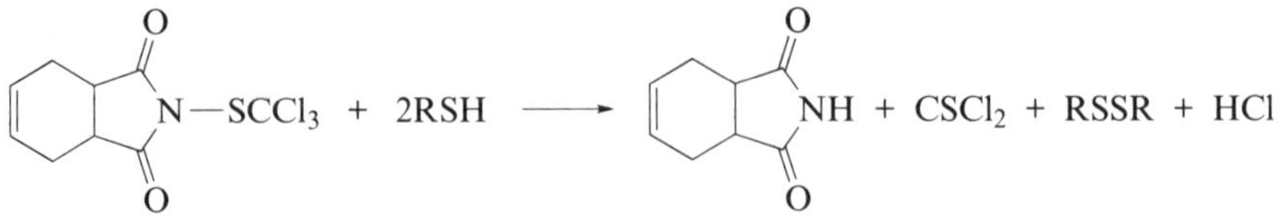

$$(72)\ \text{N}-SCCl_3 + 2RSH \longrightarrow \text{NH} + CSCl_2 + RSSR + HCl$$

(**72**)

Scheme 15

Thiophosgene would poison the fungus by combination with vital thiol-, amino- or hydroxy-containing enzymes. Systemic fungicides, e.g. compounds (**73**)–(**75**) (Figure 10), can penetrate the host plant and may therefore be able to eradicate an established fungal infection. Carboxin (**73**) owes its activity to the inhibition of respiration in the fungus, and is synthesised from α-chloroacetoacetanilide (**80**) and 2-thioethanol (**81**) (Scheme 16).

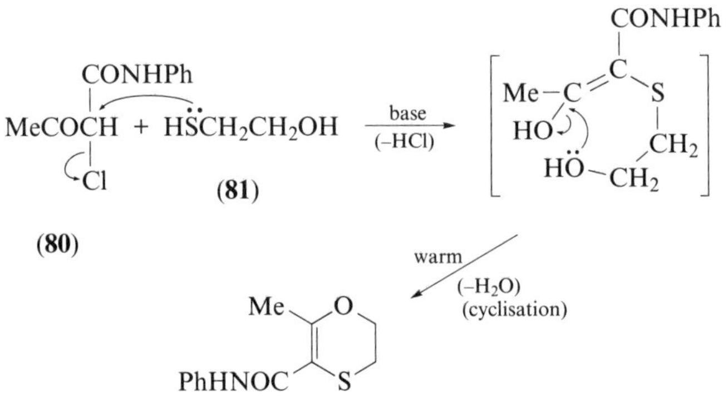

$$MeCOCH(CONHPh)Cl + HSCH_2CH_2OH \xrightarrow[(-HCl)]{base} [Me-C(HO)=C(CONHPh)-S-CH_2-CH_2-OH] \xrightarrow[(-H_2O)\ (cyclisation)]{warm} (73)$$

(**80**) (**81**)

Me, O, PhHNOC, S

(**73**)

Scheme 16

Thiophanate methyl (**74**) is, like thiabendazole (**54**), a member of the benzimidazole group of fungicides since it is metabolised *in vivo* to carbendazim (**83**), which is the active entity. Thiophanate methyl is synthesised by condensation of *o*-phenylenediamine (**82**) with potassium thiocyanate and methyl chloroformate (Scheme 17). The benzimidazoles owe their fungicidal action to the inhibition of cell division in the fungus due to interference with the microtubular assembly.

Bupirimate (**75**) is the *N*,*N*-dimethylsulfamoyl ester of the systemic fungicide ethirimol; both are members of the hydroxypyrimidine group of fungicides developed specifically to combat powdery mildew diseases. They may owe their activity to interference with purine metabolism in the fungus.

Several naturally occurring organosulfur compounds show pesticidal activity; examples include nereistoxin (**77**), isolated from the marine worm *Lumbriconereis heteropoda*, which is insecticidal, and the antifungal antibiotic gliotoxin (**78**) (Figure 10), produced by the soil fungus *Trichoderma viride*.

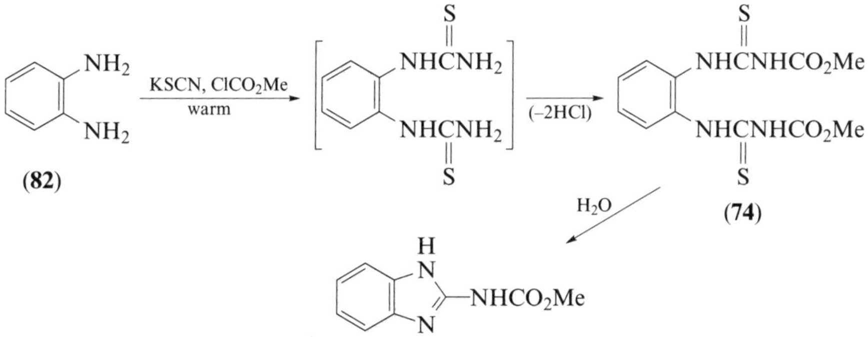

Scheme 17

Sweeteners

Several organosulfur compounds have been developed as commercial sweeteners;[12] examples are saccharin **(84)**, cyclamate **(85)** and acesulfame potassium **(86)** (Figure 11). These chemicals are usefully formulated as the sodium or potassium salts to increase the aqueous solubility.

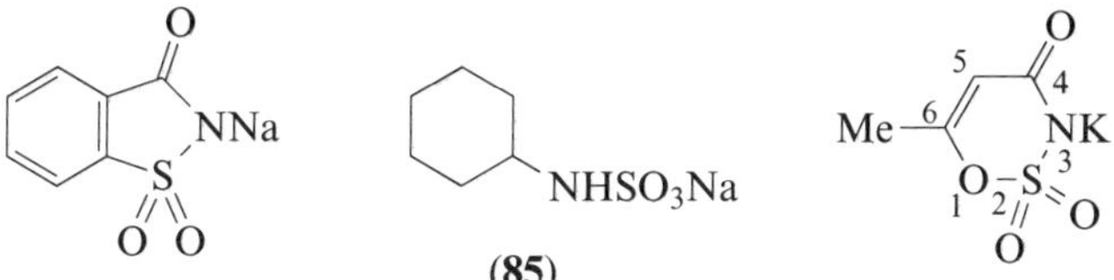

Figure 11

Saccharin **(84)** was discovered in 1878 and can be manufactured from toluene (Scheme 18). In the synthesis, the initial chlorosulfonation yields a mixture of the *o*-and *p*-toluenesulfonyl chlorides; the required *ortho*-isomer is obtained by freezing out the solid *para*-isomer to leave the liquid *o*-sulfonyl chloride behind. The imino hydrogen atom in saccharin is acidic owing to the electron-attracting (–I) properties of the adjacent carbonyl and sulfonyl moieties (Scheme 18). Consequently, saccharin readily forms the sodium salt, sodium saccharin **(84)**, on treatment with sodium hydroxide.

Saccharin is approximately 300 times as sweet as sucrose but can have a bitter after-taste in concentrated solution; it is non-calorific, does not contribute to the problem of obesity or tooth decay, and can be used by diabetics as a sugar substitute. Saccharin is stable to heat and so can be used in cooking. A number of saccharin derivatives, e.g. **(88a)**–**(88e)**, have been synthesised as potential sweetening agents (Figure 12).

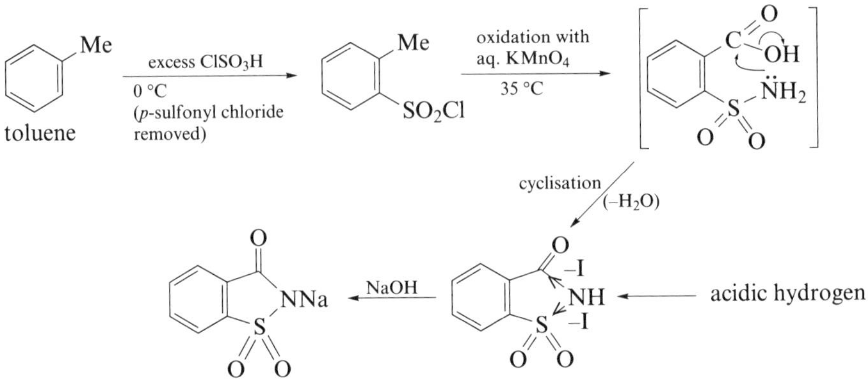

Scheme 18

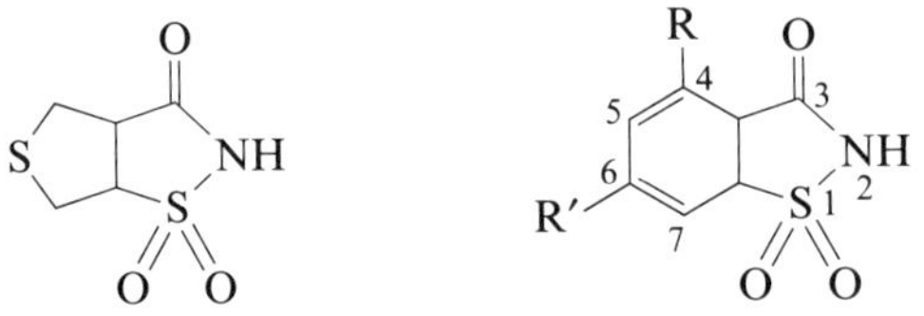

	R	R′
(88a)	NO_2	H
(88b)	NH_2	H
(88c)	H	NO_2
(88d)	H	Cl
(88e)	H	F

Figure 12

The sulfur derivative (**87**) is 1000 times as sweet as sugar and without the bitter after-taste of saccharin; however, it was discovered that *N*-alkylation of (**87**) removed the sweetness. On the other hand, in the saccharins (**88a**)–(**88e**) containing substituents in the 4-position and 6-position, sweetness was retained after *N*-alkylation. Many sulfamic acid derivatives are sweet, and there have been numerous studies of structure–taste relationships which have highlighted the importance of molecular shape and stereochemistry (see Chapter 9, p. 162). Two sulfamates which are commercial, non-nutritive sweeteners are cyclamate (**85**) and acesulfame potassium (**86**) (Figure 11). Cyclamate (**85**) is manufactured by refluxing cyclohexylamine either with triethylamine–sulfur trioxide in dichloromethane or with sulfamic acid (see Chapter 9, p. 162).

Cyclamate was discovered in 1937 and is about 30 times sweeter than sucrose, but was withdrawn from use in the USA and Canada in 1970 and later in the UK because rats fed with large doses of cyclamate developed bladder cancer. However, more recent research with laboratory animals at dosage levels up to 240 times the normal intake failed to confirm the carcinogenicity, but the USA Food and Drug Administration ban on cyclamate remains in force. The use of cyclamate is, however, permitted in many other countries. Studies have demonstrated that cyclamate (**85**) is metabolised in animals to cyclohexylamine, which causes chromosome damage. Structural analogues in which the size of the alicyclic ring is varied retain sweetness, but the *N*-alkyl derivatives are not sweet.

Acesulfame potassium (6-methyl-4-oxo-1,2,3-oxathiazine(3*H*)-2,2-dioxide) (**86**), introduced by Hoechst in 1973,[12] is approximately 130 times as sweet as sucrose; it is not metabolised and produces no pharmacological effects, but like saccharin can produce a bitter after-taste. Acesulfame potassium is prepared from the trityl derivative of acetoacetic ester (89) (Scheme 19).

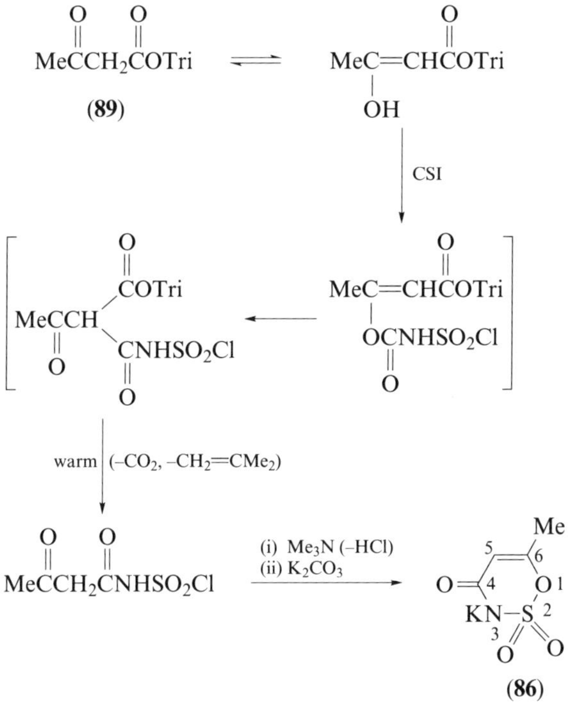

$MeC(=O)CH_2C(=O)OTri$ (**89**) ⇌ $MeC(OH)=CHC(=O)OTri$

CSI ↓

[$MeC(=O)CH[C(=O)OTri][C(=O)NHSO_2Cl]$ ← $MeC[OC(=O)NHSO_2Cl]=CHC(=O)OTri$]

warm ($-CO_2$, $-CH_2{=}CMe_2$) ↓

$MeC(=O)CH_2C(=O)NHSO_2Cl$ —(i) Me_3N ($-HCl$); (ii) K_2CO_3→ (**86**)

Scheme 19

Several analogues of acesulfame potassium (**86**) have been synthesised, e.g. compounds (**90a**)–(**90d**) (Figure 13). Several of these derivatives are sweeter than acesulfame potassium (**86**); indeed, the best one (**90b**) is almost twice as sweet. However, they do not taste as good, and accordingly (**86**) has been selected for commercial exploitation.

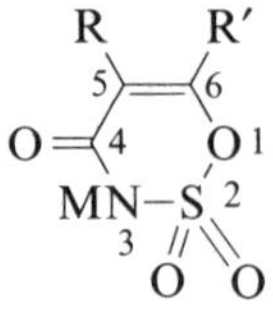

(90)

	R	R′	M	Potency*
(86)	H	Me	K	130
(90a)	H	Et	Na	150
(90b)	Et	Me	Na	250
(90c)	H	CH_2Cl	K	150
(90d)	Cl	Me	K	200

* Potency (sweetness) taken relative to sucrose as unity.

Figure 13

References

1 M. Porter, in *Organic Chemistry of Sulfur* (Ed. S. Oae), Plenum Press, New York, 1977, Chap. 3, p. 71.

2 I. L. Finar, *Organic Chemistry* 5th Edn. Vol. 1, Longman Green, London, 1967, Chap. 3, p. 821.

3 P.G. Sammes, in *Comprehensive Medicinal Chemistry* (Eds P.G. Sammes and J. B. Taylor), Vol. 2, Pergamon Press, Oxford 1990, p. 255.

4 L. A. Damani and M. Mitchard, in *Sulfur-Containing Drugs and Related Organic Compounds—Chemistry, Biochemistry and Toxicology* (Ed. L. A. Damani), Ellis Horwood, Chichester, 1989: (a) Vol. 1, Part A, p. 92; (b) B. K. Park and J. W. Coleman, Chap. 3, p. 47.

5 R. J. Cremlyn, *Agrochemicals: Preparation and Mode of Action*, Wiley, Chichester, 1991.

6 C. E. Newall and P. D. Hallam, in *Comprehensive Medicinal Chemistry* (Eds P. G. Sammes and J. B. Taylor), Vol. 2, Pergamon Press, Oxford 1990, p. 609.

7 A. Danon and Z. Ben-Zvi, in *The Chemistry of Acid Derivatives* (Ed. S. Patai), Vol. 2, Part 2, Suppl. B, Wiley, Chichester, 1992, p. 1063.

8 D. H. Rich, in *Comprehensive Medicinal Chemistry* (Eds. P.G. Sammes and J. B. Taylor), Vol. 2, Pergamon Press, Oxford, 1990, p. 391.

9 C. R. Ganellin, in *Medicinal Chemistry: The Role of Organic Chemistry in Drug Research* (Eds S. M. Roberts and B. J. Price), Academic Press, London, 1985, p. 93.

10 J. F. Kennedy and C. A. White, in *Chemistry, Biochemistry and Biology*, Ellis Horwood, Chichester, 1983, Chap. 9, p. 211.

11 K. A. Hassall, *The Biochemistry and Uses of Pesticides*, 2nd Edn, Macmillan Basingstoke, 1990.

12 *Kirk-Othner's Encyclopedia of Chemical Technology*, 3rd Edn, Vol. 22, Wiley, New York, 1983, p. 448.

INDEX

Bold type indicates the more important references.